GRID COMPUTING

Techniques and Applications

Chapman & Hall/CRC Computational Science Series

SERIES EDITOR

Horst Simon
Associate Laboratory Director, Computing Sciences
Lawrence Berkeley National Laboratory
Berkeley, California, U.S.A.

AIMS AND SCOPE

This series aims to capture new developments and applications in the field of computational science through the publication of a broad range of textbooks, reference works, and handbooks. Books in this series will provide introductory as well as advanced material on mathematical, statistical, and computational methods and techniques, and will present researchers with the latest theories and experimentation. The scope of the series includes, but is not limited to, titles in the areas of scientific computing, parallel and distributed computing, high performance computing, grid computing, cluster computing, heterogeneous computing, quantum computing, and their applications in scientific disciplines such as astrophysics, aeronautics, biology, chemistry, climate modeling, combustion, cosmology, earthquake prediction, imaging, materials, neuroscience, oil exploration, and weather forecasting.

PUBLISHED TITLES

PETASCALE COMPUTING: Algorithms and Applications
Edited by David A. Bader

PROCESS ALGEBRA FOR PARALLEL AND DISTRIBUTED PROCESSING
Edited by Michael Alexander and William Gardner

GRID COMPUTING: TECHNIQUES AND APPLICATIONS
Barry Wilkinson

GRID COMPUTING

Techniques and Applications

BARRY WILKINSON

CRC Press is an imprint of the
Taylor & Francis Group an **informa** business
A CHAPMAN & HALL BOOK

Chapman & Hall/CRC
Taylor & Francis Group
6000 Broken Sound Parkway NW, Suite 300
Boca Raton, FL 33487-2742

Printed in the United States of America on acid-free paper
10 9 8 7 6 5 4 3 2 1

International Standard Book Number: 978-1-4200-6953-2 (Hardback)

Library of Congress Cataloging-in-Publication Data

Wilkinson, Barry.
Grid computing : techniques and applications / Barry Wilkinson.
p. cm. -- (Chapman & Hall/CRC computational science series)
Includes bibliographical references and index.
ISBN 978-1-4200-6953-2 (hardcover : alk. paper)
1. Computational grids (Computer systems) I. Title. II. Series.

QA76.9.C58W55 2010
004'.36--dc22 2009030739

Visit the Taylor & Francis Web site at
http://www.taylorandfrancis.com

and the CRC Press Web site at
http://www.crcpress.com

To my wife, Wendy
and my daughter, Johanna

Contents

Preface

The purpose of this text is to introduce Grid computing techniques and form the basis for a practical senior undergraduate course or first-year graduate course on Grid computing. Grid computing uses geographically distributed computers collectively for high performance computing and resource sharing. The computers can be distributed locally, nationally, or across countries and continents. Grid computing often involves computers from multiple organizations and hence crosses organizational boundaries. Distributed teams can be formed to tackle major problems such as searching for new drugs and cures. Apart from computers, the shared resources might include expensive experimental equipment located at one site but shared by all in the team. The experimental equipment might generate huge amounts of data that need to be studied by members of the team using computers at other sites. The term *virtual organization* is used to describe the distributed team and the shared resources that are provided.

Grid computing first became viable with the widespread growth of high-speed networks and the Internet in the mid-1990s. The continued growth and performance of networks has fueled numerous Grid computing projects across the world. Very high performance Grid projects might use special dedicated high-speed networks but the general Internet also provides the backbone for many Grid computing projects. The use of the Internet enables everyone to become involved in Grid computing. A key aspect of Grid computing is the use of standard Internet protocols and techniques, which includes the basic transport protocols, but also Web services and associated technologies.

We capitalize the word Grid, not because there is only one Grid—there are many Grids that are set up locally, nationally and internationally—but because it is a common practice to do so.

Book Background. This book is the direct outcome of work done on introducing Grid computing into the undergraduate curricula by the author and Dr. Clayton Ferner from the University of North Carolina at Wilmington (UNC–Wilmington). Since Grid computing often involves computers at multiple sites on the Internet, to teach Grid computing, it is desirable to have cooperating geographically distributed sites. A Grid was set up using computing resources at several North Carolina universities crossing administrative boundaries as in many production Grids. We also took advantage of the existing state-wide televideo/teleconferencing network called North Carolina Research and Education Network (NCREN) to present the lectures to students at many universities simultaneously. NCREN is a telecommunications network that became operational in 1985 to interconnect universities, medical center, research institutions, and graduate centers in North Carolina and provides multi-way, face-to-face video and audio communications. "Teleclass" classroom facilities are provided at each site. Each student is provided with a microphone and multiple video cameras are used so that the instructor and students can hear and see each other.

With funding from the National Science Foundation (NSF) and the University of North Carolina Office of the President (UNC-OP), our Grid computing course was first offered in Fall 2004, and repeated in Fall 2005. In Fall 2004, eight institutions participated, and in Fall 2005 twelve institutions participated. The course was redesigned to use a more top-down approach with several new features such as the use of a production-style Grid portal and less reliance of centralized servers for student assignments. The new course was offered in Spring 2007 and again Fall 2008. Additional funding was received from the NSF in 2008 to incorporate a Grid computing workflow editor called GridNexus further into the course. The book is based upon this re-designed course.

Throughout the four years of development, students and faculty from a wide range of institutions were involved, including premier research universities, comprehensive state universities, private four-year colleges, minority-serving institutions, and a technical community college, fifteen institutions in total.[1] Apart from formal lectures given by instructors, internationally known guest speakers were invited to give presentations from different sites. Streaming video of the classes was provided by NCREN, which enabled students to watch the class from the Internet in real time or download the class for watching later. Depending upon the offering, computer systems were set up at between three and five universities to create a working Grid for the students. To do all of this, the instructors were assisted by many people (see Acknowledgements).

The course was designed primarily for upper-level undergraduate Computer Science students, although graduate students were accommodated by providing extra work. The course was recognized as the "Link of the Week" in the June 15, 2005

[1] Appalachian State University, Elon University, North Carolina A & T University, North Carolina Central University, North Carolina State University, University of North Carolina at Asheville, University of North Carolina Chapel Hill, University of North Carolina at Charlotte, University of North Carolina at Greensboro, University of North Carolina at Pembroke, University of North Carolina at Wilmington, Western Carolina University, Winston-Salem State University, Lenoir Rhyne College, and Wake Technical Community College.

issue of *Science Grid This Week*, and received further national attention in the feature article of *Science Grid This Week* in December 14, 2005 (repeated in *GridToday*). In addition to conference papers, a short article of the re-designed course was published in *International Science Grid This Week* in March 26, 2008. More information can be found at http://www.cs.uncc.edu/~abw/gridcourse/ including links to publications and all course materials in each offering of the course.

Structure of Materials. The book starts with Chapter 1 which is an introduction to Grid computing and its applications. Grid computing is about executing jobs on a distributed computing platform and the chapter leads onto using a Grid computing Web-based portal, which is used in real Grid computing projects. In our course, we get the students to register on a course portal after the first class and they immediately submit jobs to the Grid platform using the portal. In Chapter 2, the underlying action of job submission using a command-line interface is considered in some depth. Grid portals hide some of these details but the command-line interface is still needed to appreciate fully the underlying Grid infrastructure. We return to the portal later. Chapter 3 discusses the use of a job scheduler. Jobs usually enter a job queue and are sent to an appropriate compute resource as selected by a scheduler. Chapter 4 describes general Internet security techniques, which are the basis for Grid computing security. Chapter 5 describes the specific security mechanisms developed for Grid computing. Chapter 6 describes Web services technology. Grid computing middleware software is aligned to Web services. Chapter 7 describes how Web services are adopted for Grid computing. Chapter 8 focuses on graphical user interfaces. A user can interact with the Grid software either through a command-line interface or usually preferably through a graphical interface. A graphical interface offers several advantages. For example, it can offer scientists a domain-specific interface. It can make it easier for non-Computer Science users to access the Grid. We concentrate upon the GridSphere portal introduced in Chapter 1. Gridsphere conforms to currently agreed standards. The chapter describes how that portal can be customized to produce specific interfaces. Also in this chapter, we describe graphical workflow editors that enable the user to compose sequences of computational tasks visually using a simple drag-and-drop interface. We concentrate upon a graphical workflow editor developed at UNC–Wilmington called GridNexus which we also use in our class. The final chapter, Chapter 9, describes how to deploy applications on Grid. Although the last chapter, this chapter is very important and often not addressed in detail. Most applications can be run at one site if everything the application needs is installed at that site. However, the Grid computing platform offers much more than simply running an application at a remote site. It can also offer using multiple geographically distributed computers collectively to obtain increased speed and fault tolerance. It can offer resource discovery.

Each chapter concludes with a summary, further reading, bibliography, multiple-choice self-assessment questions, and programming assignments. The self-assessment questions are provided to check your understanding of the presented materials. The answers are given at the back of the book. Some of these programming assignments can be done on a PC once certain open-source software is installed as explained in the assignments. Other assignments do need access to a Grid computing platform.

The appendices offer useful closely related background materials, especially for doing Grid computing assignments. Appendix A covers Internet and networking basics including IP addressing and Internet protocols. Appendix B focuses on operating system environments and covers commonly used Linux and Windows commands that are helpful in assignments. Appendix C covers the XML language, which is needed for Web and Grid computing services. Appendix C could be read prior to Chapter 6 and Chapter 7 if needed. Appendix D, written by Jeremy Villalobos, provides a tutorial on the Globus installation. This appendix is intended to supplement on-line instructions on installing the Globus toolkit and related software. It provides notes on practical experiences.

The material in the book can be used in the order presented, which is top down and back, i.e., begin with a portal and user job submission, delve into the Grid infrastructure and return at end to Grid-enabling applications. However, it can also be used in a different order to obtain a more bottom-down approach by covering Web services first. We use the material more-or-less in the order given, by starting with the users submitting jobs to a portal in the first class and then delve into what is behind the portal and study the command line interface. Security (Chapters 4 and 5) is covered later in the course. Some materials may already be known, for example that on Internet security (Chapter 4), and in that case could be skipped. If XML is not known, Appendix C would become part of the course. There are some dependencies as shown in Figure P.1. Chapter 2 is written to be done before Chapter 3. Chapter 4, if not already known, has to be done before Chapter 5. Chapter 6, if not already known, has to be done before Chapter 7. There are some partial dependencies, that is, for full appreciation of the materials, some previous materials are helpful. As one can see, after the Introduction, one could do the job submission sequence, the security

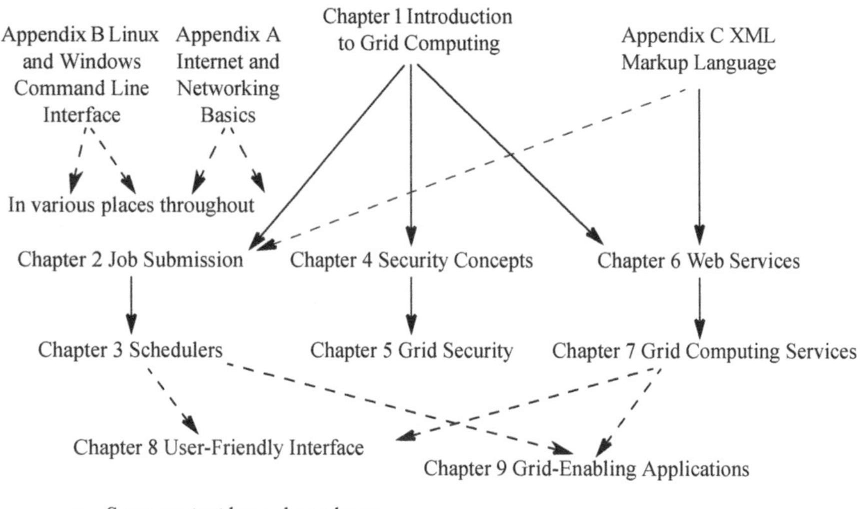

Figure P.1 Chapter dependencies.

sequence or the Web service sequence first, or interleaved combinations. Although the GridNexus workflow editor is described in a single chapter (Chapter 8), it can be introduced with schedulers (Chapter 3) for graphical job execution workflows and with Chapter 7 with graphical Web service workflows—in fact we do that in our class.

Grid Computing Platform. All the software to create a Grid computing platform is available by free download from various sites. The book concentrates on using the Globus toolkit as the central Grid computing software. Although ideally the course should use geographically distributed computers, the course can be presented to students at a single site as regular undergraduate or graduate-level Computer Science or Engineering programming course. It is usually most convenient to set aside computers/servers just for the course than attempt to use systems that are shared with other activities.

Using a centralized server for some Grid computing activities can be problematic. There are a number of educational activities that can be done on one's own personal computer (or laboratory computer) without access to a Grid platform and indeed it is much better to do so. For example, exercises using Web services (Chapters 6 and 7) can readily be done on a personal computer. Web services require a hosting environment (such as Apache Axis/Tomcat), which can be readily installed. Grid computing services can be hosted in the Globus container, a core software component of Globus. Being available in Java, it can be installed on a personal computer. Then, one's own services can be deployed and tested with local clients. This is much preferable to using a centralized server—with a single Web service hosting environment (container) or multiple hosting environments (containers). A large number of simultaneous containers is not practical because of the relatively large footprint of each container, which can cause operating system thrashing. Also, each container would need a separate port. Using a single container with multiple users is also problematic. Deploying/re-deploying services require the container to be re-started each time. Also all users would see all the deployed services and each service needs a unique name. In our early courses (Fall 2004 and Fall 2005), a script was provided that renamed students' services automatically but still continual restarting the container causes all users to be disrupted. Using one's own computer resolves all these issues and also provides users with a powerful learning experience and satisfaction in installing complex software components. Students have responded very positively to doing their software development on their own computer.

The Grid computing GUI workflow editor we use in GridNexus (Chapter 8) is freely available and can be easily installed on a personal computer to create workflows. Similarly, designing portlets within a Grid computing portal such as GridSphere can be done on a personal computer as portlet environments such as GridSphere/Tomcat can be installed there. In fact, portlets can be designed with a locally installed Grid portal that interfaces with either locally installed or remotely installed Web services (the latter with a network connection). Such activities combine materials in Chapters 6, 7, and 8, and relate to Chapter 9.

Activities relating to job submission, job scheduling, and security (Chapters 2, 3, 4, and 5) generally still need access to a Linux server or lab computer. GridNexus workflows (Chapter 8) can include workflows that execute remotely, for example remote job submission, file transfers, and Web services on different servers.

Prerequisites. The materials in this book are designed primarily for upper-level undergraduate and first-year graduate Computer Science and Engineering students with knowledge of Linux, C, and Java. Most such Computer Science and Engineering students have this knowledge or can quickly learn it sufficiently. Grid computing actually brings together topics found in other contexts. For example, job scheduling may be part of a course on operating systems. Aspects of Internet security such as certificates, certificate authorities, and secure network protocols, are found in networking courses. Web services might be found on courses on distributed computing. The underlying technology of portals and portlets such as servlets might also appear in courses on network applications. This knowledge is not assumed.

Home Page. A Web site has been provided for instructors and students at http://www.cs.uncc.edu/~abw/GridComputingBook/. This Web site includes all the instructional materials needed including slides and programming assignments. We have designed our assignments so that many could be done without access to a Grid—instead students are asked to install open-source software on their own computers or on laboratory computers to do the work. This includes software environments to deploy and test Web and Grid services, a Grid computing workflow editor to design and test workflows, and a Grid computing portal for deploying portlets. We provide step-by-step instructions for students on the home page, which supplement the description of assignments in the book. Certainly, servers are still needed for running jobs and experimenting with job schedulers and assignments are provided that do require access to servers.

Acknowledgements. Partial support for this work was provided by the NSF's Course, Curriculum, and Laboratory Improvement (CCLI) program under awards #0410667/053334 and #0737318/0737269/0737208. Funding was provided by the UNC-OP under two major 2004–2006 awards. Any opinions, findings, and conclusions or recommendations expressed in this material are those of the author and do not necessarily reflect the views of the NSF or UNC-OP. I wish to record my appreciation to the NSF and UNC-OP for providing support to enable us to develop the Grid computing courses that this book is based upon.

The course required the cooperation of *many* people at many universities. I would particularly like to thank Clayton Ferner (UNC–Wilmington) who joined me in teaching the course from the beginning and contributed throughout its development and made the Grid course a success. Mark Holliday at Western Carolina University was instrumental in starting the work. He was the co-PI on the first NSF grant and one UNC-OP grant that supported the initial development of the Grid course, and I wish to record my appreciation to him.

The following provided direct assistance at their institutions: Barry Kurtz and Rahman Tashakkori at Appalachian State University, Dave Powell and J. Hollingworth at Elon University, Yaohang Li at North Carolina A & T University, Mladen Vouk and Gary Howell at North Carolina State University, Dean Brock at the University of North Carolina at Asheville, Shan Suthaharan at the University of North Carolina at Greensboro, and Dick Hull at Lenoir Rhyne College. Dr. Mark Holliday's undergraduate students James Ruff and Jeffrey House developed several of the early assignments. James Ruff maintained the Grid computing installation at Western Carolina University. Natasha Stracener, TV Media Services Coordinator, Broadcast Communications at UNC–Charlotte managed the NCREN teleconferencing facilities at Charlotte and provided continuous support.

Appendix D, a Globus installation tutorial, was written by Jeremy Villalobos, a Ph.D. student in the Department of Computer Science at University of North Carolina at Charlotte. Jeremy is responsible for maintaining the Grid site at Charlotte for our Grid course. His Ph.D. work is on aspects of computational Grid computing, in particular in developing a high-level framework for distributed computations on Grid platform. Ramya Chaganti, a Computer Science MS student at University of North Carolina at Charlotte, was extremely helpful in assisting in the preparation of the book, including providing some screen shots in Chapter 8.

The following kindly gave guest lectures to our classes between 2004–2008: Professor Daniel A. Reed, Chancellor's Eminent Professor, Director of Renaissance Computing Institute (at the time); Wolfgang Gentzsch, Managing Director at Microelectronics Center of North Carolina (MCNC) Grid Computing and Networking Services (at the time); Chuck Kesler, Director of Grid Deployment and Data Center Services, MCNC (at the time); Jeff Schmitt, genesismolecular.com; Jim Jokl, University of Virginia; Art Vandenberg, Georgia State University; Mary Fran Yafchak, Southeastern Universities Research Association (SURA); Lavanya Ramakrishnan, Renaissance Computing Institute; Purushotham Bangalore, University of Alabama at Birmingham; Joel Hollingsworth, Elon University; Carla Hunt, MCNC; Yaohang Li, North Carolina A & T University; Rahman Tashakkori at Appalachian State University; Sammie Carter, student at North Carolina State University; Melea Williams, graduate student at the University of North Carolina at Wilmington. In addition, many students gave project presentations to the class.

Professor Ian Foster, Argonne National Laboratory and University of Chicago kindly allowed me to use in my class a recorded presentation entitled "The Grid: Beyond the Hype" he made at Duke University in November 2004. He also kindly invited Dr. Ferner and me to give a presentation describing our re-designed 2007–2008 course at *Open Source Grid & Cluster Conference*, Oakland, CA, May 12–16, 2008.

I would also like to record my appreciation to Ron Vetter at the University of North Carolina at Wilmington who involved me in his large UNC-OP grant, which led to the collaboration with Dr. Ferner at University of North Carolina at Wilmington—without this collaboration, the work would not have been possible.

This book would not have been done without a chance meeting with Alan Apt at SIGCSE 2007 and without Randi Cohen, Computer Science Acquisitions Editor, Chapman and Hall/CRC Press/Taylor and Francis Group LLC, who quickly responded to my enquiry to write a book on Grid computing.

Barry Wilkinson
University of North Carolina
Charlotte

About the Author

Barry Wilkinson is a professor of Computer Science and Computer Science MS program director at the University of North Carolina at Charlotte. He previously held faculty positions at Brighton Polytechnic, England (1984–87), State University of New York, College at New Paltz (1983–84), University College, Cardiff, Wales (1976–83), and the University of Aston, England (1973–76). From 1969 to 1970, he worked on process control computer systems at Ferranti Ltd. He has also taught at the University of Massachusetts–Boston, and Western Carolina University while on leaves of absence from the University of North Carolina at Charlotte. He is the author of *Computer Peripherals* (with D. Horrocks, Hodder and Stoughton, 1980, 2nd ed. 1987), *Digital System Design* (Prentice Hall, 1987, 2nd ed. 1992), *Computer Architecture Design and Performance* (Prentice Hall 1991, 2nd ed. 1996), *The Essence of Digital Design* (Prentice Hall, 1997), and *Parallel Programming Techniques and Applications Using Networked Workstations and Parallel Computers* (with M. Allen, Prentice Hall 1999, 2nd ed. 2005). In addition to these books, he has published many papers in major computer journals. He has been a senior member of the IEEE since 1983 and received an IEEE Computer Society Certificate of Appreciation in 2001 for his work on the IEEE Task Force on Cluster Computing (TFCC) education program. He has been supported by the National Science Foundation with five grants since 1996 for developing educational materials on cluster computing and Grid computing. He received a B.S. degree in Electrical Engineering with first-class honors from the University of Salford in 1969, and M.S. and Ph.D. degrees from the University of Manchester (Department of Computer Science), England in 1971 and 1974, respectively.

CHAPTER 1
Introduction to Grid Computing

In this introductory chapter, the concept of Grid computing and its history is outlined, to include pioneering landmark Grid computing projects. Access to Grid resources by users can be done through a command-line interface but more usually, it is done through a Web-based portal. A Grid computing course portal and registration process for users built around the established GridSphere portal toolkit is described as representative of a production-level Grid computing portal. This is a starting point for users and for a hands-on Grid computing course. How the Grid computing infrastructure works behind the portal is described in subsequent chapters.

1.1 GRID COMPUTING CONCEPT

Grid computing uses distributed interconnected computers and resources collectively to achieve higher performance computing and resource sharing. It was developed in the mid-1990s with the growth of high-speed networks and the Internet that allowed distributed computer systems to be readily interconnected. Grid computing has become one of the most important techniques in high performance computing by providing resource sharing in science, technology, engineering, and business. By taking advantage of the Internet and high-speed networks, geographically distributed computers can be used collectively for collaborative problem solving. In Grid computing, different organizations can supply the resources and personnel, and the Grid infrastructure can cross organizational and institutional boundaries. This concept has many benefits, including:

- Problems that could not be solved previously for humanity because of limited computing resources can now be tackled. Examples include understanding the human genome and searching for new drugs.
- Interdisciplinary teams can be formed across different institutions and organizations to tackle problems that require the expertise of multiple disciplines.
- Specialized experimental equipment can be accessed remotely and collectively within a Grid infrastructure.
- Large collective databases can be created to hold vast amounts of data.
- Unused compute cycles can be harnessed at remote sites, achieving more efficient use of computers.
- Business processes can be re-implemented using Grid technology for dramatic cost saving.

Collaboration. Perhaps the most important and differentiating feature of Grid computing is the ability to conduct collaborative computing. Grid computing is about collaboration and resource sharing as much as it is about high performance computing. Certainly distributed computing existed before Grid computing as will be reviewed, but the easy prospect of developing teams of geographically distributed workers—a hallmark of Grid computing—became a reality with the development of the Internet. It is common practice to use the word Grid as a proper noun (i.e., G is capitalized) although that does not refer to a universal Grid. Without qualification, the word Grid refers to Grid computing infrastructures in general. There are many Grid infrastructures. The first letter of Web is capitalized although in that case, one could legitimately say that is one universal Web.

Although very high performance Grid projects employ their own dedicated high-speed interconnection networks, using the Internet to interconnect the distributed computers really makes Grid computing possible to all. The original driving force behind Grid computing was the same as behind the early development of networks that became the Internet—connecting computers at distributed sites for high performance computing. Grid computing came from the recognition that the Internet and Internet-type interconnections provide a unique opportunity for implementing a geographically distributing computing system. Some Grid projects involve computers spread across the globe, while others are more localized depending upon the goals of the project. For example, a project close to one extreme for vast geographical distances that I participated in is a Grid computing demonstration at the Supercomputing 2003 conference called the Global Data-Intensive Grid Collaboration in which 21 countries and hundreds of computer systems were represented. A project close to the opposite extreme in terms of interconnectivity is the VisualGrid project at UNC–Charlotte that involved forming a Grid at two sites, UNC–Charlotte and UNC–Asheville. These North Carolina universities are separated by about 130 miles.

The word "Grid" in Grid computing is often compared to the word "Grid" in national electrical grids that supply electricity across a country and make electricity immediately available at outlets. The vision in using this word is that Grid computing will make high performance computing as easy to access. Some companies, such as IBM, took on this vision in the early 2000s by offering *(Grid) on-demand computing*, that is, providing access to Grid computing resources when requested by clients and paid for by the clients when used. The term *utility computing* is also used for using Grid resources in a similar way as utilities such as electrical, gas, and water are metered. Customers expect no interruption in electrical, gas, and water service (except through acts of God) and similarly on-demand or utility computing should provide high resilience to faults and security attacks. There are Grid computing research projects that focus on accounting, quality of service, service level agreements, such as the GridBus project. Providing computing on demand for pay has led to *cloud computing* in which companies offer a "cloud" of services as a business model. However, cloud computing is regarded as a distinct approach that is separate from Grid computing. We shall come back to the distinction later.

Grid computing is a form of distributed computing and builds upon earlier concepts in distributed computing, but it also introduces new important aspects. As mentioned, the most important aspects are the involvement of collaborative teams and resource sharing. Grid computing often involves computers from multiple organizations and crosses organizational boundaries. The term *virtual organization* has been coined to describe groups of people, both geographically and organizationally distributed, working together on a problem, sharing computers and other resources such as databases and experimental equipment. Sometimes, the term virtual organization just refers to the people, but we will use the term to include the physical resources.

Crossing multiple administrative domains is another hallmark of larger Grid computing projects and introduces challenging technical and socio-political issues. The resources being shared are owned either by members of the virtual organization or donated by others. They may also be used by others outside the virtual organization. There could be limited times of availability for a virtual organization. There could be multiple virtual organizations. Grid networks such as SURAGrid and TeraGrid have been established to support multiple Grid projects. In such Grid infrastructures, resources are assigned to the Grid by their owners for the common good although obviously strict usage policies are usually in place. Multiple virtual organizations can be formed that could use a subset of the resources. The key aspect of a virtual organization is its formation for a specific purpose.

A distributed team may be involved in a scientific experiment requiring data collection. Not only are computers shared by the virtual organization formed for the project, but also the experimental equipment and the data that comes from the experimental equipment. Members of the team contribute to the overall mission of the project. A well-known example of a Grid computing project that involves experimental equipment and geographically distributed Grid is at CERN (Conseil Européen pour la Recherche Nucléaire), the European Center for Nuclear Research, which straggles the Swiss/French border near Geneva Switzerland. The work there has centered around the search for new fundamental particles and understanding the universe for many years using state-of-the-art particle colliders. The data from their experimental Large Hadron Collider facility is sent to researchers around the world for analysis and collective research using Grid technology. This project involves a massive amount of data created at ultra high speed (PBytes/sec).

An extremely important aspect of Grid computing that has emerged is the use of open agreed standards. It is critically important to have agreed standards so that software can be developed by different groups and interoperate and encourage widespread adoption. Grid computing uses standard Internet standards and technology such as HTTP and SOAP. However, standards for Grid computing have to go much further than just the network protocols. Agreements must be in place for many aspects of Grid computing including security, data management, resource discovery, job submission, etc. Grid computing builds upon existing distributed computing techniques so let us begin with the history of distributed computing.

1.2 HISTORY OF DISTRIBUTED COMPUTING

Certainly one can go back a long way to trace the history of distributed computing. Types of distributed computing existed in the 1960s. Many people were interested in connecting computers together for high performance computing in the 1970s and in particular forming multicomputer or multiprocessor systems (including the author). From connecting processors and computers together locally that began in earnest in the 1960s and 1970s, distributed computing now extends to connecting computers that are geographically distant.

The distributed computing technologies that underpin Grid computing were developed concurrently and rely upon each other. There are three concurrent interrelated paths:

- Networks
- Computing platforms
- Software techniques

Networks. Grid computing relies on high performance computer networks. The history of such networks began in the 1960s with the development of packet switched networks. The most important and ground-breaking geographically distributed packet-switched network was the DoD-funded ARPNET network with a design speed of 50 Kbits/sec. ARPNET became operational with four nodes (University of California at Los Angeles, Stanford Research Institute, University of California at Santa Barbara, and University of Utah) in 1969. TCP (Transmission Control Protocol) was conceived in 1974 and became TCP/IP (Transmission Control Protocol/Internet Protocol) in 1978. TCP/IP became universally adopted. TCP provided a protocol for reliable communication while IP provided for network routing. Important concepts including IP addresses to identify hosts on the Internet and ports that identify end points (processes) for communication purposes. The Ethernet was also developed in the early 1970s and became the principal way of interconnecting computers on local networks. It initially enabled multiple computers to share a single Ethernet cable and handled communication collisions with a retry protocol although nowadays this collision detection is usually not needed as separate Ethernet cables are used for each computer, with Ethernet switches to make connections. Each Ethernet interface has a unique physical address to identify it for communication purpose, which is mapped to the host IP address. More details of underlying network technologies (ports, network addresses, etc.) are given in Appendix A.

The Internet began to be formed in early 1980s using the TCP/IP protocol. During the 1980s, the Internet grew at a phenomenal rate. Networks continued to improve and became more pervasive throughout the world. In the 1990s, the Internet developed into the World-Wide Web. The browser and the HTML markup language was introduced. (Markup languages had been conceived earlier notably as a way of making documents machine-readable.) The global network enables computers to be interconnected virtually anywhere in the world.

Computing Platforms. Computing systems began as single processor systems. It was soon recognized that increased speed could potentially be obtained by having more than one processor inside a single computer system and the term *parallel computer* was coined to describe such systems. Parallel computers were limited to applications that required computers with the highest computational speed. It was also recognized that one could connect a collection of individual computer systems together quite easily to form a multicomputer system for higher performance. There were many projects in the 1970s and 1980s with this goal, especially with the advent of low cost microprocessors.

In the 1990s, it was recognized that commodity computers (PCs) provided the ideal cost-effective solution for constructing multicomputers, and the term *cluster computing* emerged. In cluster computing, a group of computers are connected through a network switch as illustrated in Figure 1.1. Specialized high-speed interconnections were developed for cluster computing. However, many chose to use commodity Ethernet as a cost-effective solution although Ethernet was not developed for cluster computing applications and incurs a higher latency. The term *Beowulf* cluster was coined to describe a cluster using off-the-shelf computers and other commodity components and software, named after the Beowulf project at the NASA

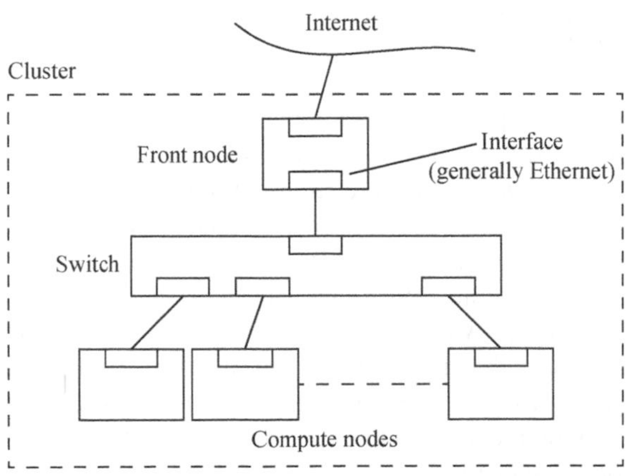

Figure 1.1 Typical cluster computing configuration.

Goodard Space Flight Center started in 1993 (Sterling 2002). The original Beowulf project used Intel 486 processors, the free Linux operating system, and dual 10 Mbits/sec Ethernet connections.

As clusters were being constructed, work was done on how to program them. The dominant programming paradigm for cluster computing was and still is message passing in which information is passed between processes running on the computers in the form of messages. These messages are specified by the programmer using message-passing routines. The most notable library of message-passing routines was PVM (Parallel Virtual Machine) (Sunderam 1990), which was started in the late 1980s and became the de facto standard in the early-mid 1990s. PVM included the implementation of message-passing routines. Subsequently, a standard definition for message passing libraries called MPI (Message Passing Interface) was established (Snir et al. 1998), which laid down what the routines do and how they are invoked but not the implementation. Several implementations were developed. Both PVM (now historical) and MPI routines could be called from C/C++ or Fortran programs for message passing and related activities.

Several projects began in the 1980s and 1990s to take advantage of networked computers in laboratories for high performance computing. A very important project in relation to Grid computing is called Condor, which started in the mid-1980s with the goal to harness "unused" cycles of networked computers for high performance computing. In the Condor project, a collection of computers could be given over to remote access automatically when they were not being used locally. The collection of computers (called a Condor pool) then formed a high-performance multicomputer. Multiple users could use such physically distributed computer systems. Some very important ideas were employed in Condor including matching the job with the available resources automatically using a description of the job and a description of the available resources. A job workflow could be described in which the output of one job could automatically be fed into another job. Condor has become mature and is widely used as a job scheduler for clusters in addition to its original purpose of using laboratory computers collectively. In Condor, the distributed computers need only be networked and could be geographically distributed. Condor can be used to share campus-wide computing resources. Chapter 3 considers Condor with other job schedulers in the light of Grid computing.

Software Techniques. Apart from the development of distributed computing platforms, software techniques were being developed to harness truly distributed systems. The *remote procedure call* (RPC) was conceived in the mid-1980s as a way of invoking a procedure on a remote computer, as an extension of executing procedures locally. The remote procedure call was subsequently developed into object-oriented versions in the 1990s; one was CORBA (Common Request Broker Architecture) and another was the Java Method Invocation (RMI). The remote procedure call introduced the important concept of a *service registry* to locate remote services. We shall describe service registries in relation to discovering services in a Grid computing environment in Chapter 6. This includes the mechanism of discovering their method of invocation.

During the early development of the World-Wide Web, the HyperText Markup Language (HTML) was conceived to provide a way of displaying Web pages and connecting to other pages through now very familiar hypertext links. Soon, a Web page became more than simply displaying information; it became an interactive tool whereby information could be entered and processed at either the client side or the server side. The programming language JavaScript was introduced in 1995, mostly for causing actions to take place specified in code at the client (client-side processing), whereas other technologies were being developed for causing actions to take place at the server (server-side processing) such as ASP first released in 1996.

In 2000, a very significant concept for distributed Internet-based computing called a *Web service* was introduced. Web services have their roots in remote procedure calls and provide remote actions but are invoked through standard protocols and Internet addressing. They also use XML (eXtensible Markup Language), which was also introduced in 2000. The Web service interface is defined in a language-neutral manner by the XML language WSDL. Web services were adopted into Grid computing soon after their introduction as a flexible interoperable way of implementing the Grid infrastructure and were potentially useful for Grid applications. Web services are described in detail in Chapter 6. Their application to Grid computing infrastructure is described in Chapter 7. XML is also used in job description languages in Chapter 2 and elsewhere. The XML markup language is covered in Appendix C.

Grid Computing. The first large-scale Grid computing demonstration that involved geographically distributed computers and the start of Grid computing proper was the Information Wide-Area Year (I-WAY) demonstration at the Supercomputing 1995 Conference (SC'95). Seventeen supercomputer sites were involved including five DOE supercomputer centers, four NSF supercomputer centers, three NASA supercomputer sites, and other large computing sites. Ten existing ATM networks were interconnected with the assistance of several major network service providers. Over 60 applications demonstrated in areas including astronomy and astrophysics, atmospheric science, biochemistry, molecular biology and structural biology, biological and medical imaging, chemistry, distributed computing, earth science, education, engineering, geometric modeling, material science, mathematics, microphysics and macrophysics, neuroscience, performance analysis, plasma physics, teleoperations/telepresence, and visualization (DeFanti 1996). One focus was on virtual reality environments. Virtual reality components included an immersive 3D environment (CAVE™ Automatic Virtual Environment). Separate papers in the 1996 special issue of *International Journal of Supercomputer Applications* described nine of the I-Way applications.

I-Way was perhaps the largest collection of networked computing resources ever assembled for such a significant demonstration purpose at that time. It explored many of the aspects now regarded as central to Grid computing, such as security, job submission and distributed resource scheduling. It came face-to-face with the "political and technical constraints" that made it infeasible to provide single scheduler (DeFanti 1996). Each site had its own job scheduler, which had to be married together. The I-Way project also marked the start of the Globus project (Globus

Project), which developed de facto software for Grid computing. The Globus Project is led by Ian Foster, a co-developer of the I-Way demonstration, and a founder of the Grid computing concept. The Globus Project developed a toolkit of middleware software components for Grid computing infrastructure including for basic job submission, security, and resource management. Globus has evolved through several implementation versions to the present time as standards have evolved although the basic structural components have remained essentially the same (security, data management, execution management, information services, and run time environment). We will describe Globus in a little more detail later.

Although the Globus software has been widely adopted and is the basis of the coursework described in this book, there are other software infrastructure projects. The Legion project also envisioned a distributed Grid computing environment. Legion was conceived in 1993 although work on the Legion software did not begin in 1996 (Legion WorldWide Virtual Computer). Legion used an object-based approach to Grid computing. Users could create objects in distant locations. The first public release of Legion was at the Supercomputing 97 conference in November 1997. The work led to the Grid computing company and software called Avaki in 1999. The company was subsequently taken over by Sybase Inc.

In the same period, a European Grid computing project called UNICORE (UNiform Interface to COmputing REsources) began, initially funded by the German Ministry for Education and Research (BMBF) and continued with other European funding. UNICORE is the basis of several of the European efforts in Grid computing and elsewhere, including in Japan. It has many similarities to Globus for example in its security model (X509 certificates and certificate authorities, see Chapter 5) and a service based OGSA standard (see Chapter 7) but is a more complete solution than Globus and includes a graphical interface. An example project using UNICORE is EUROGRID, a Grid computing testbed developed in the period of 2000–2004. (EUROGRID) A EUROGRID application project is OpenMolGRID Open Computing GRID for Molecular Science and Engineering developed during the period of 2002–2005 to "speed up, automatize, and standardize the drug-design using Grid technology" (OpenMolGRID).

The term *e-Science* was coined by John Taylor, the Director General of the United Kingdom's Office of Science and Technology, in 1999 to describe conducting scientific research using distributed networks and resources of a Grid computing infrastructure. Another more recent European term is *e-Infrastructure*, which refers to creating a Grid-like research infrastructure.

With the development of Grid computing tools such as Globus and UNICORE, a growing number of Grid projects began to develop applications. Originally, these focused on computational applications. They can be categorized as:

- Computationally intensive
- Data intensive
- Experimental collaborative projects

The computationally intensive category is traditional high performance computing addressing large problems. Sometimes, it is not necessarily one big problem but a problem that has to be solved repeatedly with different parameters (parameter sweep problems) to get to the solution. The data intensive category includes computational problems but with the emphasis on large amounts of data to store and process. Experimental collaborative projects often require collecting data from experimental apparatus and very large amounts of data to study.

The potential of Grid computing was soon recognized by the business community for so-called *e-Business* applications to improve business models and practices, sharing corporate computing resources and databases and commercialization of the technology for business applications. For e-Business applications, the driving motive was reduction of costs whereas for e-Science applications, the driving motive was obtaining research results. That is not to say cost was not a factor in e-Science Grid computing. Large-scale research has very high costs and Grid computing offers distributed efforts and cost sharing of resources. There are projects that are concerned with accounting such as GridBus mentioned earlier.

Figure 1.2 shows the time lines for computing platforms, underlying software techniques, and networks discussed. Some see Grid computing as an extension of cluster computing and it is true in the development of high performance computing, Grid computing has followed on from cluster computing in connecting computers together to form a multicomputer platform but Grid computing offers much more. The term *cluster computing* is limited to using computers that are interconnected locally to form a computing resource. Programming is done mostly using explicit message passing. Grid computing involves geographically distributed sites and invokes some different techniques. There is certainly a fine line in the continuum of interconnected computers from locally interconnected computers in a small room, through interconnected systems in a large computer room, then in multiple rooms and in different departments within a company, through to computers interconnected on the Internet in one area, in one country and across the world. The early hype of Grid computing and marketing ploys in the late 1990s and early 2000s caused some to call configurations Grid computing when they were just large computational clusters or they were laboratory computers whose idle cycles are being used.

One classification that embodies the collaborative feature of Grid computing is:

- Enterprise Grids — Grids formed within an organization for collaboration
- Partner Grids — Grids set up between collaborative organizations or institutions

Enterprise Grid still might cross administrative domains of departments and requires departments to share their resources. Some of the key features that are indicative of Grid computing are:

- Shared multi-owner computing resources
- Used Grid computing software such as Globus, with security and cross-management mechanisms in place

Grid computing software such as Globus provides the tools for individuals and teams to use geographically distributed computers owned by others collectively.

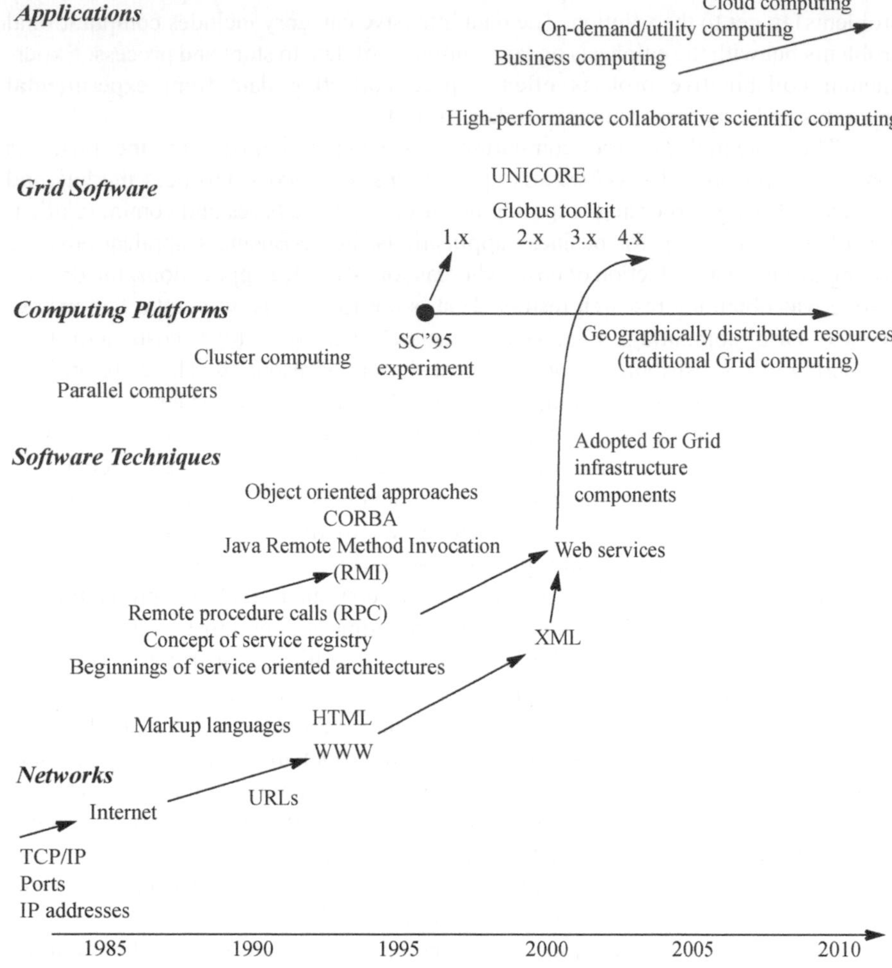

Figure 1.2 Key concepts in the history of Grid computing.

Foster's Check List. Ian Foster is credited for the development of Grid computing, and sometimes called the father of Grid computing. He proposed a simple checklist of aspects that are common to most true Grids (Foster 2002):

- No centralized control
- Standard open protocols
- Non-trivial quality of service (QoS)

Grid Computing verse Cluster Computing. It is important not to think of Grid computing simply as a large cluster because the potential and challenges are different. Courses on Grid computing and on cluster computing are quite different. In cluster computing, one learns about message-passing programming using tools such as MPI. Also shared memory programming is considered using threads and OpenMP, given that most computers in a cluster today are now also multicore shared memory systems. In cluster computing, network security is not a big issue that usually concerns the user directly. Usually an `ssh` connection to the front-end code of cluster is sufficient. The internal compute nodes are reached from there. Clusters are usually Linux clusters and in those often an NFS (Network File System) shared file system installed across the compute resources. Accounts need to be present on all systems in the cluster and it may be that NIS (Network Information System) is used to provide consistent configuration information on all systems, but not necessary so. NIS can increase the local network traffic and slow the start of applications. (A description on setting up clusters can be found in Sterling 2002.)

In Grid computing, one looks at how to manage and use the geographically distributed sites (distributed resources). Users need accounts on all resources but generally a shared file system is not present. Each site is typically a high performance cluster. Being a distributed environment, one looks at distributing computing techniques such as Web services and Internet protocols and network security as well as how to actually take advantage of the distributed resource. Security is very important because the project may use confidential information and the distributed nature of the environment opens up a much higher probability of a security breach.

Of course, there are things in common with both Grid computing and cluster computing. Both involve using multiple compute resources collectively. Both require job schedulers to place jobs onto the best platform. In cluster computing, a single job scheduler will allocate jobs onto the local compute resources. In Grid computing, a Grid computing scheduler has to manage the geographically disturbed resources owned by others and typically interacts with local cluster job schedulers found on local clusters.

Grid Computing versus Cloud Computing. Commercialization of Grid computing is driven by a business model that will make profits. The first widely publicized attempt was on-demand and utility computing in the early 2000s, which attempted to sell computer time on a Grid platform constructed using Grid technologies such as Globus. More recently, *cloud computing* is a business model in which services are provided on servers that can be accessed through the Internet.

The common thread between Grid computing and cloud computing is the use of the Internet to access the resources. Cloud computing is driven by the widespread access that the Internet and Internet technologies provide. However, cloud computing is quite distinct from the original purpose of Grid computing. Whereas Grid computing focuses on collaborative and distributed shared resources, cloud computing concentrates upon placing resources for paying users to access and share. The technology for cloud computing emphasizes the use of services (*software as a service*, SaaS), and possibly the use of *virtualization* (the process of separating the particular user's software environment from the underlying hardware, providing users with their own environment abstracted from the hardware). Cloud computing is illustrated in Figure 1.3.

A number of companies entered the cloud computing space in the mid-late 2000s. IBM was an early promoter of on-demand Grid computing in the early 2000s and moved into cloud computing in a significant way, opening a cloud computing

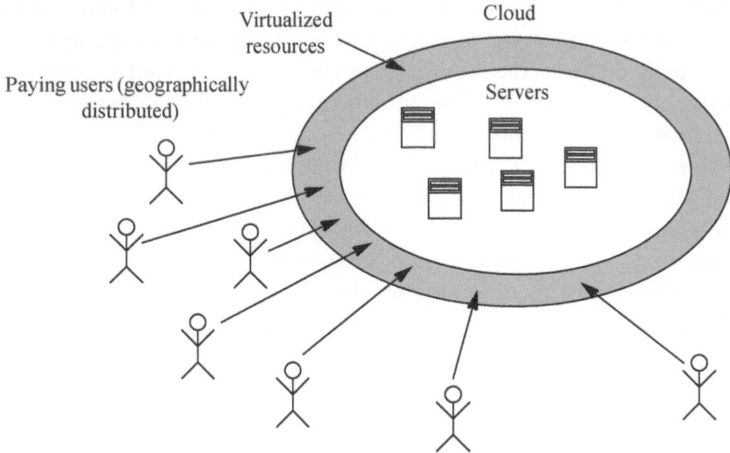

Figure 1.3 Cloud computing using virtualized resources.

center in Ireland in March 2008 (Dublin), and subsequently in the Netherlands (Amsterdam), China (Beijing), and South Africa (Johannesburg) in June 2008. Other major cloud computing players include Amazon and Google who utilize their massive number of servers. Amazon has the Amazon Elastic Compute Cloud (Amazon E2) project for users to buy time and resources through Web services and virtualization. The cloud computing business model is one step further than hosting companies simply renting servers they provide at their location, which became popular in the early-mid 2000s with many start-up companies and continues to date.

1.3 COMPUTATIONAL GRID APPLICATIONS

By the early 2000s, there were a very large number of computational Grid projects around the world, many more than can be described here. They span all science and engineering disciplines. There are Grid projects in areas such as:

- Biomedical research
- Industrial research
- Engineering research (electrical, mechanical, ...)
- High energy physics
- Chemistry
- Medical
- Bioinformatics

A representative large-scale Grid computing project involving experimental equipment mentioned earlier is the Large Hadron Collider experimental facility constructed at CERN for research into particle physics. As an aside, the CERN scientist Tim Berners-Lee invented the Web in 1990 driven by the desire for information sharing between scientists across the world. Another representative large-scale Grid computing project involving experimental equipment is the NSF Network for Earthquake Engineering Simulation (NEES) for research into reducing the effects of earthquakes. This large-scale collaborative project brought together experimental networked facilities across the US interconnected using very high-speed Internet2 connections, funded by the NSF for ten years (2004–2014), and involves 15 universities. Sites were provided with new or enhanced experimental apparatus for research into earthquakes (e.g., shaking tables, tsunami wave basin, geotechnical centrifuges), and included field monitoring. The work included the unusual aspect of merging the measurements from physical apparatus such as a shaking table at one site with computing models performed elsewhere to study the effects of earthquakes where the shaking table or computer model alone cannot represent the complete environment. More information on this project can be found at the NEES home page.

Another project focusing upon earth sciences is the Earth System Grid (ESG) for climate modeling and research. This project is concerned with providing the climate research community access to climate data and as such can be classified as a data-intensive grid application. The project was funded by US Department of Energy. The first phase of this project (Earth System Grid I) in the 2000–2001 period was pilot project, which was continued with a follow-up 5-year project called Earth System Grid II project (2001–2006). It involved Argonne National Laboratory, Lawrence Berkeley National Laboratory, Los Alamos National Laboratory, National Center for Atmospheric Research, Oak Ridge National Laboratory, and University of South California (Earth System Grid). The ESG II data Grid connected various very large climate databases and developed Grid computing solutions to access to these databases to enable researchers to perform climate research. The Grid technologies including identifying, transferring at very high speed or replicating where appropriate very large amounts of data, security aspects, and ease of access through a Grid portal. These Grid technologies are discussed later in the book. The project managed terabytes of data, in a discipline where the amount of data is growing at a very fast pace with the data coming from Earth observing satellites and other sources. Details of the project can be found in the ESG II final report (Earth System Grid II).

A Grid computing project in the medical area is the UK eDiaMoND project conducted over the period of 2002–2005 (EDiaMoNd Grid computing project). The objective of this project was to build a national database of mammographic images to aid screening and diagnosis of breast cancer. The project could be categorized as data-intensive in that medical images are stored and transferred to sites, but as most data-intensive Grid projects, it also includes experimental equipment for obtaining the data (images in this case). The collaborating organizations were Churchill Hospital Oxford, Edinburgh Ardmillan Hospital, Guy's Hospital, London, IBM, King's College London, Mirada Solutions, Oxford e-Science Centre, School of Informatics (University of Edinburgh), St Georges Hospital, University College London, and University of Oxford. This project has computing resources, data storage, and experimental equipment. Aspects of this project, in common with other medical Grid computing projects involving human data, are legal and ethical considerations of maintaining confidentiality of personnel records.

There are many more Grid projects. A key aspect of many of the computational Grid projects is:

- State-of-the-art interconnection networks
- Sharing resources
- Community of scientists

Resources can be much more than just computers. They can be:

- Storage
- Sensors for experiments at particular sites in the Grid
- Application software
- Databases

Let us reiterate that Grid computing is about collaborating and resource sharing as much as it is about high performance computing.

1.4 GRID COMPUTING INFRASTRUCTURE DEVELOPMENT

Grid computing projects can create their own Grid network for their particular application. But just as the Internet provides a universal communication platform, Grid networks can be created to support many projects (present and future). The Grid network then is not tied to a particular project or application. Such Grid networks have been set up at a local level, national level, and international level throughout the world to promote scientific Grid computing research.

1.4.1 Large-Scale U.S. Grids

One of the most notable high-speed networks constructed for Grid computing is the TeraGrid, which was funded by the NSF in 2001 initially to link supercomputer centers. Hubs were established at Chicago and Los Angeles and interconnected using a 40 Gigabit/sec optical backplane network. The five centers were connected to one of the hubs using 30 Gigabit/sec connections:

- Argonne National Laboratory (ANL) (Chicago hub)
- National Center for Supercomputing Applications (NCSA) (Chicago hub)
- Pittsburgh Supercomputing Center (PSC) (Chicago hub)
- San Diego Supercomputer Center (SDSC) (LA hub)
- California Institute of Technology (LA hub)

State-of-the-art optical lines could reach 10 Gigabit/sec in the early 2000s, and four such lines were used to achieve 40 Gigabit/sec. Three lines were used to achieve 30 Gigabit/sec.

TeraGrid was further funded by the NSF for the period of 2005–2010 and has developed into a platform for a wide range of Grid applications and is described as "the world's largest, most comprehensive distributed cyberinfrastructure for open scientific research" (TeraGrid). As of 2008, TeraGrid consists of eleven resource provider sites, each with very significant compute data storage resources: Indiana University, the Louisiana Optical Network Initiative, the National Center for Atmospheric Research, the National Center for Supercomputing Applications, the National Institute for Computational Sciences, Oak Ridge National Laboratory, Pittsburgh Supercomputing Center, Purdue University, San Diego Supercomputer Center, Texas Advanced Computing Center, and the University of Chicago/Argonne National Laboratory, up from nine resource provider sites in 2007. As of 2008, there were hubs at Chicago, Denver, Los Angeles, and Atlanta. Resource providers make connections to a hub with 10+ Gigabit/sec connections.

TeraGrid provides an open access for scientific research. Users make requests for an allocation reminiscent to getting access on a single supercomputer in the 1980s and 1990s, only now the users have access to a broad range of extremely powerful computer systems.

Open Science Grid (OSG) is another large-scale Grid computing infrastructure initiative, funded by the NSF and (US) Department of Energy's Office of Science (Open Science Grid). There is a very large number of participants in OSG, too long to list here. Consortium members have interests in particle and nuclear physics, astrophysics, bioinformatics, gravitational-wave science, and Computer Science aspects of Grid computing. Many virtual organizations exist across the OSG. In addition, a general-purpose virtual organization group is provided for those who want to use OSG for individual or small group research. New members can contribute resources to OSG, spreading out the size of the Grid infrastructure. OSG provides "Grid schools" for Grid computing education and training.

SURAGrid is another example of an expansive Grid. The Southeastern Universities Research Association (SURA) established the SURAGrid as a collaborative venture between universities to provide a shared Grid computing facility. Figure 1.4 shows participants as of 2009. The number of participants has been growing steadily over the last few years, mainly in the southeast. SURAGrid is not focused on any application domain and has applications that include storm surge modeling, multiple genome alignment, simulation and optimization for threat management in urban water systems, bio-electric simulator for whole body tissue, dynamic BLAST, and petroleum simulation. It also has interests in Grid computing education.

For infrastructure projects such as OSG and SURAGrid that can take on new members working on new and distinct projects, the joining mechanism and governance policies must be easy and uniform. New members need to know the required software they need (the software stack) and the hardware they might contribute. Obtaining and installing the software should be made easy. Grid computing software is notoriously difficult to maintain because of its immaturity. Accounts need to be organized in an efficient manner. Accounts usually need to be provided on every resource a user wishes to access, either individual accounts or a group account. Infrastructure projects such as SURAGrid provide a centralized database of information to simplify account set-up. Security in all Grid projects using the Internet typically use Internet security mechanisms (certificates, etc.). Such security mechanisms require certificate authorities. It is still an open research problem the best way to organize a dynamically growing Grid infrastructure with multiple organizations.

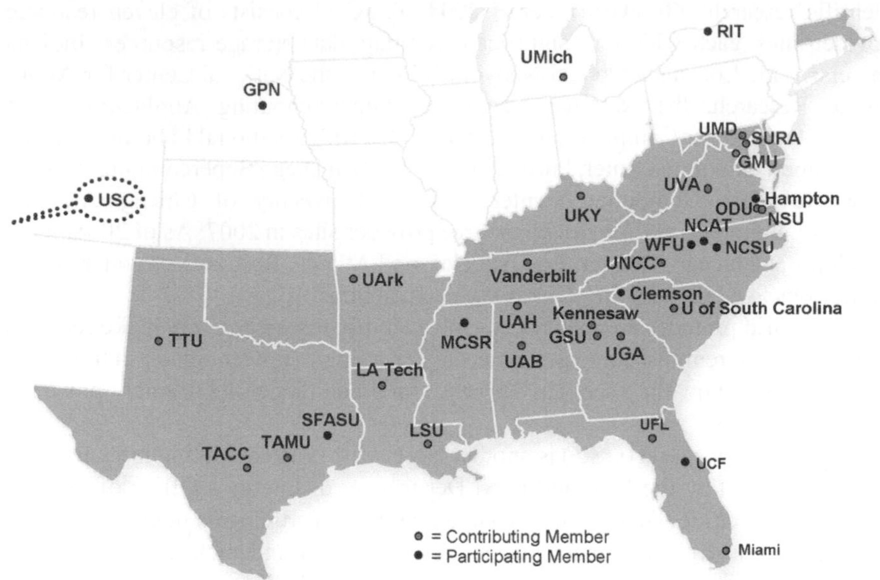

Figure 1.4 SURAGrid (Courtesy of Southeastern Universities Research Association).

SURAGrid maintains a central bridge certificate authority that is cross-certified with certificate authorities of each participating institution to provide signing authorities for users and resources. (More on security in Chapters 4 and 5.) Generally, access to Grid resources is through a Grid portal (a Web-based user interface), and this portal will display the current status of available resources. This requires all resources to communicate with the portal behind the scenes. (Chapter 8 discusses Grid portals in some detail.)

Grid infrastructures mostly use existing high-speed networks, which are being deployed everywhere for high-speed data communications and the Internet. Many states in the US have deployed state-of-the-art high-speed fiber optic networks that can be the basis for high performance Grid computing. Examples include the LONI network (Louisiana Optical Network Initiative), the Florida Lamdba Rail, and the North Carolina NCREN network, each connecting major sites across their state. LONI boasts a 870 Gigabit/sec aggregate transport capacity with connections operating between 20 Gigabit/sec and 60 Gigabit/sec in 2006.

1.4.2 National Grids

Many countries have initiated national Grid computing projects around their high-speed networks. The UK e-Science program began in November 2000 with £98 million 3-year funding for then new UK e-Science program. Funding quickly increased to £120 million, with £75 million devoted to large-scale pilot projects in science and engineering and £35 million for the so-called "core e-Science program" that focused on Grid middleware in collaboration with industry. This led to the formation of the UK e-Science Grid. Nine e-Science centers were created across the country. These centers (Southampton, London, Cardiff, Oxford, Cambridge, Manchester, Newcastle, Edinburgh/Glasgow, and Belfast) and a couple of other sites/laboratories (a site in Hinxton, and the Rutherford and Daresbury Laboratory) were connected to form original UK e-Science Grid. The project was described in a paper by Hey and Trefethen (Hey and Trefethen 2002). The network used the existing UK university network, which was upgraded to 10 Gigabit/sec by 2002. Later, seven "centers of excellence" were added to the nine regional Grid centers. The purpose of the centers of excellence was to add regional presence and add expertise and applications. Funding over the five years 2001–2006 was quoted at £250 million for more than 100 projects (highlights from the UK e-Science program). A feature of the UK e-Science program was the use of a single certificate authority for issuing certificates. This certainly greatly simplifies the process of issuing signed certificates. Such an approach has not been adopted in the US, partly because it does not scale well and partly for socio-political reasons.

A follow-up UK activity to the original UK e-Science Grid was the establishment of UK National Grid Service founded in 2004 to provide distributed access to computational and database resources, with four core sites, the universities of Manchester, Oxford and Leeds, and Rutherford Appleton Laboratory. By 2008, it had grown to 16 sites. Access is free to any academic with a legitimate need. (This is a common feature of scientific Grid infrastructure projects.)

During the period of 2000–2005, many other countries have seen the need for a national Grid. Other national Grid networks include Grid-Ireland, Scandinavian NorduGrid, DutchGrid, Polish PIONIER, and French ACI. A national Grid provides collective national computing resources to address major scientific and engineering problems and also problems of national interest such as studying or predicting earthquakes, storms, major environmental disasters, global warming, and terrorism.

1.4.3 Multi-National Grids

Also in the period of 2000–2005, several efforts were started to create Grids that spanned across many countries. There have been several initiatives for European countries to collaborate in forming Grid-like infrastructures to share compute resources funded by European programs. For example, DEISA (Distributed European Infrastructure for Supercomputing Applications) is a project to connect major supercomputing facilities across Europe. DEISA has the unique aspect of providing a global file system that eliminates the need for users to move input files to the location of the executable and moving output files back explicitly. The DEISA-1 project spanned from 2004–2008. DEISA-2 started in 2008 with funding of 12,237,000 EURO (DEISA) to continue to 2011, with the partners: Barcelona Supercomputing Centre Spain (BSC), Consortio Interuniversitario per il Calcolo Automatico Italy (CINECA), Finnish Information Technology Centre for Science Finland (CSC), University of Edinburgh and CCLRC UK (EPCC), European Centre for Medium-Range Weather Forecast UK (ECMWF), Research Centre Juelich Germany (FZJ), High Performance Computing Centre Stuttgart Germany (HLRS), Institut du Développement et des Ressources en Informatique Scientifique, CNRS France (IDRIS), Leibniz Rechenzentrum Munich Germany (LRZ), Rechenzentrum

Garching of the Max Planck Society Germany (RZG), Dutch National High Performance Computing Netherlands (SARA), Kungliga Tekniska Högskolan Sweden (KTH), Swiss National Supercomputing Centre Switzerland (CSCS), Joint Supercomputer Center of the Russian Academy of Sciences Russia (JSCC).

An example in the Asia Pacific region is ApGrid, a partnership for Grid computing in the Asia Pacific region involved Australia, Canada, China, Hong Kong, India, Japan, Malaysia, New Zealand, Philippines, Singapore, South Korea, Taiwan, Thailand, USA, and Vietnam.

The vision of a single universal international Grid in the fashion of the World Wide Web—a World Wide Grid—may never be achieved though. More likely is that Grids will connect to other Grids but will maintain their identity.

1.4.4 Campus Grids

Educational institutions also recognized the advantages of Grid technology in the early 2000s for cost-effectively sharing computing resources. Several institutions set up campus-wide Grids. Example include the University of Houston Campus Grid, Oxford University campus Grid (OxGrid), University of Texas at Austin Grid (UTGRid), University of Michigan (MGrid), University of Florida campus research Grid, and University of Virginia Campus Grid. Since most campuses nowadays provide wireless connectivity, a wireless Grid is possible for harnessing wireless resources. Note that forming a Grid on a campus faces political hurdles if the resources are not originally under one control. Resources may be controlled by individual faculty perhaps through research grants for particular purposes. Everybody has to agree in the collective advantages of sharing resources. Some campus Grids only embody those computing resources that are collectively controlled by the university's computing services.

1.5 GRID COMPUTING COURSES

Grid computing first appeared in graduate Computer Science curriculum in the late 1990s and early 2000s as special topics courses within a single university—usually at those universities that already had Grid computing research projects. Very few, if any, of these early courses involved more than one university in their teaching. The Grid computing platform used was also often limited. Grid computing entered the Computer Science undergraduate curriculum with the introduction of Grid computing courses in the 2003–2004 period. One of the first advanced undergraduate courses was a collaborative project between SUNY at Buffalo and SUNY College at Geneseo, which was funded by NSF in 2003 (Ramamurthy and Jayaraman 2003). NSF funded two further independent projects on undergraduate computing course development in 2004. In one project, investigators at the University of Arkansas and at Lewis and Clark College developed a collaborative undergraduate Grid computing course (Apon and Mache 2004). In the other project—our first Grid computing course project mentioned in the preface as the basis of this book—investigators at

UNC–Charlotte, Western Carolina University, and UNC–Wilmington developed a collaborative undergraduate Grid computing course.

Our Grid computing course was taught across the state of North Carolina starting in 2004, and repeated in 2005, 2007, and 2008 (Wilkinson and Ferner 2006, 2008). This course was one of the first at the senior undergraduate level to use both geographically distributed computers and geographically distributed students. A total of fifteen institutions across North Carolina participated in the offerings between Fall 2004 and Fall 2008. Computer systems were set up at different universities depending upon the actual participants. For example in Fall 2005, clusters were used at five sites—Appalachian State University, NC State University, UNC–Asheville, UNC–Charlotte, UNC–Wilmington, and Western Carolina University. The Microelectronics Center of North Carolina provided a cluster as a back-up facility in case there were problems with the other systems, but it was not actually needed. Each site had its own certificate authority for signing certificates of users at that site and those of users at sites without contributing computer systems.

The early undergraduate Grid computing courses in the period 2003–2006 generally took a bottom-up approach to Grid computing education starting with network protocols, client-server concepts, creating Web and Grid services, and then progressing through the underlying Grid computing middleware, security mechanisms, and job submission to a Grid platform, all using Linux a command-line interface. In 2007, our course moved away from a Linux command-line interface somewhat although for some tasks, a user/programmer might still require a command-line interface. While command-line interfaces are still used to access Grid resources in some Grid courses, it is more desirable to have a Web-based graphical user interface. The principal such interface is the Grid portal for logging on at one place and immediately having access to the distributed resources. Whether a command-line interface or a portal, the user has *single sign-on* for all Grid resources, that is, once logged in with a password, it is unnecessary to supply any passwords subsequently to reach any Grid resources, local or distant. (The actual way this is done will be described in Chapters 4 and 5 on Security.) In the next section, a typical user interface will be described, continuing with the underlying mechanisms in later chapters.

1.6 GRID COMPUTING SOFTWARE INTERFACE

The primary objective of Grid computing infrastructure (middleware) software is to make a seamless environment for users to be able to access and use distributed resources. Key aspects include:

- Secure envelope over all transactions.
- Single sign-on — being able with a single log on procedure to access all available resources and run jobs without having to supply additional passwords or account information.
- Data management tools.

- Information services providing characteristics of resources and their status (including dynamic load), and job status.
- Application programming interfaces (APIs) and services that enable applications themselves to take advantage of the Grid platform.
- Convenient user interface.

Globus. One of the most influential Grid middleware projects is the Globus project mentioned earlier. Globus has gone though several versions and bug fixes as illustrated in Figure 1.5. Version 1.0.0 was released in 1998 and was subsequently updated with version 2.0 released in early 2002. Version 2 became widely adopted, especially versions 2.2 and 2.3, which were released later.

Foster, Kesselman, Nick, and Tuecke introduced an overall Grid architecture called OGSA (Open Grid Service Architecture) in 2002, which called for a service approach to Grid components. The first Globus implementation of this architecture was Globus version 3. Version 3 had a short life because of the way these services were implemented using the now defunct OGSI (Open Grid Service Infrastructure) standard. Version 4.0.0 released in April 2005, after a number of pre-releases (3.9.x) from May 2004 onwards, implemented a new standard called WSRF (Web Service Resource Framework), which was more widely accepted. Subsequent releases include version 4.1.0 in June 2006, 4.1.1 in March 2007 and version 4.2.0 in July 2008. The Globus versioning scheme is <major>.<minor>.<incremental> where <major> is a complete redesign, <minor> might possibly include changes to APIs although infrequent, and <incremental> are generally bug fixes. Stable releases have even minor numbers.

The Globus toolkit has five major parts:

- Security — Components to provide a security envelope and secure access
- Information — Monitoring and discovery of resources and services
- Data management — Access and transfer of data

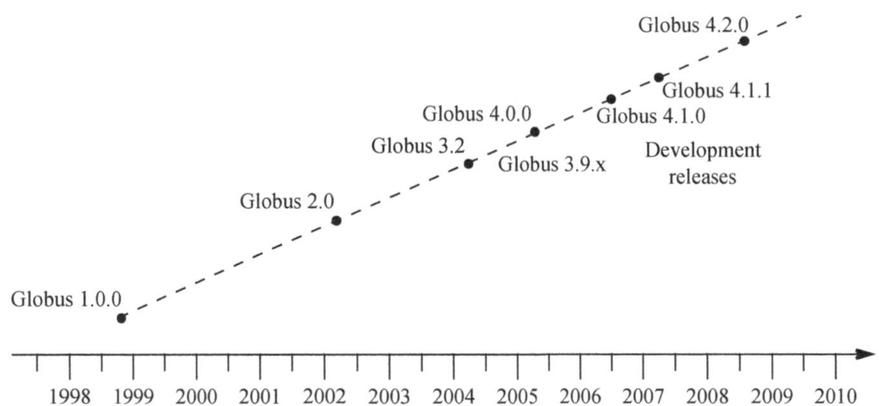

Figure 1.5 Some Globus toolkit versions (approximate time line).

- Execution management — Executing, monitoring and management of jobs
- Common run time — Libraries and core services

We shall fully describe the components of Globus and how it developed in later chapters, but let us introduce some central components here.

Security. First security is required. The distributed resources must be protected from unauthorized access. The Globus components for creating the security envelope is called GSI (Grid Security Infrastructure), which used public key cryptography. It requires each user to be authenticated (their identity vouched), which is done by each user having a digital certificate "signed" by a trusted certificate authority in a manner that is analogous to a passport or driving license being identified as true with the appropriate marks placed on it by the issuing authority. This technique is the basis of Internet security. Users will also need to be able to give their authority to Grid components to act on their behalf. This aspect is addressed in Grid computing by having special certificates called *proxy certificates,* which give resources holding them authority to act upon the issuer's behalf in a chain of trust. This is equivalent to one giving a proxy to another to vote on their behalf at a meeting. Exactly how Grid security is implemented is described in detail in Chapters 4 and 5.

Information. Next, the user often needs to know information about the available Grid resources. The basic Globus component for this is called MDS (Monitoring and Discovery System) historically, or simply information services. The users might access MDS to discover the status of the compute resources. Resource discovery is still very primitive and in the research domain, but the ideal is to be able to submit a job and the system find the best resources for that job based upon the job description and resource descriptions across the whole Grid. A Grid portal often interacts with MDS and other information services to display the current state of remote resources. It is also possible to use Globus APIs and higher-level APIs within an application to make decisions based upon the availability of resources. Chapter 9 on Grid-enabling applications touches upon this although its full treatment is beyond the scope of this book.

Executing a Job. Next, the user typically would want to submit a job. The basic Globus component for running a job is GRAM (Globus or Grid Resource Allocation Management). It may be necessary beforehand to transfer files to the resources and afterwards to transfer files to other locations including back to the user. The user might use the data management component called GridFTP for that.

The above activities are illustrated in Figure 1.6. It is important to note that Globus is a "toolkit" of components and not a complete solution for Grid computing infrastructure nor was it ever intended to be. Other higher-level components are needed in a sophisticated Grid computing infrastructure. Issues not addressed in the basic Globus toolkit include account management, job scheduling, and advanced features of security across multiple domains. Job scheduling either uses existing local job schedules such as Condor or higher-level global Grid meta-schedulers in concert with local schedulers. Chapter 3 addresses job schedulers.

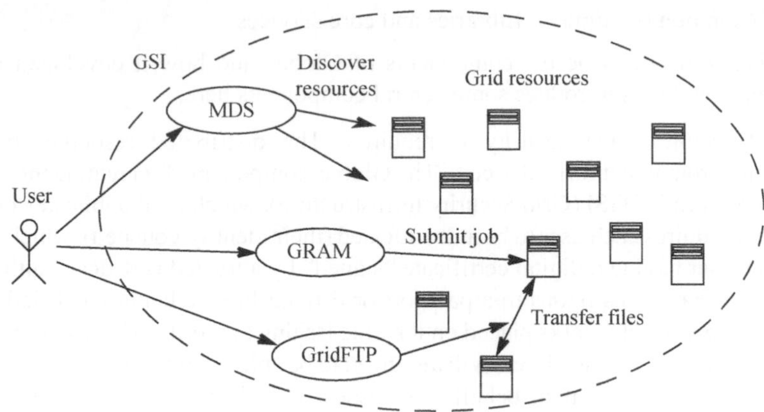

Figure 1.6 User employing Globus services and facilities.

User Interfaces. Grid computing environments are mostly Linux-based and originally accessed through the command line. Once you have established your security credentials (Chapters 4 and 5), to run a job you might issue the GRAM command:

```
globusrun-ws -submit -c prog1
```

where `prog1` is the executable of the job. The executable needs to be present on the compute resource that is to execute it. This particular command does not specify the compute resource and hence the computer executing the `globusrun-ws` command will execute the program `prog1`. If needed, transferring files to compute resources could be done with the GridFTP command such as:[1]

```
globus-url-copy  \
     gsiftp://www.coitgrid02.uncc.edu/~abw/prog1out  \
     file:///home/abw/
```

The first argument of `globus-url-copy` is the source location and the second argument is the destination location. In the above case, the file `www.coit-grid02.uncc.edu/~abw/prog1out` is transferred to `home/abw/` on the local computer. (Transferring files can be also done by specifying the file transfer in a job description language document, as described in Chapter 3.)

As one can see, the command-line interface is a very primitive way of interacting with the Grid resources. A more desirable way is to have a Web-based interface called a Grid portal or gateway. Perhaps the most successful Grid portal toolkit project is the GridSphere portal project, which is described in detail in Chapter 8 with

[1] A backslash (\) indicates the command continues on the next line. If used in practice, it must immediately precede a newline character. See Appendix B for more details on the Linux command-line interface.

other graphical user interfaces. The Grid portal used for the UNC–Charlotte/UNC–Wilmington course is based upon the GridSphere Grid portal toolkit. The login page is shown in Figure 1.7. This Grid portal is hosted on the server coit-grid02.uncc.edu:8080/gridsphere, which can be reached from anywhere on the Internet. GridSphere adheres to the JSR 168 portlet standard and can interface to the de facto standard Globus toolkit. It allows customized portlets to be created and deployed within the portal. Portlets are software components with an associated display area within the portal. Customized portlets can be created as a front-end to Grid-enabled applications. A tab on a GridSphere portlet will select a window within which there could be one or more portlet areas. The layout of a portlet is defined using HTML and JSP (Java Server Pages) or similar technologies. Creating customized portals is described in Chapter 8.

Before users can log on, they need a user name and password for the portal. Before they do anything on a Grid platform, they must have user "credentials" and accounts on the resources they wish to access. In our course portal, the PURSe (Portal-based User Registration Service) portlet is incorporated into the portal to facilitate the user setup procedures. It can be reached by selecting the "Register" tab from the main course portal page (see Figure 1.7). Figure 1.8 shows the PURSe registration portlet once the "Register" tab is selected. The user then submits the required information (name, email address, institution, etc.) This information is then forwarded to the Grid system administrator to set up accounts and credentials. A series of exchanges occur with the user by email confirming their intentions as shown in Figure 1.9. Note that communication is required with system administrators of remote resources. It is difficult to automate the process fully without communication between the user and administrators because, apart from the technical matters that need to be set up, approval is needed to use resources owned by others. A number of software projects and tools have focused on the very important matter of account management. However, account management is still often a human-centered process.

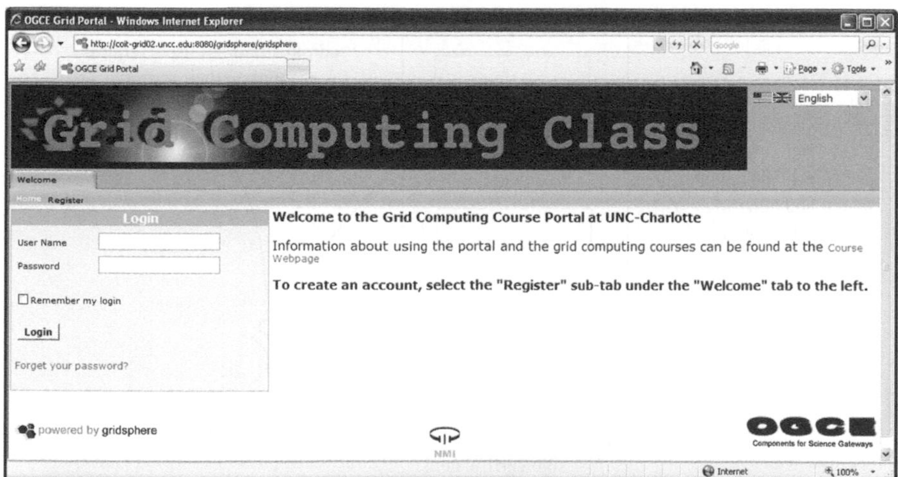

Figure 1.7 UNC–Charlotte/UNC–Wilmington Grid computing course portal (GridSphere).

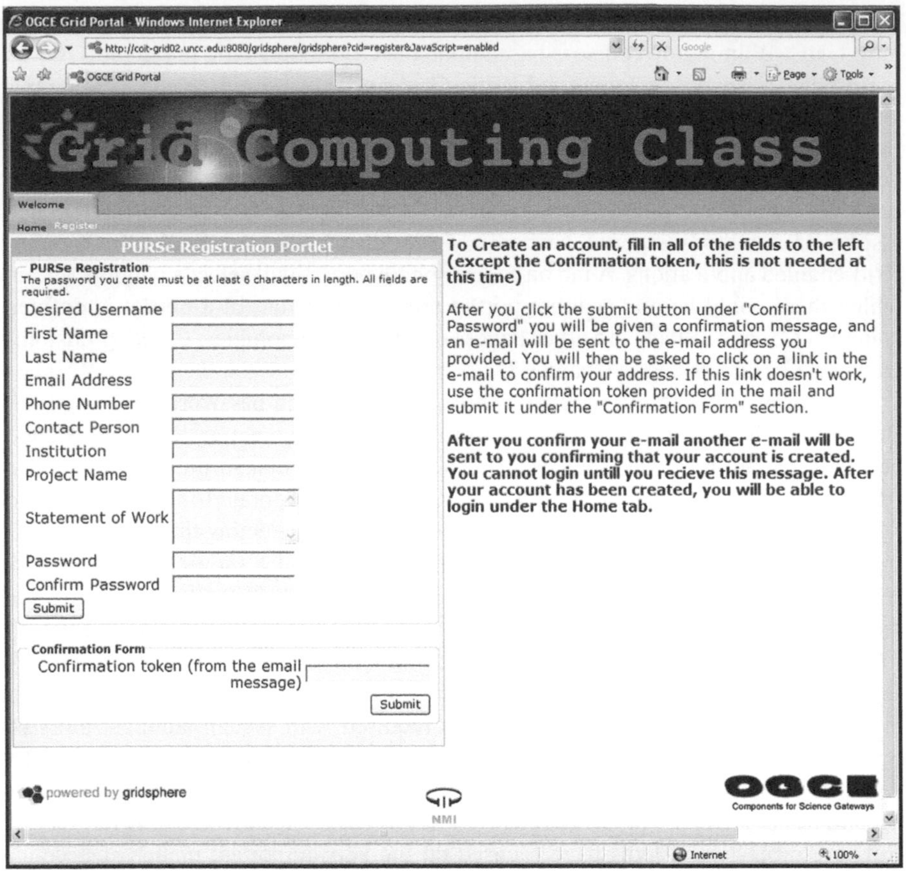

OGCE Grid Portal - Windows Internet Explorer
http://coit-grid02.uncc.edu:8080/gridsphere/gridsphere?cid=register&JavaScript=enabled
OGCE Grid Portal
Grid Computing Class
Welcome
Home Register
PURSe Registration Portlet
PURSe Registration
The password you create must be at least 6 characters in length. All fields are required.
Desired Username
First Name
Last Name
Email Address
Phone Number
Contact Person
Institution
Project Name
Statement of Work
Password
Confirm Password
Submit
Confirmation Form
Confirmation token (from the email message)
Submit
To Create an account, fill in all of the fields to the left (except the Confirmation token, this is not needed at this time)
After you click the submit button under "Confirm Password" you will be given a confirmation message, and an e-mail will be sent to the e-mail address you provided. You will then be asked to click on a link in the e-mail to confirm your address. If this link doesn't work, use the confirmation token provided in the mail and submit it under the "Confirmation Form" section.
After you confirm your e-mail another e-mail will be sent to you confirming that your account is created. You cannot login untill you recieve this message. After your account has been created, you will be able to login under the Home tab.
powered by gridsphere
NMI
OGCE
Components for Science Gateways
Internet
100%

Figure 1.8 PURSe registration portlet.

Registration portlet
Contact other Grid administrators if users request account on their resource
Fill in form – provide password and other information
Email confirmation of request, and subsequent information
New user
System administrator
Create accounts, set access control, sign certificate, …

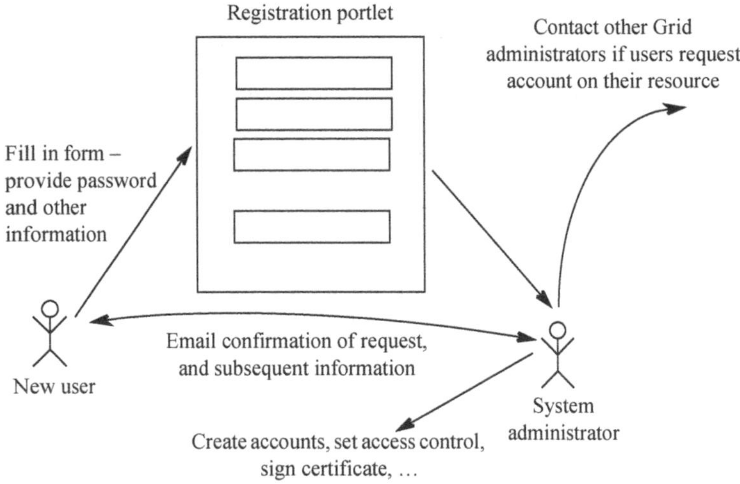

Figure 1.9 Registration activities.

In some Grid projects, it may be necessary to have face-to-face meeting with a system administrator and present a photo ID to establish identity.

Finally, once everything is in place, the user will be able to log in to the Grid portal and see a number of tabs across the top, which enable the user to perform many basic tasks. Depending upon the installed portlets, tabs typically would be for Grid information, proxy management, file management, job submission, Condor job submission, and others such as Sakai for virtual organization member communications. We shall briefly outline them now and explore the underlying software mechanisms in subsequent chapters.

Grid Information Tab. The Grid information tab will select the Grid information browser portlet and display the status of the various computing resources available (i.e., whether up or down, the current load, etc.). Figure 1.10 shows a typical Grid information portlet, in this case displaying resources at UNC–Charlotte and UNC–Wilmington. For this portlet to work, communication is needed between the resources and the portal. The GPIR (Grid Portal Information Repository) service works with the portal to gather the information into a database. Resources send an XML document in clear text (without security) at intervals to update the information. Details on exactly how GRIP does this can be found in the GPIR home page. Notice that the information can be both static information such as the amount of installed memory and dynamic information such as the amount of available memory. Grid resource can also be more than just computing resources, for example, data servers and visualization resources and information on these could be displayed.

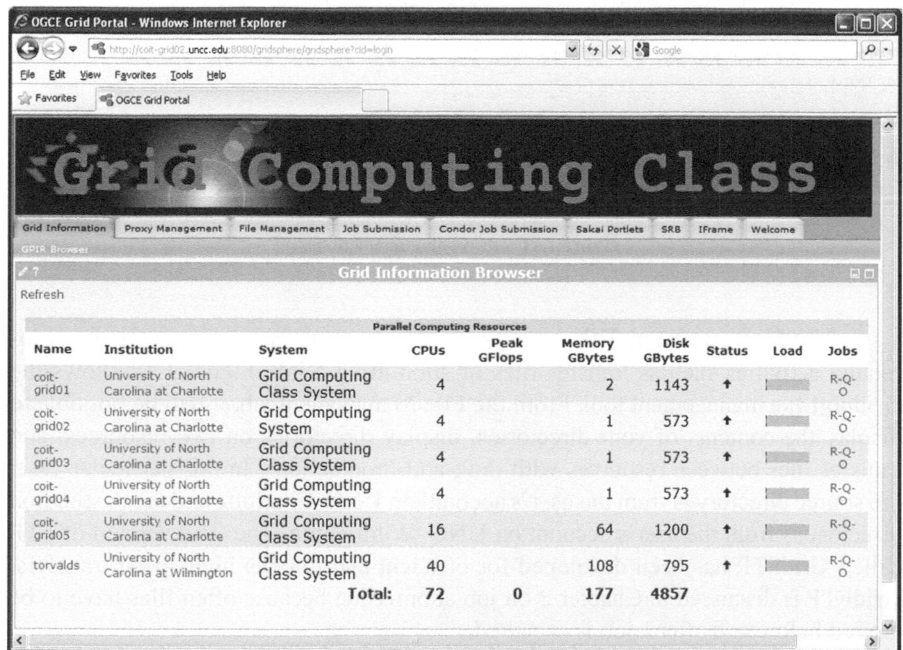

Name	Institution	System	CPUs	Peak GFlops	Memory GBytes	Disk GBytes	Status	Load	Jobs
coit-grid01	University of North Carolina at Charlotte	Grid Computing Class System	4		2	1143	↑		R-Q-O
coit-grid02	University of North Carolina at Charlotte	Grid Computing System	4		1	573	↑		R-Q-O
coit-grid03	University of North Carolina at Charlotte	Grid Computing Class System	4		1	573	↑		R-Q-O
coit-grid04	University of North Carolina at Charlotte	Grid Computing Class System	4		1	573	↑		R-Q-O
coit-grid05	University of North Carolina at Charlotte	Grid Computing Class System	16		64	1200	↑		R-Q-O
torvalds	University of North Carolina at Wilmington	Grid Computing Class System	40		108	795	↑		R-Q-O
		Total:	72	0	177	4857			

Figure 1.10 Grid information tab.

Proxy Management Tab. In order to use many services on the portal, you are required to have a proxy certificate, usually simply called a proxy. Proxies are part of the Grid security infrastructure, which will be discussed fully in Chapter 5. We have mentioned that a proxy is an electronic document that enables resources to be accessed on the user's behalf. It is very convenient to use the credential management service called MyProxy to hold proxies rather than the user holding them. Usually, GridSphere automatically obtains a proxy from the MyProxy server for you when you log in. Figure 1.11 shows a typical display when selecting the Proxy Management tab. The default lifetime of this proxy is shown as two hours. (Proxies have limited lifetimes for security reasons.) However, if you are running a job that takes longer than two hours to complete, you may need to create a proxy with a longer lifetime. One can click the "Get New Proxy" tab to do this. This will require you to provide information such as the host name of the MyProxy server. (An alternative to a longer lifetime proxy is to arrange for subsequent proxies to be retrieved as necessary during the execution of the job if that facility is available, see Chapter 3.) Note users still need to know something about the inner working of the Grid infrastructure even when interacting with a portal. Ideally, users should have to know the minimum about this.

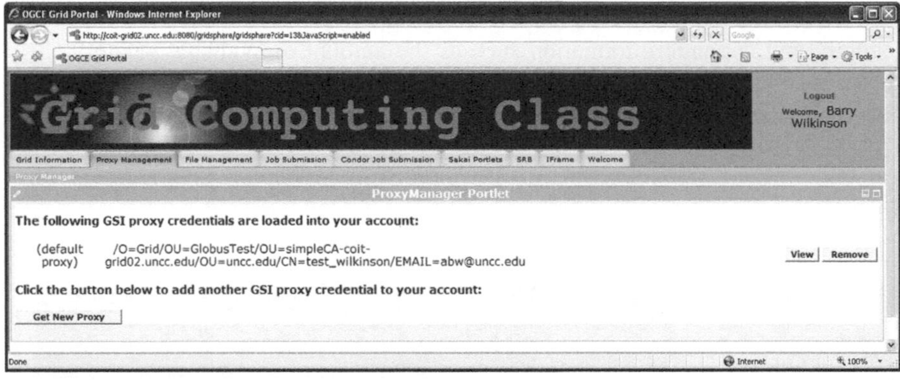

Figure 1.11 Proxy management tab.

File Management Tab. Once you have your proxy, you can do Grid-related activities such as transfer files or submitting a job. Figure 1.12 shows the Gridport file management tool. From the File Management tab, it will be possible to display the contents of your directories, display directories on two resources, and transfer files between resources with drag-and-drop actions. In this particular case, the source directory is from the user's account on UNC–Charlotte and the destination directory is from the user's account on UNC–Wilmington. The Grid version of FTP called GridFTP has been developed for efficient and reliable high-speed transfers. GridFTP is discussed in Chapter 2 on job submission because often files have to be moved before and after a job is executed.

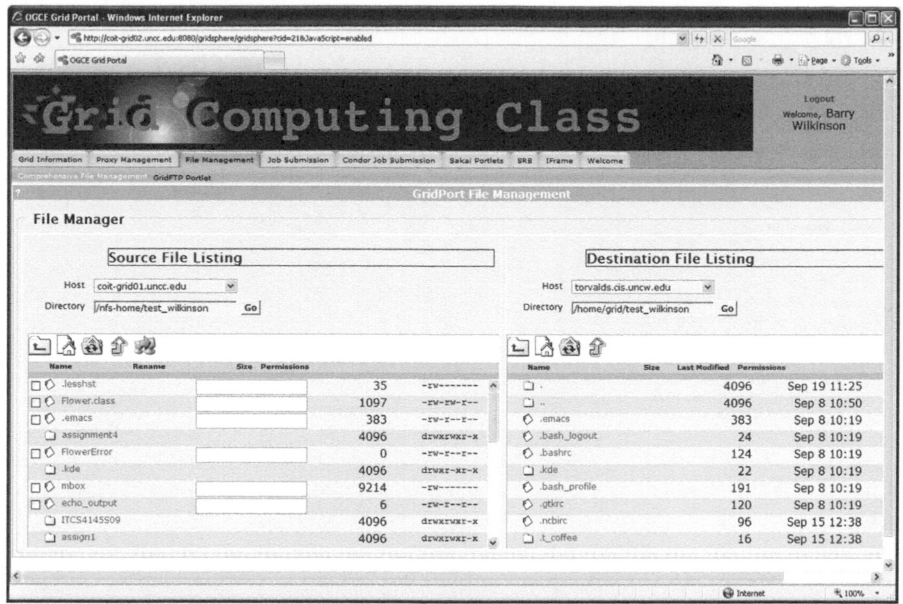

Figure 1.12 File management tab.

Job Submission Tab. Finally, one is ready to submit a job. Under the job submission tab, one has the choice of submitting an interactive job or a batch job to a Grid resource. An interactive job traditionally means that a user can interact with a program while it is running, i.e., a program that will accept keyboard input (standard input) from the user and send program output to the user's console (standard output) as it happens. In a typical Grid environment such as a Globus environment, interactive user input is not accommodated and interactive only refers to sending the program output to the user. User input has to be provided via a file. Batch submission refers to the traditional form of batch operation in which a program is sent to a compute resource for execution and control is returned to the user immediately. The job will be executed sometime in the future. Figure 1.13 shows the batch job submission portlet. Files are provided for the name of the host (from a pull-down menu), name of the executable, arguments, standard output and error files, etc. We shall see in the next chapter an alternative, more powerful way of describing the job is using a job description file.

1.7 SUMMARY

This chapter introduced the following:

- A brief history leading to Grid computing
- Grid computing infrastructure examples — local, national and international
- A brief introduction to the purpose of Grid computing infrastructure software

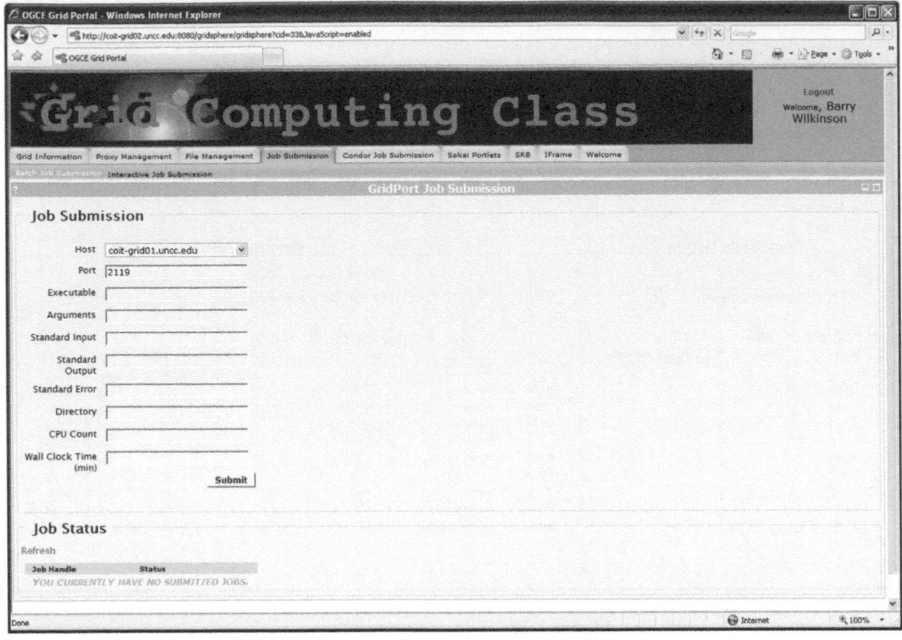

Figure 1.13 Batch job submission tab.

- Grid user interface — command line and portal interfaces

The purpose of this chapter is to set the stage for understanding how to use a Grid platform. In the next chapter, job submission will be discussed in detail.

FURTHER READING

The following is a selection of books and book chapters that are appropriate at this point:

- Chapters 1, 2, and 3, *The Grid: Blueprint for a new computing infrastructure 2nd ed.* (Foster and Kesselman, eds. 2004).
- Chapters 1, 2, 3, and 4, *Grid computing: Making the global infrastructure a reality* (Berman, Fox, and Hey, eds. 2003). Chapter 4 is devoted to describing the I-WAY experiment.
- Chapters 1 and 2, *Grid computing* (Joseph and Fellenstein 2004).
- Chapter 1, *The Grid core technologies* (Li and Baker 2005).
- Chapter 1, *Grid computing for developers* (Silva 2006).
- *Grid computing: A Practical guide to technology and applications* (Abbas 2003). A slightly less technical book on Grid computing worthy of reading.

- *Grid computing: The savvy manager's guide* (Plasczak and Wellner, Jr. 2006). Written from a Grid management perspective, but technical and definitely an excellent book to read.

von Laszewski wrote a nice paper describing the evolution of Grid computing from the perspective of a Globus team member (von Laszewski 2005).

Two important seminal papers are:

- "The anatomy of the Grid: Enabling scalable virtual organizations" (Foster, Kesselman, and Tuecke 2001) and
- "The Physiology of the Grid: An Open Grid Services Architecture for Distributed Systems Integration" (Foster, Kesselman, Nick, and Tuecke 2002).

Both papers can be found in the book by Berman, Fox, and Hey, eds. (2003). The paper "The Physiology of the Grid: An Open Grid Services Architecture for Distributed Systems Integration" describes the OGSA standardized architecture, which is the basis for the service-oriented approach now used in Grid computing. An introduction to Globus version 4 can be found in (Foster 2005).

BIBLIOGRAPHY

Abbas, A. 2003. *Grid computing: A practical guide to technology and applications*. Hingham, MA: Charles River Media, Inc.

AP Grid. http://www.apgrid.org/

Apon, A., and J. Mache. 2004. Collaborative project: Adaptation of Globus toolkit 3 tutorials for undergraduate computer science students. National Science Foundation grant, ref. DUE #0410966/0411237.

Berman, F., G. Fox, and T. Hey, eds. 2003. *Grid computing: Making the global infrastructure a reality*, Chichester Englang: John Wiley & Sons.

Condor High Throughput Computing. http://www.cs.wisc.edu/condor/

DeFanti, T., I. Foster, M. Papka, R. Stevens, and T. Kuhfuss. 1996. Overview of the I-WAY: Wide area visual supercomputing. *Int. Journal of Supercomputer Applications* 10 (2): 123–130.

DEISA. Distributed European Infrastructure for Supercomputing Applications. http://www.deisa.eu/

Earth System Grid II. Turning climate datasets into community resources. Final report. http://datagrid.ucar.edu/esg/about/docs/ESG_II_Final_Report.doc

Earth System Grid. http://www.earthsystemgrid.org/

EDiaMoNd Grid computing project. http://www.ediamond.ox.ac.uk/whatis.html

EUROGRID. Application testbed for European GRID computing. http://www.eurogrid.org/

Florida Lamdba Rail. http://www.flrnet.org/infrastructure.cfm

Foster, I. 2002. What is the Grid? A three point checklist. *Grid Today*, 1(6), July.

Foster, I. 2005. Globus toolkit version 4: Software for service-oriented systems. *IFIP International Conference on Network and Parallel Computing*, Springer-Verlag LNCS 3779, pp. 2–13.

Foster, I., and C. Kesselman. 1997. Globus: A Metacomputing Infrastructure Toolkit. I. Foster, C. Kesselman. *Intl J. Supercomputer Applications*, 11(2):115–128.

Foster, I., and C. Kesselman, eds. 2004. *The Grid: Blueprint for a new computing infrastructure,* 2nd ed. San Francisco, CA: Morgan Kaufmann.

Foster, I., C. Kesselman, J. M. Nick, and S. Tuecke. 2002. The physiology of the Grid: An Open Grid Services Architecture for distributed systems integration. Global Grid Forum. Also appears in Berman, F., G. Fox, and T. Hey, eds. 2003. *Grid Computing: Making The Global Infrastructure a Reality*, Chapter 8. Chichester England: John Wiley & Sons.

Foster, I, C. Kesselman, and S. Tuecke. 2001. The anatomy of the Grid: Enabling scalable virtual organizations. *Int. J. Supercomputer Applications* 15(3): 200–222. Also appears in Berman, F., G. Fox, and T. Hey, eds. 2003. *Grid computing: Making the global infrastructure a reality*, Chapter 6. Chichester England: John Wiley & Sons.

Global Data-Intensive Grid Collaboration. http://www.gridbus.org/sc2003/

Globus Project. http://www.globus.org

GPIR. http://gridport.net/services/gpir/index.html

GridBus project. The Grid Computing and Distributed Systems (GRIDS) laboratory, University of Melbourne. http://www.gridbus.org/

Hey, T., and A. E. Trefethen. 2002. The UK e-Science core program and the Grid. *Future Generation Computing Systems* 18:1017.

Highlights from the UK e-Science program. http://www.rcuk.ac.uk/cmsweb/downloads/rcuk/research/esci/escihighlights.pdf

Joseph, J., and C. Fellenstein. 2004. *Grid computing*. Upper Saddle River, NJ: Prentice Hall PTR.

Legion Worldwide Virtual Computer. http://legion.virginia.edu/

LONI design and capabilities. 2006. *Louisiana Tech HealthGrid Symposium*. http://www.loni.org/network/LA_Tech_HealthGrid_Symposium_2006_Presentation.pdf

Li, M., and M. Baker. 2005. *The Grid core technologies*. Chichester England: John Wiley & Sons.

MGrid. http://www.mgrid.umich.edu/

MPICH. http://www-unix.mcs.anl.gov/mpi/mpich/

MPICH-G2. http://www3.niu.edu/mpi/

MyProxy credential management service. http://grid.ncsa.uiuc.edu/myproxy/

Network for earthquake engineering simulation (NEES). http://www.nees.org/About_NEES/

North Carolina NCREN network. http://www.mcnc.org/

OGCE portal toolkit. http://www.collab-ogce.org/ogce2/

Open Science Grid. http://www.opensciencegrid.org/

OpenMolGRID Open computing GRID for molecular science and engineering. http://www.openmolgrid.org/

OxGrid. http://www.mc.manchester.ac.uk/research/seminars/past_seminars/OxGrid.pdf

Plasczak, P., and R. Wellner, Jr. 2006. *Grid computing: The savvy manager's guide*. San Francisco, CA: Morgan Kaufmann.

PURSe: Portal-based user registration service. http://www.grids-center.org/solutions/purse/

Ramamurthy, B., and B. Jayaraman. "Collaborative: A multi-tier model for adaptation of Grid technology to CS-based undergraduate curriculum," National Science Foundation grant, ref. DUE # 0311473, 2003–2006. http://www.cse.buffalo.edu/gridforce/index.htm.

Silva, V. 2006. *Grid computing for developers*. Hingham, MA: Charles River Media, Inc.

Snir, M., S. W. Otto, S. Huss-Lederman, D. W. Walker, and J. Dongarra. 1998. *MPI: The complete reference, Vol 1 The MPI core*. Cambridge, MA: MIT Press.

Sterling, T. ed. 2002. *Beowulf cluster computing with Linux*. Cambridge, MA: MIT Press, Cambridge.

Sunderam, V. 1990. PVM: A framework for parallel distributed computing. *Concurrency Practice and Experience* 2 (4): 315–339.

SURAGrid. http://www.sura.org/programs/sura_grid.html

TeraGrid. http://www.teragrid.org/about/

Unicore. http://www.unicore.eu/

University of Florida campus research Grid. http://www.hpc.ufl.edu/index.php?body=grid

University of Houston campus Grid. http://www.grid.uh.edu/

University of Virginia campus Grid. http://vcgr.cs.virginia.edu/campus_wide_grid/main.html

UTGrid. http://www.utgrid.utexas.edu/

von Laszewski, G. 2005. The Grid-idea and its evolution. *Journal of Information Technology* 47(6): 319–29. http://www.mcs.anl.gov/~gregor/papers/vonLaszewski-grid-idea.pdf.

Wikipedia E-business. http://en.wikipedia.org/wiki/E-business

Wikipedia E-science. http://en.wikipedia.org/wiki/E-Science

Wilkinson, B., and C. Ferner. 2006. Teaching Grid computing across North Carolina, Part I and Part II. *IEEE Distributed Systems Online*, 7(6–7). http://www.cs.uncc.edu/~abw/papers/DSonline6-2006.pdf, http://www.cs.uncc.edu/~abw/papers/DSonline7-2006.pdf

Wilkinson, B., and C. Ferner. 2008. Towards a top-down approach to teaching an undergraduate Grid computing course. *SIGCSE 2008 Technical Symp. on Computer Science Education* March 12-15, Portland, Oregon USA.

Worldwide LHC computing Grid. http://lcg.web.cern.ch/LCG/

SELF-ASSESSMENT QUESTIONS

The following questions are multiple-choice questions. Unless otherwise noted, there is only one correct answer for each question.

1. What is a virtual organization?

(a) An imaginary company
(b) An organization consisting of virtual resources
(c) A group of people that come together from different organizations or groups to work on a Grid project together with resources
(d) A group of people that come together to work on a virtual reality Grid project

2. What is meant by the term cloud computing?

(a) Atmospheric computing
(b) Computing using geographically distributed computers
(c) A central facility providing services and software applications
(d) A distributed facility providing services and software applications

3. In addition to computers, which of the following can be shared on a Grid? (May be more than one answer. Check all.)

(a) Storage
(b) Application software
(c) Specialized equipment (such as sensors)
(d) Databases
(e) Office space
(f) Computer accounts
(g) All of the above
(h) None of the above

4. What is meant by "single sign-on"?

(a) Allowing only one person to log onto a computer
(b) None of the other answers
(c) Not allowing a person to log onto a computer more than once in any one period
(d) A mechanism in which a user does not need to sign again to acquire additional resources

5. Which of the following is not provided for directly in the Globus version 4 software?

(a) Job submission
(b) Accounting
(c) Network security
(d) Resource discovery
(e) File transfers

PROGRAMMING ASSIGNMENTS

A suitable assignment at this stage is to register and log into a Grid platform, and perform some simple tasks through the portal. Our registering and login procedure can be found in the first assignment for our course at the textbook home page http://www.cs.uncc.edu/~abw/GridComputingBook/. The actual details of registering and logging onto the Grid platform may differ, depending upon the Grid platform you are using, but afterwards, the following are assignments:

1-1. Execute the Linux command (program) `echo` with suitable arguments from the job submission portlet, redirecting standard output to a file called `echo_output`. Go to the file transfer portlet, and find this file. Download the file to your computer and take a screenshot of its contents.

Note: The Linux program `echo` simply sends its command line arguments to standard output. This program comes with the standard Linux distribution. Its full path is `/bin/echo`.

1-2. In this assignment, you are to execute your own program rather than a pre-existing program in Assignment 1-1. The standard features of the portal may not provide a portlet specifically for compiling programs directly on a Grid resource. In this assignment, we will compile the program on the local machine and upload the program onto the remote machine.

Write and compile a C program to do the same as the Linux `echo` program on your own computer (or a laboratory computer). Upload the program onto the remote resource using the portal file transfer portlet, and execute the program there using the batch job submission portlet, redirecting standard output to a file called `myecho_output`.

1-3. In this assignment, you are to execute a Java program rather than the C program in Assignment 1-2. We will compile a Java program locally to obtain a platform-independent class file, which will be uploaded and executed on the Grid resource with a Java Virtual Machine.

Write and compile a Java program to computer 10! (factorial 10) on your own computer (or a laboratory computer). Upload the program onto the remote resource using the portal and execute the program there. The path to the Java interpreter is typically `/usr/java/jdk1.5.0_08/bin/java`. This is the executable to be specified in the portlet. The arguments will include `-classpath` flag and the classpath, and the name of the java class file. Notice that you will need to specify `CLASSPATH`, which is your home directory.

Take care that the program is compiled to match the version of the Java Virtual Machine installed on the remote resource that is to execute the program. It may be that you will have to compile to a lower version than on you local machine, which can be done by using the `-target` option with `javac`, e.g. `javac -target 1.5 YourProgramClassname.java` would compile to version 1.5. A program compiled to a higher version than installed may not run on a system. Although Java is porported to be platform independent, if a program uses features in a higher version than installed, it will not run. It may not be obvious that these features are being invoked.

1-4. In this assignment, we will compile a program on the remote resource rather than on the local machine, which will eliminate the possible problem experienced of not having compatible versions of Java Virtual Machine in Assignment 1-3. A program can be compiled through the portal interface by simply invoking the compiler as the executable and the program as the argument.

Write a Java program (class file) to do the same as the Linux `echo` program. Upload the class file onto the remote resource using the portal file transfer portlet. Then run a batch job on the portal that specifies the `javac` compiler as the executable, the Java class file program as the first argument and the string `hello` as the second argument. Re-direct output to a file called `mynewecho_output`.

CHAPTER 2
Job Submission

This chapter considers how jobs are submitted to a Grid computing environment. It focuses on Globus components and commands. The discussion of job submission is continued in Chapter 3 with the study of job schedulers. Although one cannot submit jobs to a Grid platform without security credentials and an account, this matter is deferred to Chapters 4 and 5, as its full treatment requires considerable discussion.

2.1 INTRODUCTION

In Chapter 1, we reviewed using a Grid portal, a Web-based interface, for accessing the Grid platform. Most production Grid computing platforms provide a Grid portal for their users. Now, we will look at the command-line interface that is essentially underneath the Portal. Users often find it necessary to access the Grid platform through a command-line interface. The command-line interface provides direct access to the middleware components of Grid platform such as the job manager, schedulers, security infrastructure, and data management components. In this chapter, we shall look at job submission and Globus job manager.

The preliminary purpose of a Grid computing platform is to provide distributed computing resources to run jobs on. The term *job* here describes any user-initiated program to be executed on computers on the Grid. These jobs are programs that the user might have written in C, C++, Java, or another language, or a pre-compiled application package. Perhaps one of the first tasks a user might do after having access to the Grid platform is to run simple Linux commands to establish that the account is set up correctly and there is proper access to compute resources. Then, the user might write a simple program and submit for execution. Consider programs that need to be compiled such as in C or C++. The appropriate compiler and libraries need to be available to create the executable code for the target machine. The executable code has to be stored at the compute sites that are to execute the code, as are any required input files. In a simplistic approach, a user might need to move executables and input files to remote locations by direct commands themselves. Afterwards, the user may then need to move output files back to a local site. (A Grid does not usually have a shared file system.) The user activities are illustrated in Figure 2.1. The user interface might be a Grid portal or a command-line interface.

Suppose we want to submit a Java program. The process is quite similar except the Java compiler (`javac`) creates a class file (bytecode) that is interpreted by a Java Virtual Machine (`java`). It is the Java Virtual Machine (JVM) that is the executing program and the class file is specified as an argument to JVM as illustrated in Figure 2.2. Other class files usually need to be called too, which are found in the path specified by the `CLASSPATH` variable, so this variable must be set up properly. Java programs offer more portability because the class file could be sent to any remote computer having a JVM installed.[1] The speed of execution of interpreted Java programs may be significantly less than executing fully compiled binaries and typically high performance computing uses compiled binaries from languages such as C, but that might depend upon the application. Many internal components of Grid middleware software such as Globus actually use a mixture of Java and C. Java is commonly used to create the Web service components.

So far in the description of what the user might want to do, the user is just running a job on a remote computer system. In fact, that can be easily achieved by

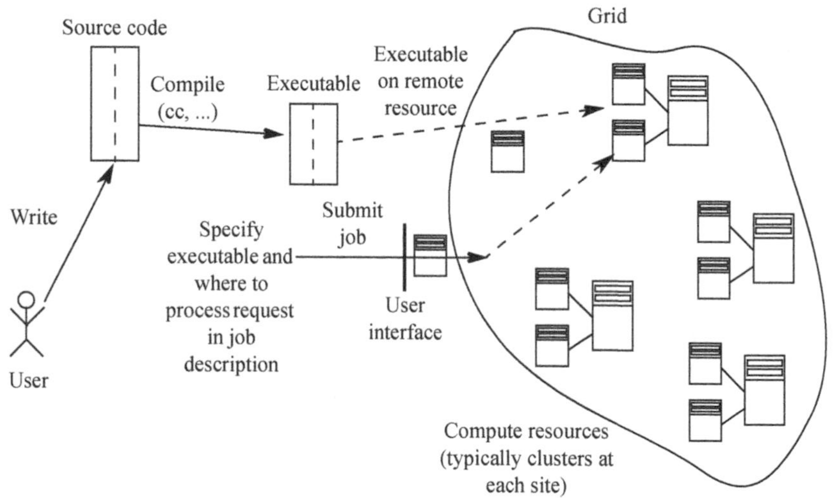

Figure 2.1 User activities using a Grid environment–compiling and submitting a C/C++ job.

[1] Although Java is purported to be platform-independent, one can have problems with different versions of Java installed on different compute resources. The safest way is to compile down user programs to the earliest installed version using the `-target` option in `javac`.

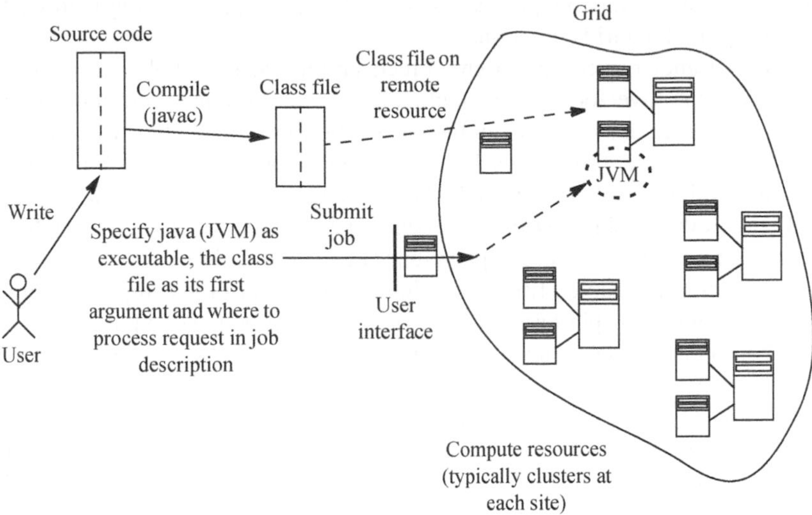

Figure 2.2 User activities using a Grid environment–compiling and submitting a Java job.

using an `ssh` connection, and also one would usually compile the code on the remote computer that is to execute the code. *But a Grid environment offers much more including a single interface and single sign-on to Grid resources, a unified view of the resources, discovering and using multiple resources, and collaborative shared activities.*

Since the Grid is a collection of computers, a user might wish to use these computers collectively to solve problems. Two ways to use multiple computers are:

- Break the problem down into a collection of tasks that need to be done to solve the problem and submit individual tasks to different computers to work on them simultaneously, known as *parallel programming*.
- Run the same job on different computers at the same time but with different input parameters, known as *parameter sweep* problems.

Parallel programming has had a very long history, at least back to 1958 (Gill 1958). The concept is fairly obvious. Dividing a problem into parts that can be done simultaneously should lead to increased execution speed on a multicomputer system. Typically, the parts will communicate with each other in the processing of the problem solution. Ideally, there should be significant computations in the parts and as little interaction as possible between the parts for the greatest performance improvement, that is, the computation/communication ratio should be as high as possible as the communication is an overhead not present when using a single computer. Some problems can be easily divided into parts with minimal communication between the parts. Such problems are called *embarrassingly parallel* problems. Other problems require significant effort to "parallelize" into an efficient solution. Parallel programming has become important even on single computer systems with the introduction of multicore processors. We will explore parallel programming techniques as applied to Grid computing in Chapter 7.

The parameter sweep approach is much simpler than parallel programming and particularly attractive for Grid computing platforms because there are no dependences between each sweep. It can be part of the job specification when the job is submitted. Of course, this approach only represents one area for high performance computing and may not be applicable. One benchmark from NAS Grid benchmark (NGB), the Embarrassingly Distributed benchmark, is designed to represent the characteristics of parameters sweep applications (Peng et al. 2004).

In the current chapter, first submitting a job to one of the computers that make up the Grid computing platform will be explored, whether the computer is local or a remote computer, in the context of a Globus environment. As mentioned in Chapter 1 (see page 20 and Figure 1.5), Globus has gone through several major versions, Globus version 1 (GT1), Globus version 2 (GT2), Globus version 3 (GT3), and Globus version 4 (GT4). Globus version 1 (GT1) appeared before Grid computing was widely used and essentially a prototype. Globus version 2 (GT2) was widely used. Globus version 3 (GT3) had limited life (2003–2004). Globus version 4 (GT4) has been widely used since its introduction in 2005. The focus here is on Globus version 4.

2.2 GLOBUS JOB SUBMISSION

2.2.1 Components

Grid Resource Allocation Management (GRAM). The software component in Globus that receives job submission requests is called *Grid Resource Allocation Management* (GRAM), which is part of Globus' execution management group. Earlier, it was called Globus Resource Allocation Management. With the introduction of Web service components in Globus version 3 onwards, GRAM became a Web service component (WS-GRAM), although version 3 and version 4 are not the same and rely on different standards (see Chapter 3). The pre-Web service component from Globus version 2 still exists in the Globus package and could be used, although here only the version 4 Web service component will be considered. From the user's perspective, the differences are only in the names of the commands to issue and control the jobs.

Job Schedulers. Figure 2.3 shows the basic job submission components on a local computing resource. In this figure, jobs are submitted to a "front-end" computer system, which in turn passes the jobs onto a cluster of compute nodes through a scheduler as in a traditional cluster. The front-end could also execute jobs. The user issues a job submit command to GRAM, which provide a unified interface to accessing Grid resources to manage jobs. GRAM, being a Globus Web service, is hosted in a Web service container provided in the so-called Globus core.

GRAM could be told to submit the job directly onto the local host ("fork") but more typically it will be told to submit the job to a job scheduler which will forward

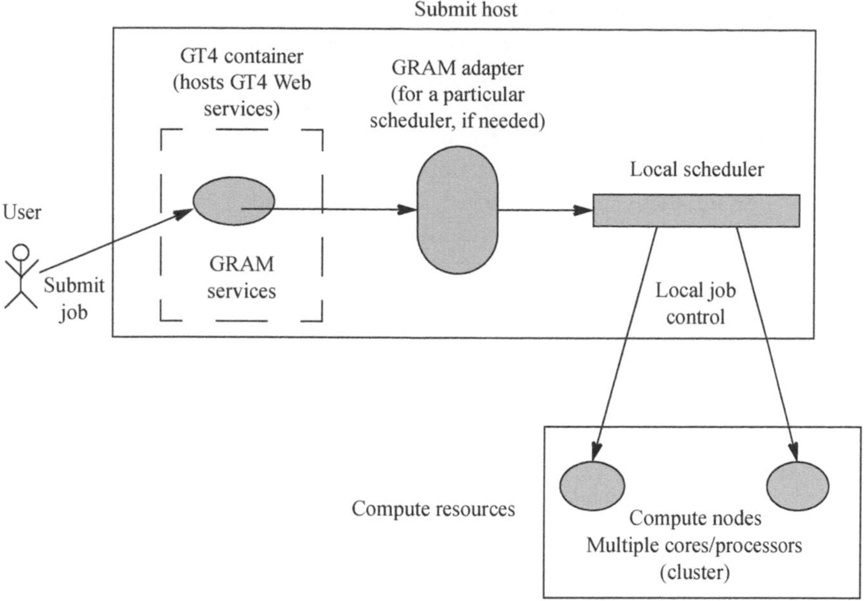

Figure 2.3 Basic job submission components.

the job to a compute cluster. GRAM has a language for specifying the job (see later) and the job specification needs to be converted into the job specification of a particular job scheduler.

Job schedulers interface to GRAM through a GRAM scheduler adapter. Adapters are provided in Globus version 4 for Portable Batch System (PBS), Load Sharing Facility (LSF), and Condor. Third-party adapters are available for Sun Grid Engine (SGE), IBM LoadLeveler, and GridWay. PBS is a traditional job scheduler that was originally developed by NASA in the early 1990s for allocating jobs on Linux computing resources such as in a cluster. There is an open-source version (openPBS) and a professional supported version (PBS Pro). LSF is a commercial job scheduler from Platform Computing. SGE is a job scheduler from SUN with an open-source version (Grid Engine) and commercially supported version (N1 Grid Engine). We touched upon Condor in Chapter 1 as an important project that started before Grid computing. Condor began its development in the 1980s as a research project at the University of Wisconsin–Madison to use idle workstations collectively. It became a very popular job scheduler for clusters and now has Grid environment version called Condor-G. We shall describe Condor and SGE later, as we have used these schedulers in Grid computing. Condor has some unique features for distributed computing environments. GridWay is an open source scheduler developed in the mid-2000s specifically for scheduling jobs across a Grid computing environment. Sometimes, the term *meta-scheduler* is used to describe scheduling jobs across distributed compute resources such as a Grid rather than just using local compute resources.

GRAM Entry Point for Submitting Jobs. GRAM in Globus version 2 and version 3 used the term *gatekeeper* as an entry point for submitting jobs on compute resources. This term vanished in Globus version 4, and the entry point is a service called `ManagedJobFactoryService` (MJFS).[2] In a Globus Grid environment, there will typically be a cluster at each site controlled by a front-end computer/server and each site would look similar to Figure 2.3, as illustrated in Figure 2.4. In this figure, there are four distributed sites, each with a front-end that has the Globus toolkit installed (specifically the GRAM services) and a local scheduler. Each scheduler could be different, e.g., one or more could use PBS, one or more could use Condor, and one or more could use SGE. (It is even possible to have multiple schedulers installed although only one can be used at a time.) The choice of which

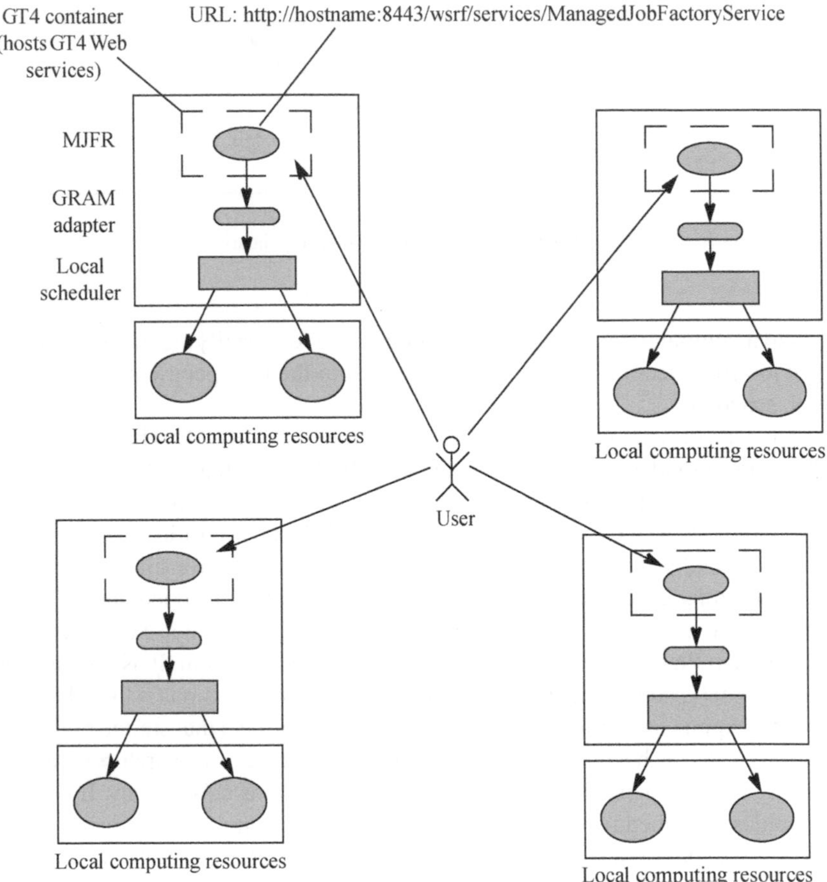

Figure 2.4 Running simple jobs across a Grid computing environment.

[2] `ManagedJobFactoryService` (MJFS) actually creates another service called `ManagedJobService` (MJS) that contacts the local scheduler to submit a job. For full details, see the Globus documentation.

scheduler to install is made by the site administrator. Clients could connect to any front-end.

GRAM operates within the security envelope of GSI. It has a number of features in addition to just submitting local jobs and passing jobs onto a scheduler and can enable users to monitor jobs and cancel them. It uses the Linux `sudo` command to gain access to user accounts as user "globus" (rather than as root in the earlier pre-Web service version of GRAM). The user globus is created during the installation procedure, see Appendix D.

2.2.2 Job Specification

The basic job submission command to execute a job in Globus version 4 is called `globusrun-ws`, which submits and monitors GRAM jobs. It replaces the earlier Globus version 3 `managed-job-globusrun` command, which was implemented in Java. `globusrun-ws` is implemented in C for faster startup and execution. It supports multiple and single job submission and handles credential management and streaming of stdout/stderr during execution.

Specifying Job. There are two basic ways a job can be specified:

- Directly by the name of the executable with required input arguments.
- With a job description file.

Directly. For very simple jobs, one can submit a single job using the `-c` option, e.g.,

```
globusrun-ws -submit -c prog1 arg1 arg2
```

which executes the program `prog1` with arguments `arg1` and `arg2` on the local host. The `-c` option actually causes `globusrun-ws` to generate a job description with the named program and arguments that follow. The `-c` option must be the last `globusrun-ws` option. An example is a simple test of using Linux `echo` program say with the argument `hello`:

```
globusrun-ws -submit -c /bin/echo hello
```

This would create the Globus job monitoring output on the command line and will indicate that the job completes. However, the output from the echo program (`hello`) is not displayed and is lost as is any standard output without further specification (see later).

With a Job Description File. Although a job might be specified on the command line as an executable with its arguments, a more flexible and powerful way is to use a job description file to describe details about the job, its inputs and outputs. Typically, a job description file would contain details such as:

- Job description
  - Name of executable

- Number of instances
- Arguments
- Input files
- Output files
- Directories
- Environment variables, paths, ...

- Resource requirements
  - Processor
    Number, cores, ...
    Type
    Speed, ...
  - Memory

The resource requirements give the opportunity to match the job with available resources.

A job description language or syntax is needed to describe the job. There are several invented, for example:

- Globus
  - Globus versions 1 and 2 used their Resource Specification Language RSL-1
  - Globus version 3 used an XML version called RSL-2
  - Globus version 4 uses a variation of RSL-2 called a JDD (Job Description Document)
- Job Submission Description Language (JSDL) — A recent standard (2005)

Resource Specification Language (RSL) Version 1—Historical. The RSL language of Globus version 1 and version 2 was a meta-language describing the job. It provided a specification for matching the job with resources using Boolean expressions using relational operators and AND, OR, and NOT logical operators. The logical operators preceded the list of Boolean expressions. For example, suppose we wished to describe the following:

Create four instances of prog1, each on a machine with at least 64 Mbytes memory.

In RSL-1, it would be written as

```
& (executable=prog1)(count=4)(minMemory=64)
```

where `&` is the logical AND operator (conjunction), and `executable`, `count`, and `minMemory` are attribute names in the RSL-1 language. The full list of GRAM attribute names is: `arguments`, `count`, `directory`, `executable`, `environment`, `jobType`, `maxTime`, `maxWallTime`, `maxCpuTime`, `gramMyjob`, `stdin`, `stdout`, `stderr`, `queue`, `project`, `dryRun`, `maxMemory`, `minMemory`, `hostCount`. Descriptions of these attributes can be found at (Globus Toolkit GT 2.4).

Complex specifications could be constructed. For example

Create ten instances of prog1 on a machine with 64 Mbytes of memory or five instances of prog1 on a machine with 128 Mbytes of memory and make the system available for a maximum of 15 minutes for each instance.

could be specified in RSL-1 by

```
& (executable=prog1)(maxCpuTime=15)
(|(&(count=10)(minMemory=64)(&(count=5)(minMemory=128)
```

where | is the logical OR operator (disjunction) and `maxCpuTime` is the maximum CPU time for a single execution of the executables.

Multiple resources and programs can be specified using the multi-request symbol +. For example

Execute two instances of prog1 with at least 64 Mbytes of memory and three instances of prog2 with at least 32 Mbytes of memory

could be specified in RSL-1 by

```
+(&(executable=prog1)(count=2)(minMemory=64))
 (&(executable=prog2)(count=3)(minMemory=32))
```

XML Job Description Languages. With the introduction of XML in the early 2000s, job description languages began to be changed to XML.[3] The RSL job description language of Globus version 1 and version 2 was replaced by an XML version of the language in Globus version 3. This XML language was first called Resource Specification Language version 2 (RSL-2). Later in Globus version 4, simplifying syntax changes were made to RSL-2 and the language renamed to a *Job Description Document* (JDD). Although RSL-2/JDD describes the job using an XML schema, it still has much in common with RSL-1 in terms of the attributes used. It has many of the same attributes. RSL-2/JDD can specify everything from executable, paths, arguments, input/output, error file, number of processes, maximum and minimum execution time, maximum and minimum memory, job type, and also a few less used features. Table 2.1 shows a list of most of the RSL-2/JDD XML elements for the attributes. This table is simplified and derived from information at (Globus Toolkit. GT 4.0 WS GRAM: Job Description Schema Document). The full details can be found at this reference. Using an XML language is much more elegant and flexible, and in keeping with systems employing Web services. One can use XML parsers and it allows more powerful mechanisms with job schedulers. Resource schedulers and brokers can apply the job specification to the local resources.

There are some differences in the RSL-2/JDD language specification as it was developed through Globus 3.2 and Globus 4.0 so that Globus 3.2 RSL-2 and Globus 4.0 JDD are not completely interchangeable! The most notable change is in simpli-

---

[3] For review of XML languages, consult Appendix C.

TABLE 2.1 SOME RSL-2/JDD ELEMENTS (PARTIAL LIST)

Element	Meaning
`argument`	A command line argument for the executable.
`count`	The number of executions of the executable. Default: 1.
`directory`	Path of default directory used for the requested job. Default: `${GLOBUS_USER_HOME}`
`environment`	Definition of environment variables in addition to default variables.
`executable`	Name of the executable file.
`factoryEndpoint`	Managed Job Factory service endpoint for submission of job.
`fileCleanUp`	Files local to the job to be removed.
`fileStageIn`	Files to be staged to nodes which will run job. Each specified as pair ("remote URL" "local file").
`fileStageOut`	Files to be staged from job. Each specified as pair ("local file" "remote URL").
`job`	Job description element. Contains elements describing job.
`jobType`	Specifies how the jobmanager should start the job: `single`, `multiple`, `mpi`, or `condor`.
`maxCpuTime`	Maximum CPU time for a single execution of executable, in minutes.
`maxMemory`	Maximum amount of memory for a single execution of executable, in Megabytes.
`minMemory`	Minimum amount of memory for a single execution of executable, in Megabytes.
`multiJob`	Multiple job element. Contains job elements.
`stderr`	Name of remote file to store standard error from job.
`stdin`	Name of file to be used as standard input for executable on remote machine.
`stdout`	Name of remote file to store standard output from job.

fying the syntax. For example, suppose we wished to describe the executable as `/bin/echo`. Originally in RSL-2 (Globus 3.2), it would be described as

```
<gram:executable>
     <rsl:path>
          <rsl:stringElement value="/bin/echo"/>
     </rsl:path>
</gram:executable>
```

where `gram` is the namespace prefix in the above. In Globus 4 JDD, one can simply write

```
<executable>/bin/echo</executable>
```

and similarly for other attributes. Hence RSL-2 (Globus 3.2) is more verbose than JDD (Globus 4.0).

To specify a simple job `/bin/echo` in JDD, we can simply write

```
<?xml version="1.0" encoding="UTF-8"?>
<job>
    <executable>/bin/echo</executable>
    <argument>Hello</argument>
    <argument>World</argument>
    <stdout>${GLOBUS_USER_HOME}/jobOut</stdout>
    <stderr>${GLOBUS_USER_HOME}/jobErr</stderr>
</job>
```

One can also write more complex JDD job descriptions to include memory and requirements just as one could in RSL-1/2.

Job Submission Description Language (JSDL). RSL-1, RSL-2, and JDD are products of the Globus team. As you can see from previously, it was an evolving design. And it was still just for the Globus toolkit. Other projects developed their own job description languages. Different schedulers also use different job description languages, so there was an urgent need to standardize. *Job Submission Description Language* (JSDL) is an attempt to provide a standard job description language that can be used widely. Version 1 was introduced in November 2005 as a GGF standard (Anjomshoaa et al. 2005) particularly focusing on computational jobs. JSDL has been adopted in several projects including UNICORE and the GridWay metascheduler. GridWay had its own existing job description language. JSDL can be used, which is then converted into GridWay's internal job description language. JSDL was introduced after Globus 4.0 was developed but tools exist to use JSDL with Globus 4.0.

The element names used in JSDL are quite recognizable. The following is not a complete list of every possible allowable element or tag. The basic structure of a JSDL document is similar to other job description language documents:

```
<JobDefinition>
  <JobDescription>
    <JobIdentification > ... </JobIdentification >
    <Application> ... </Application>
    <Resources> ... </Resources >
    <DataStaging> ... <DataStaging >
  </JobDescription>
</JobDefinition>
```

The `<jobIdentification>` element can include

```
<JobIdentification>
  <JobName> ... </JobName>
  <Description> ... </Description>
  <JobAnnotation> ... </JobAnnotation>
  <JobProject> ... </JobProject>
</JobIdentification>
```

The `<Application>` element can include

```
<Application>
  <ApplicationName > ... </ApplicationName >
  <ApplicationVersion > ... </ApplicationVersion >
  <Description > ... </Description >
</Application>
```

For executables operating in a Linux environment, one would replace `<application>` with the POSIX[4] "normative extension" `<POSIXapplication>`, which includes the elements

```
<POSIXApplication name="xsd: ... ">
  <Executable> ... </Executable>
  <Argument> ... </Argument>
  <Input> ... </Input>
  <Output> ... </Output>
  <Error> ... </Error>
  <WorkingDirectory> ... </WorkingDirectory>
    :
</POSIXApplication>
```

A different namespace would be used with `POSIXApplication` elements; see later.

The `<Resources>` element describes the requirements of resources for the job and can include

```
<Resources>
  <CandidateHosts> ... </CandidateHosts>
  <FileSystem> ... </FileSystem>
  <ExclusiveExecution> ... </Exclusive Execution>
  <Operating System>... </Operating System>
  <CPUArchitecture> ... </CPUArchitecture>
  <IndividualCPUSpeed> ... </IndividualCPUSpeed>
  <IndividualCPUTime> ... </IndividualCPUTime>
  <IndividualCPUCount> ... </IndividualCPUCount>
  <IndividualNetworkBandwidth> ... </IndividualNetworkBandwidth>
```

[4] Portable Operating System Interface, a collection of IEEE standards that define APIs, compatible with most versions of Unix/Linux.

```
    <IndividualPhysicalMemory> ... </IndividualPhysicalMemory>
    <IndividualVirtualMemory> ... </IndividualVirtualMemory>
    <IndividualDiskSpace> ... </IndividualDiskSpace>
    <TotalCPUTime> ... </TotalCPUTime>
    <TotalCPUCount> ... </TotalCPUCount>
    <TotalPhysicalMemory> ... </TotalPhysicalMemory>
    <TotalVirtualMemory> ... </TotalVirtualMemory>
    <TotalDiskSpace> ... </TotalDiskSpace>
    <TotalResourceCount> ... </TotalResourceCount>
  </Resources>
```

The `<CandidateHosts>` element above can be further described by the host names of machine that can be used, for example using the `jsdl` namespace prefix, for example

```
  <jsdl:CandidateHosts>
    <jsdl:HostName> coit-grid01.uncc.edu </jsdl:HostName>
    <jsdl:HostName> coit-grid02.uncc.edu </jsdl:HostName>
    <jsdl:HostName> coit-grid03.uncc.edu </jsdl:HostName>
    <jsdl:HostName> coit-grid04.uncc.edu </jsdl:HostName>
    <jsdl:HostName> coit-grid05.uncc.edu </jsdl:HostName>
  </jsdl:CandidateHosts>
```

Listing just one machine would force the job to run on that machine. The other elements inside `<resources>` also have several sub-elements. The details can be found in (Anjomshoaa et al. 2005). The `<totalCPUcount>` is the number of processors needed for the job.

JSDL provides for file staging (moving files to and from the executable). The `<datastaging>` element specifies what files are to be moved to the job before execution and what files are to be moved from the job after execution. A sample Linux job description is shown below:

```
<?xml version="1.0" encoding="UTF-8"?>
<jsdl:JobDefinition
  xmlns:jsdl="http://schemas.ggf.org/jsdl/2005/11/jsdl"
  xmlns:jsdl-posix="http://schemas.ggf.org/jsdl/2005/11/jsdl-posix">
<jsdl:JobDescription>
  <jsdl:Application>
    <JobName>Test Job</JobName>
    <Description>Hello world Job</Description>
    <jsdl-posix:POSIXApplication >
      <jsdl-posix:Executable>/bin/echo</jsdl-posix:Executable>
      <jsdl-posix:Argument>hello, world</jsdl-posix:Argument>
      <jsdl-posix:Output>${GLOBUS_USER_HOME}/stdout</jsdl-posix:Output>
      <jsdl-posix:Error>${GLOBUS_USER_HOME}/stderr</jsdl-posix:Error>
    </jsdl-posix:POSIXApplication>
  </jsdl:Application>
</jsdl:JobDescription>
</jsdl:JobDefinition>
```

which is very similar to our previous example using JDD. JSDL version 1 omits certain features such as specifying multiple instances of the same job, which was found in RSL/JDD. That could be achieved by re-submitting the job description file. Chapter 9 includes more powerful ways for parameter sweep.

2.2.3 Submitting a Job

Coming back to submitting a job on the command line in a Globus 4 environment. The command to use is `globusrun-ws`. Before `globusrun-ws` can be used, the user must have valid credentials and a valid proxy. A user would create a proxy with the `grid-proxy-init` command. Then, the user can issue `globusrun-ws` with parameters to specify job and how and where to execute it. Table 2.2 shows a list of options (flags) for `globusrun-ws`. This table is simplified and derived from information at (Globus Toolkit. GT 4.0 WS GRAM Command-line Reference). The full details can be found at this reference.

The most basic Globus 4 `globusrun-ws` command for running a job with a JDD job description file is

```
globusrun-ws -submit -f prog.xml
```

where `-f` specifies use a job description file and `prog.xml` is the job description file that specifies the job. (The extension `.jdd` could also be used to indicate a JDD file.) The `-submit` option causes the job to be submitted (or resubmitted) to the job host. The above command will execute on the local host immediately, that is, so-called "fork."

Output. When jobs are executed on a computer, output from these jobs are send to stdout (standard output) and stderr (standard error) streams, i.e., the user's console. However a remote computer does not have access to the user's console and the output would be lost unless redirected to a file or back to the user's console. Globus offers both redirections in the `globusrun-ws` *output modes*. There are three types of job output modes:

- Interactive
- Interactive-streaming
- Batch

The default is interactive.

Globus Output from Submitting an Interactive Job. Suppose one submits the following job:

```
globusrun-ws -submit -f prog1.xml
```

where `prog1.xml` is the job description file. No host or scheduler is specified in the above (see later) so the job will run (fork) on localhost. Each job has a unique job ID. Typically, Globus monitoring output for a successful job might be

TABLE 2.2 `globusrun-ws` **OPTIONS (FLAGS, PARTIAL LIST)**

Option	Argument (if any)	Effect
`-b`		Batch output mode
`-c`		Causes `globusrun-ws` to generate a simple job description with the named program and remaining arguments. Must appear as last of `globusrun-ws` arguments.
`-f`	Filename	Job description read from specified file.
`-F`	URL	Specifies contact for the job submission (URL is ManagedJobFactory).
`-Ff`	Filename	Specifies contact for the job submission. EPR for ManagedJobFactory read from given file.
`-Ft`	Name	Specifies contact for the job submission (named scheduler).
`-j`	Filename	EPR for ManagedJob read from specified file. EPR used as endpoint for service requests.
`-kill`	Filename	Requests immediate cancellation of job specified in file and exits.
`-monitor`		Attaches to an existing job in interactive or interactive-streaming output modes. Used with `-j`.
`-o`	Filename	ManageJob EPR created for job stored in specified file.
`-s`		Interactive streaming output mode. Standard output and error files of job monitored and data written to `globusrun-ws` output.
`-so / -se`	Filename	Appends standard output and error stream to specified file instead of to stdout/stderr.
`-status`		Reports current state of job and exits. Used with `-j`.
`-submit`		Submit a job to a job host using job description file. Three output modes: batch (`-b`), interactive-streaming (`-s`) or default, interactive.
`-term`	Time	Set an absolute termination time (mm/dd/yyyy HH:MM) or termination time relative to successful creation of job (+HH:MM).

```
Submitting job...Done.
Job ID: uuid:d23a7be0-f87c-11d9-a53b-0011115aae1f
Termination time: 07/20/2008 17:44 GMT
Current job state: Active
Current job state: CleanUp
Current job state: Done
Destroying job...Done.
```

The job submission goes through several states from submitted to done and in the interactive mode, the user's console is tied up as the job goes through the various stages. Any job output is lost unless files are specified in the job description file (`prog1.xml`) for the output.

Streaming. *Streaming* refers to sending the contents of a stream of data from one location to another location as it is generated. It is often associated with Linux and with standard output and standard error streams. In fact, for a program that creates output on remote machine, one would typically need to specify files to hold the output and error messages or redirect the output and error messages to the user console by streaming. Streaming has the advantage of seeing the messages immediately as they are generated. The `globusrun-ws` interactive-streaming output mode is a version of interactive mode that captures the program output and error messages and redirecting them to the user's console (that is, output of `globusrun-ws`). It also provided for redirection to files. It is selected with the `-s` option. For example, issuing the command

```
globusrun-ws  -submit -s -c /bin/echo hello
```

would redirect the output to the user's console. The output would show up inbetween the normal Globus status report, for example

```
Submitting job...Done.
Job ID: uuid:d23a7be0-f87c-11d9-a53b-0011115aae1f
Termination time: 07/20/2008 17:44 GMT
Current job state: Active
Current job state: CleanUp-Hold
hello
Current job state: CleanUp
Current job state: Done
Destroying job...Done.
```

Alternatively if redirection to files is desired, then one use `-s` option with the `-so` and `-se` options

```
globusrun-ws  -submit -s -so outfile -se errorfile  \
              -c /bin/echo hello
```

where the files `outfile` and `errorfile` hold the output and error messages, respectively. For streaming to work, the Globus GridFTP file transfer component (see later) must be operational and be able to capture the output for redirection.

A job description file can specify where `stdout` and `stderr` streams go, as we saw earlier, e.g., in JDD

```
<job>
    <executable>/bin/echo</executable>
    <argument>Hello</argument>
```

```
    <argument>World</argument>
    <stdout>jobOut</stdout>
    <stderr>jobErr</stderr>
</job>
```

In this case, the streaming option would not be present. One could simply issue the command such as

```
globusrun-ws -submit -f prog2.xml
```

where `prog2.xml` including the streaming information.

Batch Submission. The term *batch* is a long-standing Computer Science term from the early days of computing where jobs are submitted to a system in a group (a batch) and wait their turn to be executed sometime in the future. It originally appeared when programs were submitted by punched cards to a time-shared system, perhaps to be run overnight. (I remember those days with frustration.) The context for the term batch here is not much different. Jobs will be queued for execution by a scheduler. Batch submission is really part of a scheduling approach, but in `globusrun-ws`, it is selected separately to the actual scheduler being used. The term batch in `globusrun-ws` is referred to as an output mode because of the way output is generated. In the batch output mode, once the job is submitted, control is returned to the command line, and one will need to query the system to find out the status of the job.

For example, suppose a job was run in the interactive mode, say the `sleep` command

```
globusrun-ws -submit /bin/sleep 100
```

it would return when the program (`sleep` command in this case) completes. The `sleep` command waits the period given by the argument (default is in seconds) and provides a delay of that time. Afterwards, the normal `globusrun-ws` output would be obtained, such as

```
Submitting job...Done.
Job ID: uuid:d23a7be0-f87c-11d9-a53b-0011115aae1f
Termination time: 07/20/2005 17:44 GMT
Current job state: Active
Current job state: CleanUp
Current job state: Done
Destroying job...Done.
```

only each line would appear as the process moves to the next state.

Alternatively, `sleep` could be executed in batch output mode (`-b` option) with the command

```
globusrun-ws -submit -b /bin/sleep 100
```

The output would immediately appear of the form

```
Submitting job...Done
JoB ID: uuid:f9544174-60c5-11d9-97e3-0002a5ad41e5
Termination time: 01/08/2005 16:05 GMT
```

and displays the job ID. Technically, the job ID is actually an identifier called the ManagedJob service end-point-reference for the job (ManagedJob EPR), see Chapter 7 for a discussion on end-point references but for now we can simply consider it as a number that identifies the job. Control will then be returned to the command line. The program (in this case, `sleep`) may not have finished, and in this case will not for 100 seconds.

There are `globusrun-ws` command options to query the status of the job. The job has to be identified, i.e., the ManagedJob EPR is needed, which could be obtained from original batch output. Commands to query the status of a job require the ManagedJob EPR to be held in a file, so the most convenient approach is to run the job with the `-o` option, which stores the ManagedJob EPR of the newly created job in a specified file, for example

```
globusrun-ws -submit -b -o jobEPR /bin/sleep 100
```

where `jobEPR` holds the ManagedJob EPR.

To watch the status of a submitted job, one can "attach" interactive monitoring with the `-monitor` option. The ManagedJob EPR needs to be provided with the `-j` option, for example

```
globusrun-ws -monitor -j jobEPR
```

where `jobEPA` holds ManagedJob EPR. Then, we can see the stages that the job goes through with interactive output immediately, such as

```
job state: Active
Current job state: CleanUp
Current job state: Done
Requesting original job description...Done.
Destroying job...Done
```

although the job itself is still a batch output job. Other options are the `-status` option to display the current state of the job and the `-kill` option to request that the job be cancelled immediately. These options also need a file holding the ManagedJob EPR, for example

```
globusrun-ws -submit -b -o jobEPR /bin/sleep 100
Submitting job...Done
JoB ID: uuid:f9544174-60c5-11d9-97e3-0002a5ad41e5
Termination time: 01/08/2005 16:05 GMT
```

```
globusrun-ws -status -j jobEPR
job state: Active

globusrun-ws -status -j jobEPR
Current job state: CleanUp

globusrun-ws -status -j jobEPR
Current job state: Done

globusrun-ws -kill -j jobEPR
Requesting original job description...Done.
Destroying job...Done
```

The interactive mode corresponds to the default mode on a Linux command line in which commands are executed as submitted and control is only returned to the command line when the command has been completed. The batch mode corresponds to running a process in the background in Linux using the `&` option. For example, the Linux command[5] on `coit-grid02`

```
[abw@coit-grid02 ~]$ sleep 100
```

waits 100 seconds before executing another command, i.e., it will take 100 seconds to get back to the prompt

```
[abw@coit-grid02 ~]$
```

Alternatively, running in the background

```
[abw@coit-grid02 ~]$ sleep 100 &
```

will immediately return with the process ID, for example

```
[1] 31480
```

but the process will still be running in the background. Its status can be queried using the `ps` command with the process ID:

```
[abw@coit-grid02 ~]$ ps 31480

  PID TTY      STAT   TIME COMMAND
31480 pts/2    S      0:00 sleep 100
[abw@coit-grid02 ~]$ ps 31480
```

i.e., sleeping.

[5] Basic Linux commands are given in Appendix B.

```
  PID TTY      STAT   TIME COMMAND
[1]+  Done                    sleep 100
[abw@coit-grid02 ~]$
```

i.e., finished.

Specifying Where Job Is Submitted. The request to run a job is processed by a "factory" service called `ManagedJobFactoryService`, which is part of the Globus package and exists in the Globus container. The default URL for `ManagedJobFactoryService` is `https://localhost:8443/wsrf/services/ManagedJobFactoryService`. To run a job on the localhost using the default URL for factory service, one could simply issue the command such as

```
globusrun-ws -submit -s -c /bin/echo hello
```

The `globusrun-ws -F` option is provided to specify the "contact" for the job submission (factory information). For example, suppose the port was 8440 rather than 8443. We could issue the command

```
globusrun-ws -submit -F http://localhost:8440 -f prog1.xml
```

The factory service is still located at `wsrf/services/ManagedJobFactoryService`.

For remote host, one could simply give the URL or IP address if the Managed Job Factory service is present in the default location and port at the remote site, i.e.,

```
globusrun-ws -submit -F http://coit-grid01.uncc.edu   \
     -f prog1.xml
```

An example of selecting a different host and port with a full specification is

```
globusrun-ws -submit   \
     -F http://coit-grid01.uncc.edu:8440/   \
     wsrf/services/ManagedJobFactoryService -f prog1.xml
```

A variation of the `-F` option is `-Ff` to specify that a file is provided that gives the EPR of the ManagedJobFactory service. This way is recommended for interoperability with other implementations of WS-GRAM.

Selecting a Scheduler. So far, a job scheduler is not specified. As mentioned in Section 2.2.1, a job scheduler is usually present. GRAM submits its jobs to a scheduler if a scheduler is specified, otherwise the GT 4 "fork" job manager attempts to execute the job immediately on the specified host. In `globusrun-ws`, the scheduler is selected by name using the `-Ft` option (i.e., factory type). Common schedulers are Condor, LSF, PFB, and SGE. Examples of commands are

```
globusrun-ws -submit -Ft Condor -f prog1.xml
globusrun-ws -submit -Ft SGE -f prog1.xml
```

Clearly, the selected scheduler has to be present. Different schedulers have different features and modes of operation. The discussion of schedulers will be continued in Chapter 3 where two common schedulers, SGE and Condor, are described in detail.

2.3 TRANSFERRING FILES

Before a job can be run on a resource, both the executable and any input files have to be accessible on the compute resource used to execute the program. Any generated output files need to be accessible by the user. Most Grid platforms do not have shared network file system (NFS) across geographically distributed sites for performance reasons and so it is necessary to transfer the files to and from the compute resource explicitly.[6]

So far, the job submission command

```
globusrun-ws -submit -F http://coit-grid01.uncc.edu -c prog1
```

requires `prog1` to be existing on the remote machine in the default directory (`${GLOBUS_USER_HOME}`). It is up to the user to ensure the executable is in place. Moving input is called *input staging*. Moving output files is *output staging*. Staging can be specified in the job description but first let us review simply moving any file using a command line.

2.3.1 Command-Line File Transfers

File movement in a Grid environment requires the use of Grid data transfer services. It may be that several files need to be transferred together and it may be that the files are very large and need very high-speed transfers. Globus provides several components for such data management. The Globus component GridFTP provides for large data transfers, secure transfers, fast transfers, reliable transfers, and *third-party transfers*. A third-party transfer describes transferring a file from one remote location to another remote location controlled by a party at another location (the third party). Third-party transfers have actually already been seen in Chapter 1, in a Grid portal at the file management portlet (Figure 1.12). There, the user can initiate a transfer between two locations from the portal running on a third system. As we shall see shortly, the same actions can be achieved on the command line.

GridFTP uses control and data channels that one finds in FTP, but it operates in the Grid Security Infrastructure (GSI) environment for a secure transfer. Third-party transfers using GridFTP is shown in Figure 2.5. The client first establishes a control channel with each of the servers, in this case, *A* and *B*. Then, the client can set the parameters for the transfer and requests a data channel creation. Once the data channel is established, the client sends transfer commands over control channels to initiate the third-party transfers through the data channel. Note that the ports that GridFTP uses have to be open for GridFTP to operate. (The servers' control channel

---

[6] The DEISA project uses a global shared file system.

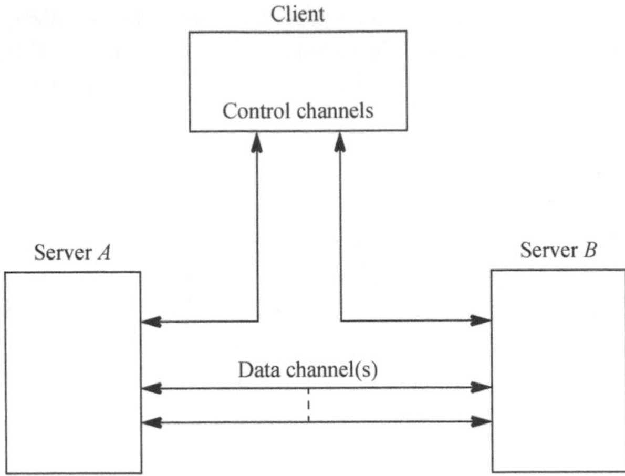

Figure 2.5 GridFTP third-party transfers.

ports are 2811, other ports are selectable.) The control channel is encrypted by design and does not require high bandwidth. The high bandwidth data channel is not encrypted by default because of performance considerations, but can be encrypted.

GridFTP offers features for very high-speed transfers, including parallel transfers using multiple virtual or physical channels. The term *parallel transfer* is used with GridFTP when employing multiple virtual channels sharing a single physical network connection. Some speed improvement can be made using multiple virtual channels across one network connection, but is ultimately limited by the network card and connection. The term *striping* is used when employing multiple physical channels requiring multiple hardware interfaces.

GridFTP is not a Web service. Another Globus component called ReliableFile-Transfer (RFT) service provides a service interface and additional features for reliable file transfers (retry capabilities, etc.). RFT uses GridFTP servers to effect the actual transfer. This combination, RFT and GridFTP, replaces the older Globus Access to Secondary Storage (GASS) data transfer server, which also provided a means of transferring between HTTP, FTP (and GASS) servers.

Coming back to user commands, the user can issue a Globus file transfer command

```
globus-url-copy [options] Source_URL Destination_URL
```

to move files or complete directories from a source location to a destination location, both identified by URLs, using one of several supported protocols including GridFTP (`gsiftp`) as specified in the protocol part of the URL. For example

```
globus-url-copy    \
    gsiftp://www.coit-grid02.uncc.edu/~abw/hello    \
    file:///home/abw/
```

copies the file hello from `coit-grid02.uncc.edu` to the local machine using GridFTP. Of course, it is necessary to have valid security credentials (a certificate and proxy) as globus-url-copy operates within the GSI security envelope. Note the general form of file URL is `file://host/path`. If host omitted, it is assumed to be localhost; in that case, one needs three /'s, i.e., `file:///`.

2.3.2 Staging

File staging refers to arranging that complete files are moved to where they are needed. *Input staging* refers to input files, that is, files needed as input for a program and *output staging* refers to output files, that is, files produced as output by a program. Input files need to be moved to where the program is located, and similarly output files generated by a program need to be moved back to where the user is, or in some situations to other locations as input to programs or to other users. Note that this is different to *input and output streaming*, which refers to bringing into a program a series of data items as a stream from another location, or sending the contents of a stream of data generated from a program to another location as it happens, as illustrated in Figure 2.6. Many applications require staging and it can be achieved using the file movement commands given previously, but staging can be thought of as part of the job description and can be specified in the job description document.

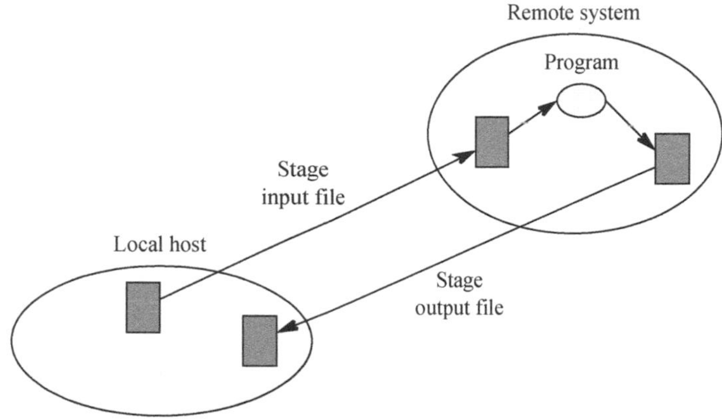

Figure 2.6 File staging.

Staging in JDD. Staging can be specified in the JDD job description file using `<fileStageIn>` and `<fileStageOut>` elements (see Table 2.1). These elements (attributes) have sub-elements for further specification: `<transfer>`, `<rftOptions>`, and `<allOrNone>`. The `<transfer>` element specifies the source and destinations as a pair, identified by URLs. An example of a JDD description of staging a file to another server is

```
<job>
   ...
   <fileStageOut>
```

```
    <transfer>
      <sourceUrl>
        file:///prog1Out
      </sourceUrl>
      <destinationUrl>
        gsiftp://coit-grid05.uncc.edu:2811/prog1Out
      </destinationUrl>
    </transfer>
  </fileStageOut>
  ...
</job>
```

Note that the source file here is on the machine executing the job, which is regarded as the local machine for the job but the file still has to be specified as a URL (`file:///`). Given the destination URL as `gsiftp://` ... , the transfer will be done using GridFTP. Staging can be done the opposite way using `<fileStageIn>`. The transfer will be done using GridFTP.

Staging in JSDL. The JSDL job description language has similar staging features to JDD and includes being able to specify whether files are deleted after use and whether output files are appended to existing files. There are subtle differences in the schema. With input staging in JSDL, for example

```
<jsdl:DataStaging>
  <jsdl:FileName>/inputfiles/prog1Input</jsdl:FileName>
  <jsdl:CreationFlag>overwrite</jsdl:CreationFlag>
  <jsdl:Source>
    <jsdl:URI>
      gsiftp://coit-grid05.uncc.edu:2811/prog1Input
    </jsdl:URI>
  </jsdl:Source>
</jsdl:DataStaging>
```

a URL is not used for the location on the destination resource. A path and file name is given where to place the incoming file. The source is given as a URL with the `gsiftp` protocol to establish that GridFTP will be used in the transfer. In the above, the incoming file overwrites an existing file of the same name.

For output staging in JSDL, one might have

```
<jsdl:DataStaging>
  <jsdl:FileName>/prog1Out</jsdl:FileName>
  <jsdl:CreationFlag>overwrite</jsdl:CreationFlag>
  <jsdl:Target>
    <jsdl:URI>
      gsiftp://coit-grid05.uncc.edu:2811/prog1Output
    </jsdl:URI>
  </jsdl:Target>
</jsdl:DataStaging>
```

2.4 SUMMARY

This chapter introduced the activity of submitting jobs to compute resources in a Grid environment, focusing on a Globus environment although the concepts apply generally. It introduced the following concepts:

- Job description languages with examples (RSL-1, RSL-2/JDD, and JSDL)
- Staging data
- Streaming data

The chapter described various options in the Globus `globusrun-ws` command, including for submitting a job with and without a job description file, and for streaming data and monitoring the state of the job. The Globus command `globus-url-copy` is described to moving files and also how file staging can be specified in the job description file. Ideally, one can use this information to move files and submit jobs on the command line and compare this to submitting jobs through a portal as described in Chapter 1, to give an insight into what happens behind the scenes with a portal interface.

FURTHER READING

Full details on the Globus commands can be found at the Globus toolkit home page (Globus Toolkit).

BIBLIOGRAPHY

Anjomshoaa, A., F. Brisard, M. Drescher, D. Fellows, A. Ly, S. McGough, D. Pulsipher, and A. Savva (editor). 2005. Job submission description language (JSDL) specification, version 1.0. http://www.gridforum.org/documents/GFD.56.pdf

Globus Toolkit. http://www.globus.org/toolkit/

Globus Consortium. 2006. Globus toolkit tutorial. http://www.globusconsortium.org/tutorial/

Globus Toolkit. GT 2.4: The Globus resource specification language RSL v1.0. http://www.globus.org/toolkit/docs/2.4/gram/rsl_spec1.html

Globus Toolkit. GT 4.0. http://www.globus.org/toolkit/docs/4.0/

Gill, S. 1958. Parallel programming. *The Computer Journal*, 1:2–10.

Peng, L., S. See, J. Song, A. Stoelwinder, and H. K. Neo. 2004. Benchmark performance on cluster Grid with NGB. *18th Int. Parallel and Distributed Processing Symp.* (IPDPS'04) – Workshop 17.

Wilkinson, B., and M. Allen. 2005. *Parallel programming: Techniques and application using networked workstations and parallel computers*. 2nd ed. Upper Saddle River NJ: Prentice Hall.

SELF-ASSESSMENT QUESTIONS

The following questions are multiple choice. Unless otherwise noted, there is only one correct answer for each question.

1. GRAM is:

(a) A job scheduler
(b) A service that can communicate a job to a job scheduler
(c) Grid Random Access Memory
(d) Globus Random Access Memory

2. Which Globus 4 command submits a job?

(a) `ant deploy`
(b) `submit-globus-job`
(c) `globusrun-ws`
(d) None of the other answers

3. When one issues the Globus 4 command: `globusrun-ws -c /bin/echo hello`, what is `hello`?

(a) The argument for the program to run
(b) An xml file containing the description of the job to be run
(c) The executable to run in Globus
(d) A Java class file

4. When one issues the Globus 4 command: `globusrun-ws -submit -F localhost:8440 -s -so hello1 -c /bin/echo hello`, is the order of the options important, and if so why?

(a) Not important
(b) Important: `-c` must be last as it precedes the executable, which will take the remaining as arguments
(c) Important: `-s` must be before `-so`
(d) Important: `-F` must be first

5. When one issues the Globus 4 command: `globusrun-ws -submit -F localhost:8440 -s -so hello1 -c /bin/echo hello` what is `localhost`?

(a) The server executing `globusrun-ws`
(b) The computer you are using to log into the server
(c) None of the other answers

6. What does the tag `<count>` specify in an RSL-2/JDD file?

(a) The number of arguments

(b) The number of different jobs submitted
(c) The number of identical jobs submitted
(d) The number of computers to use

7. Suppose you are asked to write a Java program called `Test.java`. What is specified as the executable in the Globus 4.0 RSL2/JDD job description file?

(a) The Java compiler, `javac`
(b) The Java interpreter, `java`
(c) The class file `Test.class`
(d) The Java program `Test.java`

8. When you run a job using the Globus command: `globusrun-ws -submit -s -c prog1`, where does the output of the program appear?

(a) Standard Output (`stdout`) on the machine executing the program
(b) Output is lost
(c) On the command line window where you are issuing the Globus command
(d) In a file called `prog1.stdout`

9. What is the basic difference RSL (Resource Specification Language) version 1 and RSL version 2?

(a) Nothing
(b) RSL-2 corrected errors in RSL-1
(c) RSL-1 is an XML schema
(d) RSL-2 is an XML schema

10. To execute job, one first needs to issue the Globus command `grid-proxy-init`. Why?

(a) To create a proxy certificate so that resources can be accessed on your behalf
(b) To initialize the Globus environment
(c) To start job submission services
(d) To initialize the proxy server

11. What is the file `echo.xml` in the command `globusrun-ws -submit -f echo.xml`?

(a) The XML file containing the schema for globusrun-ws
(b) The executable to run
(c) An RSL file containing a description of the job to run
(d) None of the other answers

12. What is a third-party transfer?

(a) A file transfer from one computer to another computer through a third-party computer
(b) A file transfer from one computer to another computer initiated by a third computer
(c) A file transfer paid for by a third party
(d) A file transfer in which one party is unknown

PROGRAMMING ASSIGNMENTS

Notes: A suitable assignment at this stage is to log in to a Grid platform on the command line, and perform some simple tasks. The actual details of logging onto the Grid platform may differ, depending upon the Grid platform and software you are using. In the following, it is assumed that you are using the Globus toolkit. Assignment 2-1 is a preliminary assignment to establish that you can submit jobs and needs to be completed before any of the subsequent assignments. These preliminary tasks are given in a step-by-step fashion to enable it to be done before fully assimilating the material on security in Chapters 4 and 5.

2-1. *Preliminary tasks:* On the command line of your computer, make remote connection to the entry point to your Grid resources. Typically this would be made using an `ssh` client such as Putty. Once an `ssh` client is installed (see Appendix B), you can type in the command such as

```
ssh coit-grid01.uncc.edu
```

In this case, the machine you are connecting to is `coit-grid01.uncc.edu`. Putty has the facility to save host names. At the command line prompts, enter your username and password to establish a connection to your home account.

Before you can issue any Grid (Globus) commands, you must have your security credentials set up, that is, your user certificate signed by a trusted certificate authority, as described in Chapter 5. Issue the command

```
grid-cert-request
```

to create your private key and a certificate request. This command requires a password to complete. It is important to choose a strong password that is different from your login password and remember the password as this password is used to submit jobs subsequently. The files created by this command are `usercert_request.pem`, `usercert.pem` (an empty file which will become the signed certificate), and `userkey.pem` all in your `.globus` directory, as described in Chapter 5. The request (`usercert_request.pem`) file is sent to the certificate authority for it to return a signed certificate. The process of doing that is dependent upon the system administration. Follow the procedures at your site to obtain a signed certificate. Once you have your signed certificate, you will need a proxy to run jobs. You can create a proxy with the command

```
grid-proxy-init
```

The following tasks may now be done.

2-2. Use the `globusrun-ws` command with suitable options to submit a job to execute the Linux `echo` command with the arguments `Hello World`, where the output is streamed back to the user console. This task is to be done without a job description file.

2-3. Repeat Assignment 2-2 but arrange that the output is placed in a file called `echo_output` in your home directory. Confirm that the output is correct.

2-4. Repeat Assignment 2-3, but use a job description file to specify the executable, arguments, standard output, and standard error.

2-5. Write a C/C++ program to perform the same actions as the Linux `echo` command. Call your version of the `echo` program `my_echo`. Compile to an executable and test your program in your home directory. Then, submit a Grid job to execute the `my_echo` program using a Globus command. Use a job description file.

2-6. Write a Java program to perform the same actions as the Linux `echo` command. Call your version of the `echo` program `myEcho`. Compile to a class file and test your program in your home directory. Then, submit a Grid job to execute the `myEcho` program using a Globus command. Use a job description file. You will need to provide the full path to the Java Virtual Machine interpreter (`java`) and to the `myEcho` class file. You can find the full path to the Java Virtual Machine with the command

```
which java
```

which returns the path to `java`, say `/usr/java/jdk1.5.0_08/bin/java`. There may be multiple versions (or copies) of `java` installed which can be identified by adding the `-a` option to `which` to get all versions or copies installed. The full path to your class file is also needed if not in your home directory of the Grid resource you are using. Note that you do not include the `.class` extension as the argument to `java`.

2-7. Perform Assignment 2-5 but arrange that `myEcho` is executed 10 times in the job description file.

Notes: In the following, you are asked to run a job on a different machine in the Grid platform to that you have logged onto. This will depend upon you having an account and access to the remote machine. You should first establish that your remote account is set up properly. You will need to know not only the name of the remote machine, but also the port assigned to the Globus container, if not the default port, and also the location of the `managedJobFactoryService` if not in the default location (very unlikely).

2-8. Execute the `globusrun-ws` command with an accessible remote Grid resource and the Linux `hostname` command as the program, streaming the output to the console.

2-9. Execute the `globusrun-ws` command with an accessible remote Grid resource and the Linux `hostname` command as the program, streaming the output to a file called `hostname_output` in your home directory. Confirm that the output is correct.

2-10. Repeat Assignment 2-9 but with your own program (for example, the `my_echo` program of Assignment 2-5 or the `myEcho` program from Assignment 2-6). This task will require the executable of your program to reside on the remote system.

2-11. Use the `globus-url-copy` command to transfer a file from one available Grid resource to another, and back again.

CHAPTER 3
Schedulers

This chapter is a continuation of Chapter 2 on job submission and focuses on scheduling jobs. The chapter starts with job schedulers that schedule jobs across compute resources in a cluster at one location. Job schedulers were developed for this purpose prior to Grid computing. A Grid computing environment still uses such schedulers for local job scheduling, but in addition has further requirements of scheduling jobs across distributed sites and scheduling multiple simultaneous resources to bring to bear on a problem. Later in the chapter, meta-schedulers that schedule jobs across distributed sites are explored.

3.1 SCHEDULER FEATURES

3.1.1 Scheduling

Schedulers assign tasks (jobs) to compute resources to meet specified requirements within the constraints of the available resources and their characteristics. Scheduling is an optimization problem. The objective is usually to maximize throughput of jobs. Each job will have certain static and dynamic characteristics, and resource requirements. The static job characteristics, such as code size, can be deduced prior to execution, whereas dynamic job characteristics refer to when the job is being executed and can change during execution. Jobs may also have user requirements such as maximum time to completion. Each compute resource will also have certain static and dynamic characteristics and will be affected as jobs are executed upon them. The static resource characteristics refer to fixed characteristics such as processor type or installed memory whereas dynamic resource characteristics refer to when the resource is operating and can change under varying conditions such as current load.

Different schedulers may have different approaches for allocating jobs to resources but typically jobs are entered into a queue as illustrated in Figure 3.1. There may be multiple queues separating jobs by priority. Jobs are taken from a queue and sent to a particular compute resource, based upon a scheduling algorithm that takes into account both the job and resource characteristics. A component called a *dispatcher* performs this operation.

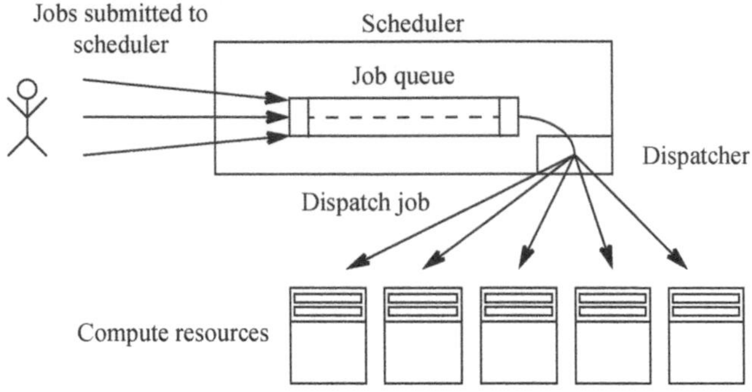

Figure 3.1 Job scheduler.

Scheduling Policies. The following are some traditional scheduling policies:

- First-in, first-out (longest waiting job)
- Job requiring smallest (or largest) memory first
- Short (or long) running job first
- Job with highest priority first

Scheduling is often based upon job priority. If a high priority job cannot be run because all the necessary resources for the job are not available, scheduling may have a *backfilling* mechanism in which lower priority jobs are run while waiting to run the high priority job, to utilize the compute resource fully.

Job schedulers can also include:

- Job and resource matching
- Dynamic scheduling based upon dynamic characteristics such as load
- Preemptive scheduling with process migration

which we shall discuss. These features are particularly relevant for a distributed heterogeneous computing platform such as a Grid platform. Job schedulers described in this chapter have job and resource matching, which requires both the characteristics of the job and the resources to be specified. For matching in the face of dynamically changing characteristics such as resource load, a mechanism for reporting these dynamic characteristics is necessary.

Type of Jobs. Schedulers expect the job to name an executable that can execute on the target resources, possibly with named input and output files. Linux commands are executables immediately available on resources and might be used for testing. The job may be a script, which can also be executed immediately. If it is source code in languages that need to be compiled such as C and C++, it will need to be compiled to an executable. If it is a Java program to be interpreted by the Java Virtual Machine, it will need to be compiled to a class file and provided as an argument to the Java Virtual Machine, which is the executable, as described in Chapter 2. It is possible to submit a job that actually compiles the program as the job on the Grid resource, i.e., the compiler is the executable and the argument is the source file to be compiled.

There can be multiple executables specified or multiple instances of the same job executable. A group of instances of the same job with different arguments is sometimes called an *array* job. Most schedulers have facilities for submitting array jobs, as array jobs are extremely efficient to execute multiple resources simultaneously, and also they appear in scientific applications such as simulation studies (parameter sweep applications).

It may be that a series of jobs have to be executed to arrive at the final solution to a problem. The results of one or more jobs might be needed for another job to be able to start. There might be a complex relationship between jobs. The jobs might be complete applications that have to be interfaced together. Job schedulers, for the most part, offer some facilities to enable jobs to be submitted with dependencies in a *workflow*. Usually only very basic control-flow dependencies can be handled by the scheduler, that is, a job must be complete before another job can start. Such dependencies are given in the job description file sent to the scheduler. Additional tools are available for handling more complex dependencies and for providing drag-and-drop interfaces for building workflows graphically. Chapter 8 will describe graphical workflow editors as separate components.

Most jobs are expected to be batch jobs. One of the expected types of jobs are long running unattended batch jobs. Standard output and standard error streams are often redirected to files. Standard input would be from a file. Schedulers may not provide for true interactive jobs where the user can respond on the keyboard to programs while they are running on remote machines. Occasionally, true two-way interactive operation is possible but is done by essentially bypassing the scheduler and making a direct `ssh` connection to a remote computer.

Schedulers should be able to handle jobs that operate in specialized environments for particular types of programs such as parallel programs. For MPI message-passing parallel programs (see Chapter 9), the MPI environment with compatible MPI libraries must be present.

Types of Compute Resources. Usually, the compute resources at each site consist of a number of individual computers, sometimes hundreds of computers, connected together in a cluster. Clusters have been around for many years and job schedulers are designed that can handle cluster configurations. A typical cluster configuration is a group of computers connected through a local switch and accessed from front-end node as shown in Figure 3.2. In this configuration, jobs are received by the front-end node, which uses a local job scheduler such as Condor or SGE to route jobs to internal compute nodes.

The specific architecture of each computer system will have an effect on scheduling the jobs. Each computer system could have multiple processors sharing a common main memory, for example dual-processor or quad-processor systems. The shared common main memory enables variables and data to be shared and makes it more efficient for processors to communicate between themselves. That aspect needs to be taken into account in the scheduling algorithm.

Apart from systems having multiple physical processors, processors could have hardware support for time-sharing threads on a single processor. The general term for this technique is called *multithreading*. Intel calls their patented version *hyper-threading technology*, which was introduced in the early 2000s. Hyper-threaded Intel processors can operate as two virtual processors. Hyper-threaded processors also have to be scheduled accordingly to make best use of the hyper-threading.

Processors containing multiple execution cores were introduced in the mid 2000s when it became clear that increases in clock speed to obtain increased performance could no longer be maintained. Each core in a *multi-core processor* is capable of independent execution of processes/threads. Again, the scheduling algorithm needs to take this processor design into account.

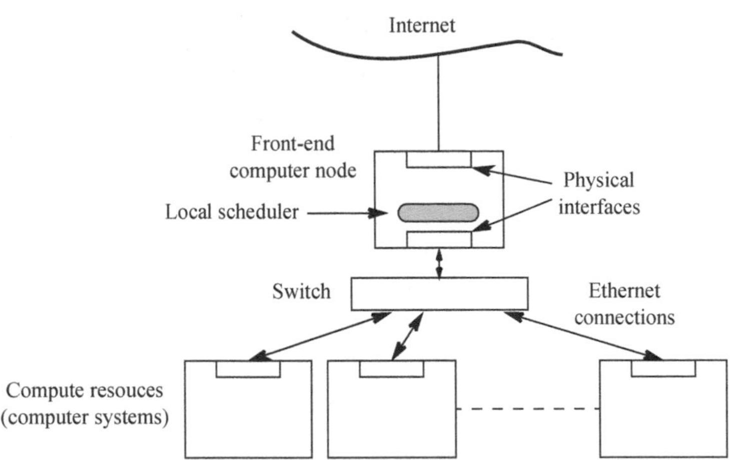

Figure 3.2 Typical computer cluster configuration.

Scheduling Compute Resources. Resource characteristics that the scheduler will consider include:

- Static characteristics of machines
  - Processor type
  - Number of cores, threads
  - Speed
  - Main memory, cache memory, disk memory, ...

- Dynamic machine conditions including:
  - Load on machine
  - Available disk storage
  - Network load
- User preferred/required compute resources
- Network connections and characteristics
- Characteristics of job including:
  - Code size, data size, ...
  - Expected execution time
  - Memory requirements, ...
  - Location of input files and input staging
  - Location required for output files and output staging

among other factors. The input data has to be arranged in the right place (staging). Data may need replication and/or movement as appropriate.

Job Matching. Job matching implies that more than one resource might be suitable for the job and some resources are more suitable than others. Hence, a selection process is necessary, typically by computing a figure of merit for a particular job-resource matching using attributes of the job and resources, and choosing the highest figure of merit. The figure of merit of a particular matching will use an algebraic expression with the job and resource attributes as terms. Sometimes that expression can be defined by the user in the job description. Usually, scheduling occurs without the user having to be involved (*automatic job scheduling*), except for providing information about the job and its requirements. It may be possible to guide the scheduling or say which computer system(s) should or must run the job in the job description.

3.1.2 Monitoring Job Progress

One of the central features of schedulers is that they monitor the progress of a job and report back to the user. Typically, a job exists in one of various states as it goes through processing, such as:

- Submitted/pending
- Running
- Completed
- Stopped/aborted
- Hold

although each scheduler may use different terms and more states. For example, there may be an additional state for failed jobs as opposed to jobs that have been stopped for intentional purposes. Figure 3.3 shows a very simple state diagram. When a user submits a job, the job is placed in a queue and enters the *pending* state. When the job is dispatched and begins execution, it enters the *running* state. When the job finishes, it enters the *completed* status. The job may be interrupted during the running state and then it enters the *stopped* state and has to be rescheduled before it can be run again, i.e., it goes back to the *pending* state before entering the *running* state again. A job might be removed from the queue (*pending* state) by the user to prevent it from being scheduled for execution. In that case, the job goes into the *hold* state. It might also be possible for a job to enter the *hold* state from the *running* state. Users should be able to query schedulers to report the job status (state) and errors that occur during job execution.

Figure 3.3 is only a representative state diagram for a job scheduler. It can differ depending upon the scheduler's internal operation. In practice, the state diagram can be more complicated, for example with the introduction of other states to show prologue and epilog states (states for pre-processing and post-processing). In a Globus environment, the job schedulers will interface to the Globus GRAM job manager. GRAM has nine external states for its Managed Executable Job service (MEJS) and 41 internal states in a highly complex state diagram (see GT 4.0 WS GRAM: Developer's Guide).

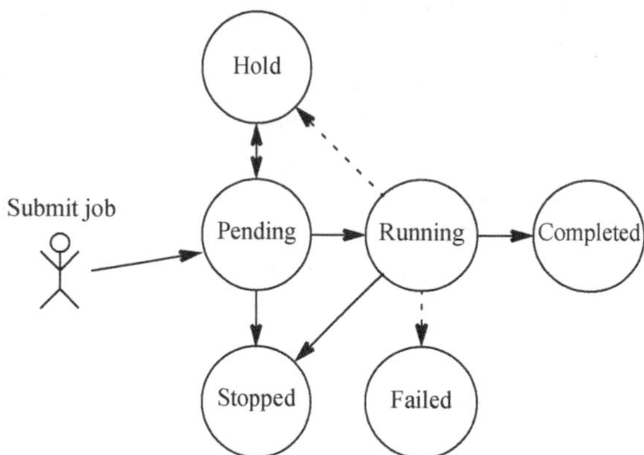

Figure 3.3 Simple local job state diagram.

3.1.3 Additional Scheduler Features

Schedulers may optionally offer such features as:

- Data placement components
- Co-scheduling
- Advance reservation
- Dynamic scheduling
- Fault tolerance and checkpointing

- Cost as a requirement/constraint
- QoS (quality of service) as a requirement/constraint
- Security

Data Placement. In Grid computing, moving input and output data (data placement or staging) is common for a job. Components for data placement may be separate to the job scheduler as illustrated in Figure 3.4. An example of a component that is concerned with data placement is the Stork "data-aware" batch scheduler (Kosar 2006). Stork can be used in conjunction with existing schedulers, for example with Condor.

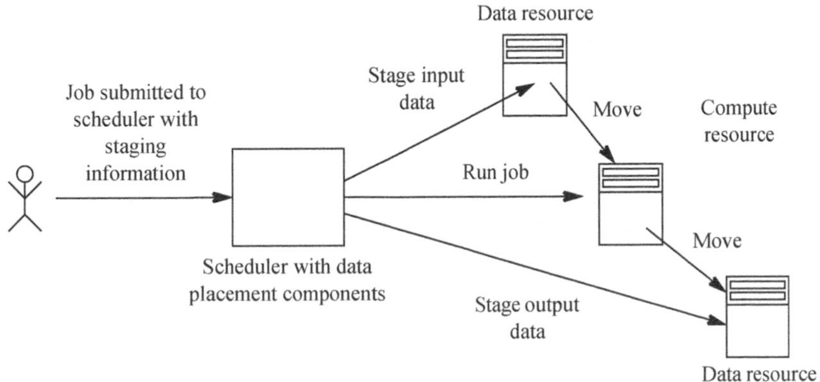

Figure 3.4 Staging input and output.

Co-Scheduling. So far, scheduling has been described as allocating a compute resource to run a job. But Grid computing offers multiple resources and a fully fledged Grid scheduler should be able to allocate multiple resources for a job where the resources are geographically distributed. Terms to describe this include *co-reservation*, *co-allocation* and *co-scheduling*. Of course, the job needs to be able to use multiple resources. The most common situation requiring co-reservation is in parallel programming (Chapter 9) in which multiple communicating processes are executed to solve a problem. The communicating processes usually have to be running at the same time to communicate, and would need to be co-scheduled together. This leads on to the next closely related feature—advance reservation.

Advance Reservation. *Advance reservation*[1] is the term used for requesting actions to occur at times in the future, in this context, requesting a job to start at some time in the future. Advance reservation has appeared in other contexts, most notably in the reservation of network connections to achieve specific quality of service, for example to achieve a certain speed of interconnections, or for user par-

[1] Also called *advanced reservation*, but that term is imprecise English as it does not mean reservation that is advanced. One could have advanced advance reservation!

ticipation in video conferencing when all participants agree to meet at a certain time. In Grid computing, both computing resources and network resources are involved. The network connection is usually not dedicated and not reservable as most use the Internet, although there are examples of network bandwidth and time being reserved on networks that are not general-purpose. Job schedulers schedule jobs according to a job description provided to it. Advance reservation can be specified by the user in the job description file, i.e., by including a statement that says that the job should be executed at a particular time.

There are several reasons one might want advance reservation in Grid computing, for example

- Reserved time is chosen to reduce network or resource contention. Resources may be less expensive or better to use at a certain time (c.f. old batch processing where jobs are submitted at nighttime so as not to affect daytime interactive users.)
- Resources are not physically available except at certain times. Resources may be shared and limited access is through an agreement.
- Jobs require access to a collection of resources simultaneously, for example data generated by experimental equipment in real time.
- There is a deadline for results of the work. Using advance reservation would lead to a specified start time given by end time minus the worse case execution time.
- In parallel programming, jobs usually need to communicate between themselves during execution. Sometimes, a specific start time is not the issue—it is that all processes that make up the parallel program on the different compute resources are operating concurrently.
- In workflows, when a job completes, often another should start. Jobs can reserve a time to wait for execution to ensure efficient completion of the workflow.

Without advance reservation, schedulers will schedule jobs from a queue with no guarantee when they actually would be scheduled to run. Since advance reservation is based upon time, it is critical that the clocks of all distributed resources are synchronized, i.e., they all "see" the same time, which can be achieved by running a Network Time Protocol (NTP) daemon that synchronizes time with a public time server.

Most schedulers did not have an advanced reservation feature initially, which led to stand-alone advance reservation components being developed elsewhere. Version 6.2 of the Sun Grid Engine (SGE) scheduler considered later included advance reservation although earlier versions did not.

Static and Dynamic Scheduling. *Static scheduling* assigns jobs to a compute resource before execution using known or predicted characteristics of the job and the job's assigned location is fixed. Most scheduling is done this way as it simplifies job management. *Dynamic job scheduling* applies the scheduling while the jobs are running and takes into account dynamic characteristics of the compute resources (dynamic load, etc.). Dynamic scheduling can include *preemptive scheduling*, in which jobs are interrupted and moved after they have started. If the load becomes too large, the job is moved to another computer resource to achieve load balancing. Preemptive scheduling will transfer running jobs to achieve a scheduling goal while *non-preemptive scheduling* only transfers jobs prior to running the jobs.

Fault Tolerance. Fault tolerance implies that a system can operate in the face of hardware failures. It usually requires redundant hardware components. Most Grid resources are not designed to be fault tolerant in that manner (except perhaps the disk drives). However, it is important to be able to recover from a temporary system failure and continue later. It is particularly important for a long running job that is interrupted by a system failure to be able to restart it without having to go to the beginning of the program. The basic way to achieve that is through *checkpointing* whereby information about the status of the execution of the job is stored at regular intervals during the execution of the program, as illustrated in Figure 3.5. The program can then be restarted at the last checkpoint. Some schedulers, such as Condor and SGE, do provide checkpoint abilities if selected by the user.

Related to fault tolerance is *job migration*. Job migration refers to moving a job from one compute resource to another, possibly after the job has started. This is particularly relevant for Grid computing and may be done for

- Recovery after a resource failure, or
- Distributing load dynamically to achieve a certain level of performance.

Checkpointing provides a mechanism to restart in both cases. Some schedulers provide checkpointing for process migration, but the user has to move the job physically (by user-initiated commands) and then restart elsewhere. More powerful schedulers also provide for automatic process migration for fault tolerance and load balancing.

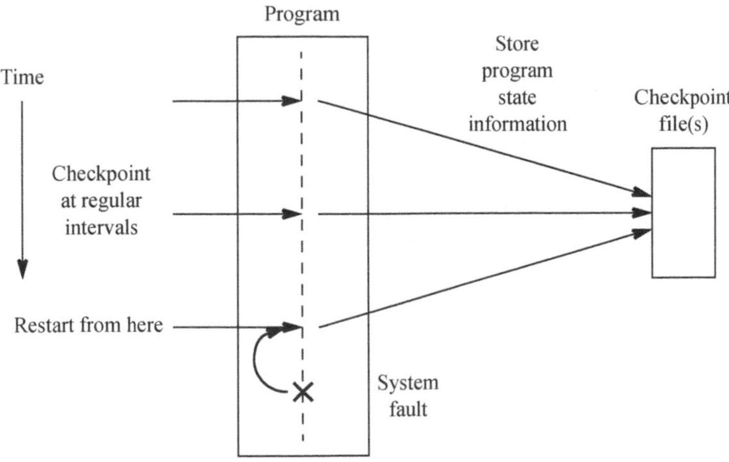

Figure 3.5 Checkpointing concept.

An advanced form of job migration, especially in Grid computing, is where the job itself can control its migration during execution. This might be done if the job demands more resources during execution, for example more memory than is available.

Cost. (Financial) cost is somewhat a unique factor for schedulers and would be significant if users were charged for computer time and that needed to be incorporated into the scheduling decisions. Cost is specifically addressed by the Nimrod-G scheduler.

Quality of Service. Quality of service (QoS) is related to cost and implies that a formal agreement has been made to provide computing facilities to an agreed level of performance for the job. Of course, there is always an informal expectation from users that reasonable QoS is provided for all computing resources. QoS is often associated with network performance (speed, delay, etc.) but in context of schedulers, it is extended to include all equipment and software involved in using the computing resources. Quality of service becomes extremely important in a Grid computing environment where different compute resources are provided by different organizations to create the collective Grid computing environment for users. A user may have a choice of which resources to use. Agreements need to be in place on policies and availability.

The job description may have implied QoS conditions. Advance reservation is a service-level agreement between the user and the scheduler stating that the scheduler will arrange that the job will be executed at a certain time. The scheduler should honor this agreement. There is the possibility that the scheduler cannot honor the request due to scheduling and resource conflicts.

Resource Broker. The term *resource broker* is used for a component that can negotiate for resources, although the term is also used simply as a powerful scheduler that can optimize resource performance. But the word *broker* is best used to indicate that it is some form of intermediary between the user and the resources that brokers an agreement. In some special cases, that agreement includes a financial aspect (paying for use of a compute resource). With that definition, a resource broker would be a higher level component that interfaces with information services and schedulers as illustrated in Figure 3.6. Resource brokers exist that have a portal/GUI interface through which users can specify their requirements.

Agreements can be described using WS-Agreement documents. The WS-Agreement specification is an XML language and protocol developed by GRAAP (Grid Resource Allocation and Agreement Protocol) Working Group of the Global Grid Forum for describing resource capabilities and forming agreements between service providers and service consumers.

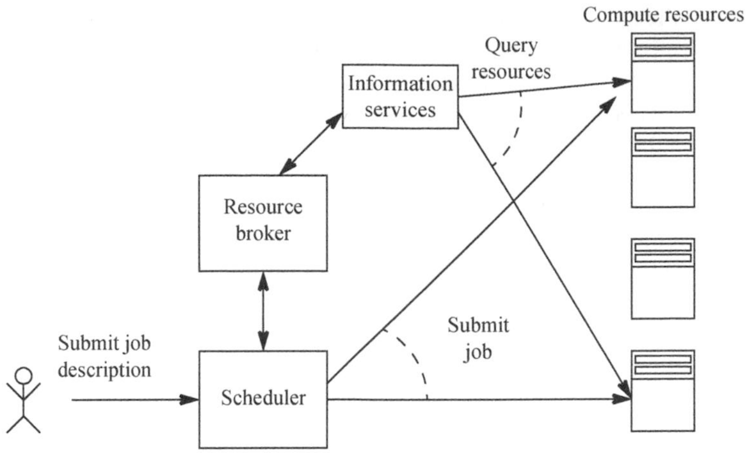

Figure 3.6 Resource broker.

3.2 SCHEDULER EXAMPLES

As described in Chapter 2 (Section 2.2.1), the Globus toolkit was not provided with a job scheduler. The GRAM job manager is able to "fork" a job locally and interface to common local job schedulers. It is assumed that clusters of compute resources would be provided with a local scheduler and that is the case for most systems. In this section, two commonly used local schedulers will be described—SGE and Condor. Both of these have been used in our Grid computing course. SGE is representative of a traditional job scheduler. Condor began somewhat differently as a research project in the 1980s and has some interesting features for matching jobs to resources and creating workflow. Historically, Condor predates SGE, but we will begin with SGE as the more traditional job scheduler. Both SGE and Condor were not originally designed for Grid computing environments, although Condor now has a Grid version called Condor-G and SGE now has advance reservation, useful for Grid computing. A scheduler that was designed for a Grid computing environment and for performing meta-scheduling across geographical sites is GridWay, which is now part of the Globus distribution. GridWay will be described in Section 3.3 under meta-schedulers.

3.2.1 Sun Grid Engine

Sun Grid Engine (SGE) is a job scheduler provided by SUN Microsystems for distributed compute resources. It was originally developed by Gridware Inc, a Germany company founded by Wolfgang Gentzsch. Gridware was acquired by SUN in 2000. The supported commercial version of the SGE scheduler is called *N1 Grid Engine (N1GE)*. An open source version is available, called simply *Grid Engine*. The name of the scheduler, (SUN) Grid Engine, is somewhat misleading as it implies that it is primarily a Grid meta-scheduler meant for geographically distributed Grid computing applications. However, SUN documentation defines three levels of Grid, a *cluster grid*, a *campus grid*, and a *global grid*. Grid Engine has the target of cluster and campus grids but an upward transition path to global grids (Sun Microsystems 2007). A SUN cluster grid is simply a traditional local cluster contained in one location, typified by Figure 3.2 earlier, which we have not described as a true Grid. The SUN campus grid describes a group of compute resources and clusters at different locations with different owners but within a single organization, which comes under our term of a distributed Grid as an enterprise Grid. A SUN global Grid is a group of geographically distributed compute resources and clusters in the traditional Grid sense.

Grid Engine has all the usual features of a job scheduler including:

- Various scheduling algorithms
- Job and machine matching
- Multiple job queues
- Checkpointing for both fault tolerance and job migration
- Multiple array jobs
- Parallel job support
- Advance reservation (from Grid Engine version 6.2 onwards)
- Accounting
- Command line and GUI interfaces
- DRMAA interface (see later)

Machines within a Grid Engine environment are classified according to their duties and can have multiple classifications (duties). *Submit hosts* can submit jobs. *Execute hosts* can execute jobs. An *administrative host* performs system administration work. A single *master host* controls the cluster activity handling job scheduling, and by default is also an administrative host and a submit host. (Condor also has a somewhat similar structure with submit hosts, execute hosts and a master host, see later.)

Grid Engine has both a command-line interface (CLI) and a graphical user interface (GUI).

Command-Line Interface. Grid Engine (version 6.2) has 18 user commands available for submitting and controlling jobs. Remote execution of both batch and interactive jobs are supported. Each Grid Engine command starts with the letter *q*. Some common commands are shown in Table 7.1. A user might start with issuing `qhost`, which lists the machines available. Sample output from `qhost` is:

```
HOSTNAME                 ARCH         NCPU  LOAD  MEMTOT  MEMUSE  SWAPTO  SWAPUS
--------------------------------------------------------------------------------
global                   -               -     -       -       -       -       -
coit-grid01              lx24-x86        4  0.00    1.9G  821.7M    1.9G     0.0
      .
      .
      .
```

TABLE 3.1 SOME GRID ENGINE COMMANDS (NOT ALL OPTIONS GIVEN)

Command	Purpose
`qhost <options>`	Display status of hosts, queues, and jobs
`qstat <options>`	Display status of jobs and optionally queues
`qsub <options> <command>` `<command arguments>`	Submit a batch job. `<command>` specifies either a script or binary. If binary, include option `-b` yes
`qdel <options>`	Delete job(s) from queues Options: `<job list>` list of jobs to delete
`qrsh <options> <command>` `<command arguments>`	Submit interactive `rsh` session
`qmon`	Start X-Windows GUI interface

Jobs are submitted using the `qsub` command. A job can be specified as a (shell) script or as named executable (the latter with the `-b yes` option). For example, for a simple test, one might run

```
qsub -b y uptime
```

The immediate output will be of the form

```
Your job 238 ("uptime") has been submitted.
```

The `qstat` command can be used to display the status of jobs that are in the system (not completed). For example, `qstat` issued after the above might produce the display

```
job-ID  prior   name     user     state submit/start at      queue slots ja-task-ID
-----------------------------------------------------------------------------------
    238 0.00000 uptime   abw       qw    06/03/2008 13:34:52    1
```

The status `qw` indicates waiting in queue. Once the job is completed, `qstat` will display nothing. By default, standard input and standard output is redirected to files named `<job_name>.o<job-ID>` and `<job_name>.e<job_ID>`, that is, for the above `uptime.o238` and `uptime.e238` will be created in the user's current directory.

Job described by a script can be run with simply

```
qsub my_script.sh
```

where `my_script.sh` is the script. Any `qsub` options (flags) can be placed in the script by starting the line with `#$`.

Graphical User Interface. Everything a typical user might want to do can be achieved using the Grid Engine graphical user interface shown in Figure 3.7. This GUI is started with the `qmon` command. Buttons on the GUI will take you to other GUI interfaces for submitting jobs and checking their status. Figure 3.8 shows

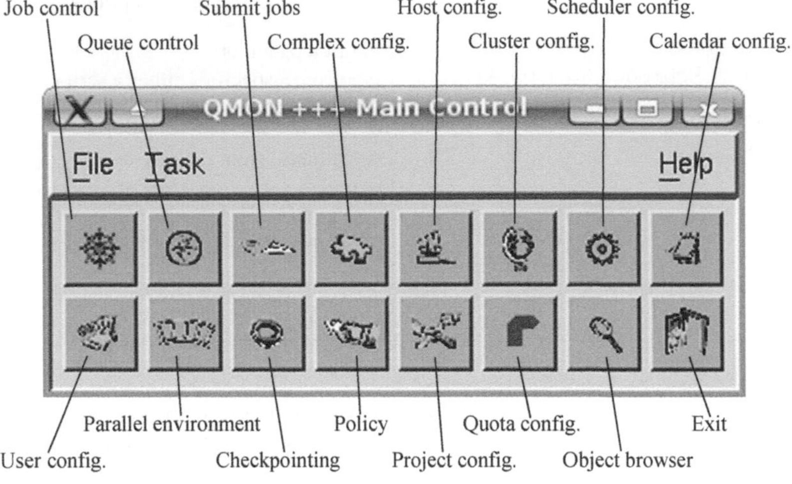

Figure 3.7 Representative Grid Engine graphical user interface – main control.

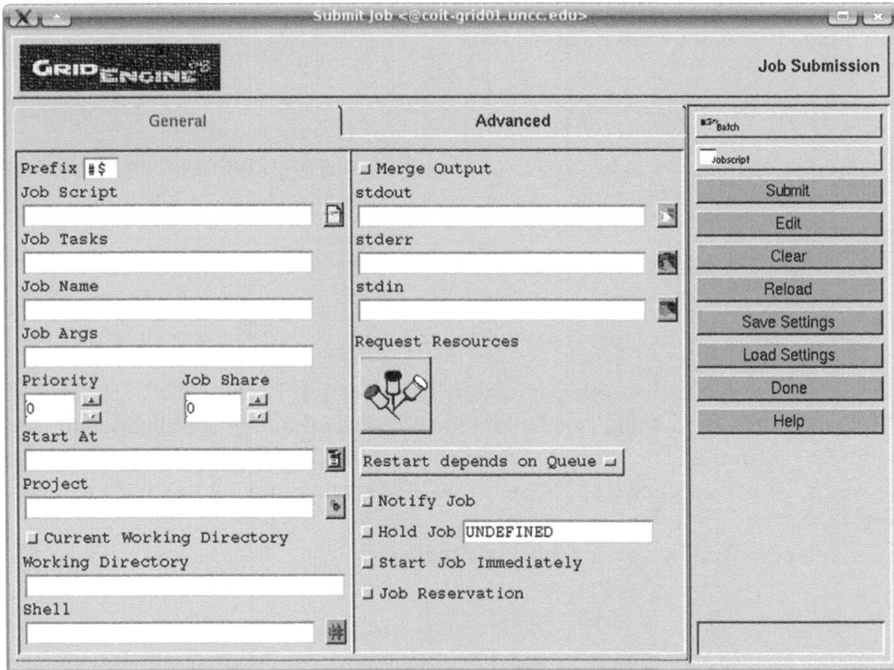

Figure 3.8 Grid Engine graphical user interface – job submission interface.

the job submission interface where one can enter the details of the job. On the advanced section shown in Figure 3.9, one can arrange for an email to be sent when the job starts, when it ends, or if the job is aborted or is suspended, and other things can be set such as checkpointing. Finally, one would then click on "submit" button to submit the job. On a job control window, one can monitor the progress of jobs (pending jobs, running jobs, finished jobs). The qmon GUI interface can be customized.

A script offers the possibility of compiling the program on the actual machine that you want to execute the program. For example, one could compile and run the Java program prog1 with a simple script that includes

```
#!/bin/bash
javac prog1.java
java prog1 $1
```

#! states which shell should be run (bash in this case), $1 refers to the first command-line argument after the script.

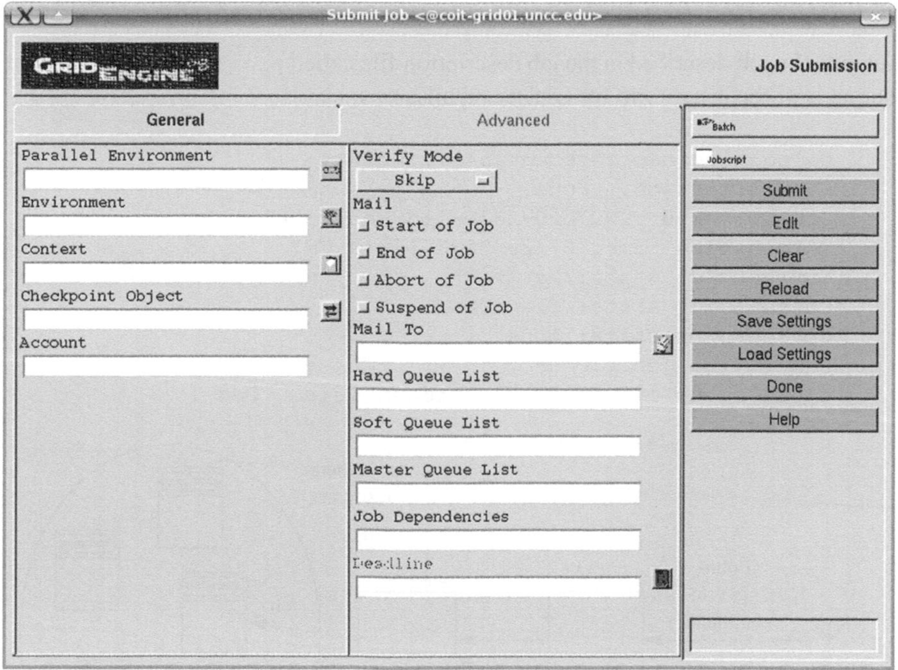

Figure 3.9 Grid Engine job submission interface – advanced section.

Interactive Session. Grid Engine offers the ability to have an interactive session, that is, directly communicate with a machine in the cluster and run jobs with standard input from the console redirected to the program and standard output redirected to the console. There are several commands for that, depending upon the type of connection one wants to use. Each command has the same command line arguments as qsub. The qrsh command makes rsh connection. Without any arguments, it simply makes connection with a machine chosen by Grid Engine. An example of an interactive session would be

```
qrsh hostname
```

which will return the hostname of an available machine in the cluster, as chosen by the scheduler. This may change as the load on machines change.

Running a Globus Job with the SGE Scheduler. In a Globus Grid environment, GRAM is the front-end of the scheduler. Jobs are submitted to GRAM, with the globusrun-ws command instead of directly to the scheduler with the qsub command and GRAM forwards the job to the scheduler as illustrated in Figure 3.10. Using the globusrun-ws command, the scheduler is selected by name with the -Ft option (i.e., factory type), as mentioned in Chapter 2. The name for Sun Grid Engine is SGE (obviously). Hence, the command

```
globusrun-ws -submit -Ft SGE -f prog1.xml
```

submits the job described in the job description file called prog1.xml. The resultant output is of the usual form for Globus submissions, i.e.,

```
Delegating user credentials...Done.
Submitting job...Done.
Job ID: uuid:d23a7be0-f87c-11d9-a53b-0011115aae1f
Termination time: 07/20/2008 17:44 GMT
Current job state: Active
Current job state: CleanUp
Current job state: Done
Destroying job...Done.
Cleaning up any delegated credentials...Done.
```

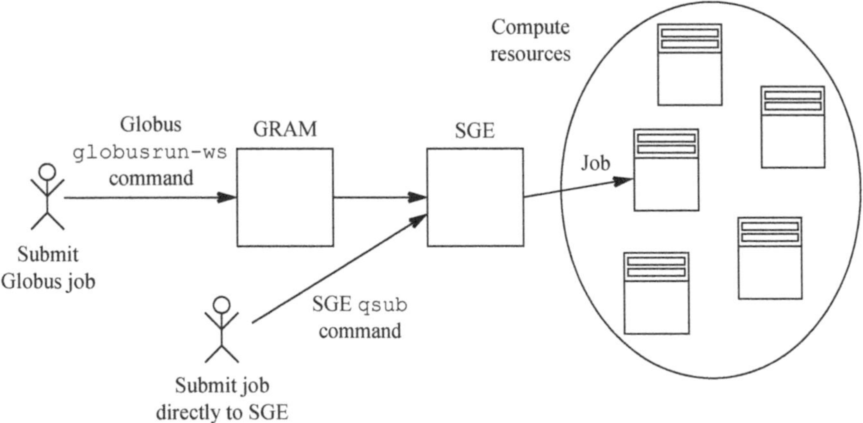

Figure 3.10 Submitting a job through GRAM and through an SGE scheduler.

This output shows that the user credentials have to be delegated. The scheduler will choose the actual machine that the job is run on, which can vary for each job. Hence

```
globusrun-ws -submit -s -Ft SGE -c /bin/hostname
```

submits the executable `hostname` to the SGE scheduler in streaming mode redirecting output to the console, with the usual Globus output. The hostname displayed will be that of machine running the job and may vary.

Specifying Submit Host. The previous commands specify the scheduler but not the submit host and hence will submit to the machine running the Globus command as default, i.e., through `https://localhost:8443/wsrf/services/ManagedJobFactoryService`. A specific location for the host and/or factory service can be specified by using the `-F` option (see Chapter 2). The command could get somewhat long, e.g.,

```
globusrun-ws -submit -s     \
     -F http://coit-grid03.uncc.edu:8440    \
     -Ft SGE -c /bin/hostname
```

or even longer if output is to be streamed to specific files.

3.2.2 Condor

Objectives. Condor was developed at the University of Wisconsin–Madison in the mid-1980s to enable networked computers to be used for high performance computing when the computers are idle (*cycle scavenging*). It was recognized that many laboratory computers were not being used continuously by users. It was clear that computers were not used at all in the nighttime, as well as during the daytime, users are not fully occupying them with tasks. The original premise is still valid. It is still the case that many office and laboratory computers are under-utilized, yet the power of these computers continues to increase. It is not uncommon for processors in desktop computers to have similar performance to those in servers and computers in clusters. With the advent of multicore processors, desktop computers are multicore, yet for the most part they are used for simple office tasks such as accessing the Internet and word processing. The goal of Condor is to harness the unused computing power of computers by making a collection of computers accessible as a high-performance cluster. The only requirement is that the computers are networked.

The Condor project has been hugely successful. Many institutions now operate Condor clusters. The Condor software is essentially a job scheduler where the jobs are scheduled in the background on distributed computers, but without the user needing an account on the individual computers. Users need to compile their programs for the types of computers Condor is going to use. To use some of Condor's features, Condor libraries need to be included. A compile script is provided, `condor-compile`. The job is described in a job description file. Condor then ships the job off to the appropriate computers. The classical ideal use case for Condor is for executing a long-running job multiple times with different parameters (*parameter-*

sweep problem), where there is no communication between jobs and each job can be scheduled independently. If a single parameter sweep takes n hours on a single computer. With p sweeps, it would take np hours. With m computers, it would take np/m hours, where p is a multiple of m, as illustrated in Figure 3.11.

Condor includes a number of interesting features. It is a scheduler with job queues, but it is able to find necessary resources for the job and send the job to that resource. Condor is designed with the view of long running batch jobs where the user may leave after submitting the jobs. It provides for restarting jobs that are interrupted, through checkpointing if the user should enable that feature. It is able to migrate processes from one location to another (*job migration*). It can handle machine crashes or disk space becoming exhausted, or software not being installed. The machines can be remotely located and owned by others (assuming permission is given to use them as Condor resources). Computers can be heterogeneous. Condor is available for a range of platforms including various versions of Linux, Mac OS X, and Windows, although not all features are implemented for Windows.

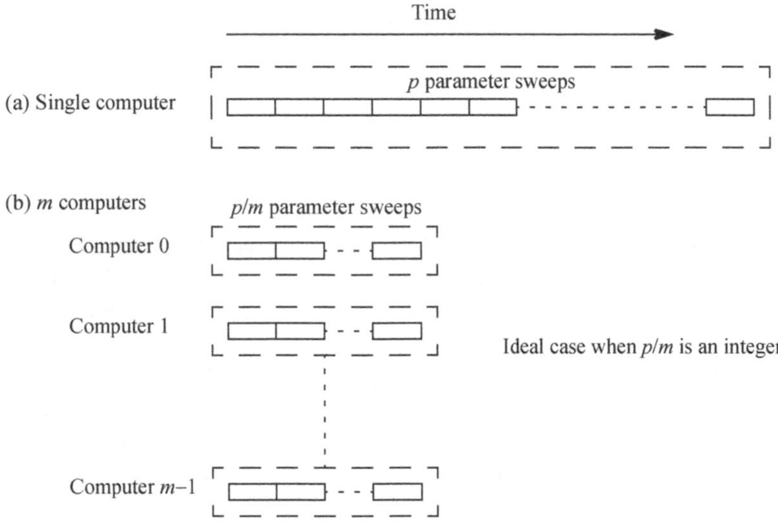

Figure 3.11 Submitting multiple parameter sweeps across m computers.

Physical Structure. Condor is structured as a collection of machines called a Condor *pool*. There are four roles for machines in a Condor pool: a *central manager*, *submit machine*, *execution machine*, and *checkpoint server*. The central manager is the resource broker for the pool and negotiates with compute resources, checking the status of the resources and running jobs. There can be only one central manager in a pool. The submit machine submits jobs to the pool. Clearly, there has to be at least one submit machine in a pool, but there can be and usually is more than one submit machine. Users access the submit machine to submit their jobs, usually through `ssh`. The execute machine runs the jobs that are submitted. Again there has to be at least one execute machine in a pool, but there usually is more than one execute machine. A computer can serve as both a submit machine and an execute machine. The checkpoint server stores checkpoint files produced by jobs for which checkpointing is enabled. There can be only one checkpoint server in a pool. It is optional to have a checkpoint server.

A general configuration would have single central manager, a single checkpoint server, computers that are submit machines, computers that are execute machines, and computers that are both submit and execute machines, as shown in Figure 3.12. Another possibility is to have a single central manager that also operates a single checkpoint server when needed, and make all other computers both submit and execute machines. This configuration might suit the situation when a group of lab computers is being used submit/execute machines. Users submit their jobs by logging onto a submit host or lab machine. One configuration would be comparable to a cluster with a single front end node as illustrated previously in Figure 3.2. The front-end would be the submit machine and could also be the central manager. The compute nodes could be execute machines. Obviously, there are other possibilities.

It may be desirable to separate functions to individual computers for workload and scalability. Multiple submit machines reduce bottlenecks that might occur if there were only one submit machine. Communication is needed between all types of machines. The submit machine will communicate with the central manager to determine where to submit the job, and communicate with the execute machine to submit the job. The central manager will communicate with the execute machines to find out their status and the status of jobs. Each machine will run a Condor daemon for inter-machine communication. The steps performed to run a job are as follows:

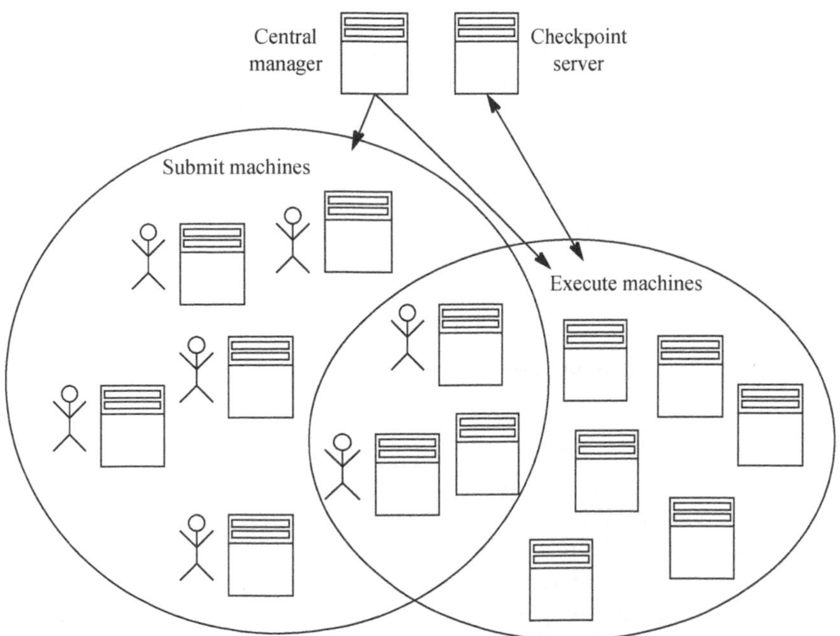

Figure 3.12 General Condor configuration.

1. User submits job description to submit machine.
2. Submit machine tells central manager of job description and requirements.
3. Central manager matches job with suitable execution units. (At intervals, central manager receives from execute machines descriptions and availability of the execution machines.)
4. Central manager informs execute machine that it has been claimed for the job and gives it a *ticket*.
5. Central manager gives matching ticket to submit machine.
6. Submit machine contacts execution machine and provides ticket and copies job executable and any necessary data files to execution machine.
7. Execute machine executes job.
8. Execute machine returns results to submit machine.

These steps are shown in Figure 3.13. In this scenario, the job executable is physically transferred to the execute machine automatically by Condor. If there is a shared file system, the executable host can access the files itself. Condor does not require a shared file system but can use one if available. If a connection is broken, the job will be resubmitted from the beginning unless checkpointing is available and enabled. Then, the job can restart from the last checkpoint.

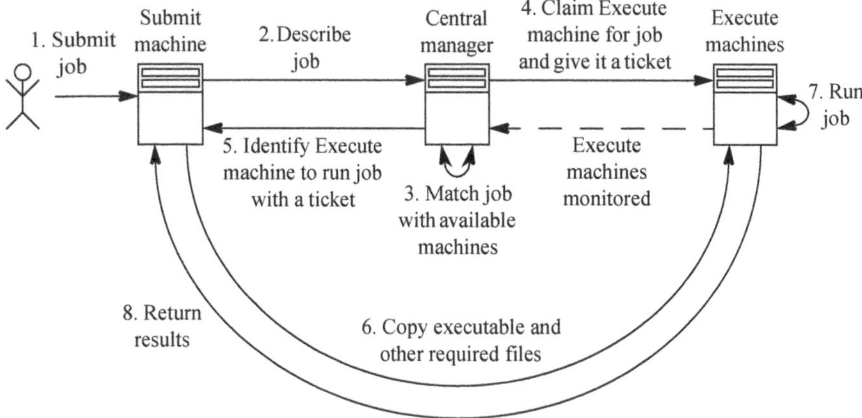

Figure 3.13 Internal steps to execute a job in Condor.

Types of Jobs. An important aspect of Condor is that the jobs are run on the execute machines as background batch jobs and so one cannot run interactive jobs where there is real-time interaction with the user. Such jobs will need to be reformulated so that input can be read from a file and output is also redirected to a file. Condor defines so-called *universes,* which are environments that the job will run in and in particular provide the appropriate support and Condor libraries for the type of job. The number of universes has grown and has been refined over the years. Condor version 7 has nine universes:

1. Vanilla
2. Standard
3. Java
4. MPI
5. Parallel
6. Grid
7. Scheduler
8. Local
9. Virtual Machine (VM)

but that includes at least one universe for legacy purposes—the MPI universe for MPI parallel programs. MPI programs can also be executed in the newer Parallel universe. The Scheduler universe may also be subsumed into the Local universe. The Local universe runs jobs on the submit machine where they are submitted. The Vanilla, Standard, Java, Parallel, and Grid universes will be discussed further here. The previously named Globus universe is now incorporated into a universe called Grid.

Vanilla Universe. Straightforward computational jobs written in regular compiled languages such as C or C++ or pre-compiled applications can use the Vanilla universe (the default universe from Condor 7.2). The Vanilla universe is also used for shell scripts and Windows batch files. The universe does not provide for features such as checkpointing or remote system calls and may be less efficient than under other universes, but it only requires the executable. Programs are not compiled with Condor libraries. Jobs that cannot be compiled with these libraries would use the Vanilla universe. There are certain code restrictions such as on the use of system calls.

Let us first take an example of a program already compiled, say the `uptime` Linux command, which displays the following information: the current time, how long the system has been running, how many users are currently logged on, and the system load averages for the past 1, 5, and 15 minutes. The job has to be described in a Condor job description file. When Condor was first developed, a standard job description language such as JDSL (see Chapter 2) did not exist. Even now, there are many non-XML proprietary job description languages. Condor has its own format for the job description consisting of a list of parameters and their values, very much like the RSL-1 language described in Chapter 2. The Condor documentation uses the term *submit description file*. The principal parameters in this file are the name of the universe (unless already specified with a system configuration variable), the name of the executable, the names of the arguments, the names of the input files, and the name of directories. The Condor submit description file is constructed as a list of commands in a script. There are a very large number of Condor submit description commands. A macro feature enables substitution of values, which is especially useful for complex multi-job situations. The reader is referred to the on-line Condor manual for full details of the commands, restrictions, and features. Here, few important commands will be introduced.

The Condor submit description for the simple `uptime` job could be

```
# This is a comment condor submit file for uptime job
universe = vanilla
executable = /usr/bin/uptime
arguments = 23 45
output = uptime.out
error = uptime.error
log = uptime.log
queue
```

Notice that the full path is given for `uptime`. Alternatively, directories could be specified separately. The log file named `uptime.log` record events that may be helpful such as when a job begins running, checkpoints in the standard universe (see later), completes, etc. The log file is not necessary but is recommended.

The command `queue` tells Condor to submit the job to the scheduling queue. To execute the same job multiple times, an integer specifying the number of times would follow the `queue` command. To execute the same job multiple times each with different arguments, one would have single `executable` command specifying the executable but multiple `argument` and other commands relating to the instance of the job, each group of commands terminated with a `queue` command.

The Condor command to submit the job is `condor_submit`, which takes the submit description file as an argument. For example, if the submit description file is called `uptime.sdl`, the command would be

```
condor_submit prog1.sdl
```

Without any other specification, Condor will attempt to find a suitable executable machine from all available. As mentioned, Condor works with and without a shared file system. The default setting for a Linux system specifies a shared file system and the default setting for a Windows system specifies that there is not a shared file system. Most local clusters, such as illustrated in Figure 3.2, will be set up with a shared file system and Condor will not need to transfer the files explicitly. In the case when it is necessary to explicitly tell Condor to transfer files, additional parameters are included in the submit description file, as in

```
# This is a comment, condor submit file for uptime
# with explicit file transfers
universe = vanilla
executable = uptime
output = uptime.out
error = uptime.error
log = uptime.log
should_transfer_files = YES
when_to_transfer_output = ON_EXIT
queue
```

Some basic commands available are shown in Table 3.2. The command `Should_transfer_files = YES` says that Condor is to transfer the executable

TABLE 3.2 SOME BASIC SUBMIT DESCRIPTION FILE COMMANDS (Derived from the Condor Manual (Condor team 2009).)

<table>
<tr><th>Command</th><th>Provides or action</th></tr>
<tr><td>universe = <vanilla | standard | scheduler | local | grid | mpi | parallel | java | vm></td><td>Condor Universe to use when running job.</td></tr>
<tr><td>environment = <parameter_list></td><td>List of environment variables each defined by <name>=<value>.[a]</td></tr>
<tr><td>arguments = <argument_list></td><td>Program command line arguments. For Java, universe, the class file.[b]</td></tr>
<tr><td>executable = <pathname></td><td>Name of executable file (required). Full path or relative path from current directory.</td></tr>
<tr><td>input = <pathname></td><td>Name of file containing keyboard input (stdin).</td></tr>
<tr><td>output = <pathname></td><td>Output file to hold redirected program screen output (stdout).</td></tr>
<tr><td>error = <pathname></td><td>Path and file name to hold redirected error messages (stderr).</td></tr>
<tr><td>log = <pathname></td><td>Name of file Condor will use to record job events.</td></tr>
<tr><td>should_transfer_files = <YES | NO | IF_NEEDED ></td><td>Define whether to transfer files to and from machine where job runs. Not used in Grid universe.</td></tr>
<tr><td>when_to_transfer_output = < ON_EXIT | ON_EXIT_OR_EVICT ></td><td>Defines when to transfer files if transfer specified by should_transfer_files.</td></tr>
<tr><td>transfer_input_files = < file1,file2,file... ></td><td>Files to be transferred into local working directory for job before it is started on remote machine.</td></tr>
<tr><td>notification = <Always | Complete | Error | Never></td><td>Owners notified by e-mail when job events occur.</td></tr>
<tr><td>notify_user = <email-address></td><td>E-mail address Condor uses to send e-mails.</td></tr>
<tr><td>queue [number-of-procs]</td><td>Places one or more copies of previously described job in Condor queue. Possible to have multiple queue commands, see text.</td></tr>
</table>

a. Delimiter is a semicolon (;) for Unix, a vertical bar (|). New syntax uses space delimiters with whole list surrounded by double quotes.
b. Delimiter is a white space. For whitespace in list item, use double quotes around list and single quotes around item with space(s).

and input files to the remote machine to execute the job and transfer all files generated by the job back to the submit machine. (The alternatives are NO and IF_NEEDED, the latter transfer does not take place if a shared file is present.) when_to_transfer_output says when the files generated by the job are to be transferred back to the submit machine. In this example (ON_EXIT), and usually, the transfer is at the end of the job execution. The alternative is ON_EXIT_OR_EVICT, see the Condor manual for more information. Condor has the command transfer_input_files to transfer files prior to execution. It also has the command transfer_output_files for transferring specific files back to the submit machine, but mostly Condor will do this automatically, transferring all files created by the job. In some specific situations, for example running Java programs that require additional class files and packages, it may be necessary to explicitly transfer files necessary for the job in additional to the executable and named input files.

As with SGE, a useful feature is that Condor can be instructed to send you an email regarding job events (completion, if errors occurs, ...). If you do not specify the email address, Condor will use a default one based upon the user ID, so it is best to specify the email address that you actually want to use.

After submitting the condor_submit command, there will be a message indicating that the job has been submitted to Condor, such as:

```
Submitting job(s).
Logging submit event(s).
1 job(s) submitted to cluster 662.
```

One can query the status of the Condor queue with the condor_q command. The output will be of form

```
Queue

-- Submitter: coit-grid02.uncc.edu : <152.15.98.25:32821> :
ID      OWNER            SUBMITTED     RUN_TIME ST PRI SIZE CMD
    :
662.0   abw              5/23 17:36   0+00:00:00 I  0   9.8  uptime

16 jobs; 1 idle, 0 running, 15 held
```

Status can be H (hold), R (running), I (idle, waiting for machine), C (Completed), U (unexpanded, never being run) or X (removed), where applicable. Once the job is completed, it will vanish from the queue. The log file will show details, for example

```
cat uptime.log
000 (662.000.000) 05/23 17:36:10 Job submitted from host: <152.15.98.25:32821>
...
001 (662.000.000) 05/23 17:36:43 Job executing on host: <152.15.98.25:32822>
...
```

```
005 (662.000.000) 05/23 17:36:43 Job terminated.
        (1) Normal termination (return value 0)
                Usr 0 00:00:00, Sys 0 00:00:00  -  Run Remote Usage
                Usr 0 00:00:00, Sys 0 00:00:00  -  Run Local Usage
                Usr 0 00:00:00, Sys 0 00:00:00  -  Total Remote Usage
                Usr 0 00:00:00, Sys 0 00:00:00  -  Total Local Usage
        71  -  Run Bytes Sent By Job
        7928  -  Run Bytes Received By Job
        71  -  Total Bytes Sent By Job
        7928  -  Total Bytes Received By Job
...
```

Condor provides input and output streaming in the Vanilla universe, selected with the commands given in Table 3.3 and also in the Grid universe by default. Should the job terminate before finishing for some reason, the input and output streams will also be disrupted without a permanent record.

TABLE 3.3 CONDOR STREAMING COMMANDS (Derived from the Condor Manual (Condor team 2009).)

Method	Effect if TRUE
`stream_error = <True | False>`	`stdin` is streamed from the execute machine back to the submit machine.
`stream_input = <True | False>`	`stdout` is streamed from the submit machine to the execute machine (Vanilla and Java universes only).
`stream_output = <True | False>`	`stdout` is streamed from the execute machine back to the submit machine.

Standard Universe. The Standard universe provides for checkpointing and remote system calls in a Linux environment. (The Standard universe is not available in Windows as of Condor version 7.0.) One would use the Standard universe if one wanted the ability to roll back and continue the execution of a program after a system failure. If the checkpointed job is interrupted by a machine failure, it can resume at the last checkpointed state on a different machine rather than have to start from the beginning. Generally, there is no change to source code, but one has to link Condor's standard universe support libraries. This is done by using the Condor `condor_compile` command, which is placed before whatever command you would normally use to compile and link the program. (It will even accept make and shell scripts.) The complete command to compile `prog1.c` and link Condor libraries would be

```
condor_compile cc -o prog1 prog1.c
```

A simple submit description would be

```
#Simple condor submit file for prog1 job
universe = standard
executable = prog1
queue
```

Without file transfer commands in the submit description file, all standard input (`stdin`), output (`stdout`) and standard error (`stderr`) would be lost, or in Linux jargon, are redirected to `/dev/null`.

The Standard universe provides for remote system calls. It causes system calls in the executing program to be caught and executed on the submit machine rather than on the execute machine executing the program. The result of the call is returned to the execute machine. For example, a file can be opened on the submit machine and sent to the execute machine. There are some limitations including not being able to use certain system calls such as `fork()`, `alarm()`, and `sleep()`. The program must be single threaded. The full list of restrictions can be found on the Condor documentation (Condor team 2009). Checkpointing does carry the overhead of storing a checkpoint image of the job at regular intervals, and would only be needed for a long running job. Checkpointing can be disabled in the Standard universe by including the command

```
+WantCheckpoint = False
```

in the submit description file before `queue`. (Commands are not case-sensitive.)

Java Universe. The Java universe is for submitting Java programs that will be executed by the Java Virtual Machine. In a Globus environment, a Java program can be executed from the Globus job submit command, `globusrun-ws`, by specifying the Java Virtual Machine as the executable, i.e., `java` (usually with the full path) and the Java program class file as the first argument (without the class extension as Java expects a class file). Certainly that approach could be used in Condor in the Vanilla and Standard universes, assuming the remote machines have the Java Virtual Machine installed. However, Condor provides a special universe called the Java universe just for Java programs with some job description commands and features to help. In the Java universe, the executable specified in the submit description file is the Java class file, as this universe will invoke the Java Virtual Machine automatically. One might wish to first make sure that the remote machines have the Java (Virtual Machine) installed, by issuing the command `condor_status -java`, which will list those machines with the Java installed. This will include the Java versions, which can be important. To ensure complete compatibility, Java programs should be compiled for the lowest version number (`javac -target` option).

Suppose one wished to run the Java program, `prog1.java`. First, it would be converted into bytecode (a class file) with

```
javac prog1.java
```

The resulting class file is `prog1.class`. The submit description file might be

```
# This is a comment condor submit file for java job
universe = java
executable = prog1.class
arguments = prog1
output = prog1.out
error = prog1.error
log = prog1.log
should_transfer_files = IF_NEEDED
when_to_transfer_output = ON_EXIT
queue
```

In the Java universe, the first argument must be the name of the main class. More complicated Java programs can be handled including those requiring multiple specific class files (transferred using the `transfer_input_files` command) and jar files using `jar_files` command.

Parallel Universe. The Parallel universe is for executing parallel programs, that is, programs that execute together and collectively to solve a problem using multiple processors, usually for the purpose of increased speed. Parallel programming in the context of Grid computing will be discussed in Chapter 9. A key aspect of parallel programming is that the parts usually communicate between themselves during execution and need to execute at the same time on the different computers. Condor provides a mechanism for that to happen, co-scheduling the parts (basically co-advance reservation).

With networked computers as the target platform, the most common way to write parallel programs is to use message-passing routines that send messages between the parts. In the late 1980s, a very popular library of message-passing routines and environment was developed called PVM (Parallel Virtual Machine). PVM was superseded by a standard set of library APIs for message passing called Message Passing Interface (MPI) in the mid-1990s. Condor did provide specific universes for both PVM and MPI but later substituted a new universe called *parallel* to encompass all parallel programming. It discontinued the PVM universe altogether in version 7.0.0 onwards, but left in the MPI universe for legacy purposes.

Grid Universe. Condor can be used as the environment for Grid computing without Grid middleware such as Globus or alternatively it can integrated with the Globus toolkit.

Stand-alone: A Condor pool is a group of interconnected computers generally under one administrative domain. Grid computing generally involves multiple administrative domains. Condor pools can be joined together in a process called *flocking*. Pools have to be configured to enable flocking. Flocking is Condor's own way of create a Grid computing environment and enables jobs to migrate from one computer to another even when the computers are in different pools and under different administrative domains. Migration will only occur if a suitable computer is not available in the original pool. A relatively recent addition to Condor is Condor-C, in which jobs can move from one computer's job queue to another.

Another feature of Condor that is particularly applicable to Grid computing is the Condor *glidein* mechanism. The glidein mechanism enables computers to join a Condor pool temporarily. The Condor command for achieving this is `condor_glidein <contact argument>` where `<contact argument>` in the most simple situation is a hostname but it can also specify a job manager/scheduler or Globus resource. Various options enable more information to be passed such as the resource architecture and number of computers.

Interfaced to Globus Environment: Condor can be used as a front-end for Globus resources and in that case it is called Condor-G. We shall discuss Condor-G later in Section 3.3.2 under Grid meta-schedulers.

Condor's ClassAd Mechanism. So far, the features of Condor described allow jobs to be submitted to a group of computers (a Condor pool) in much the same way as other job schedulers. A job (submit) description file is created that describes the job. In addition, Condor has an optional feature called the *ClassAd* mechanism that enables jobs to be matched with resources (machines) according to job and machine characteristics. ClassAd is based upon the notion that jobs and resources advertise themselves in classified advertisements, which include their characteristics and requirements. The job ClassAd is then matched against the resource ClassAd. For example, a user might advertise

> *My job needs an Intel Core 2 processor a speed of at least 2 MHz (or equivalent Intel-compatible processor)*[2] *and with at least 6 GB of main memory and 1 TB of working disk space.*

Compute resources advertise their capabilities, for example

> *I am a computer with an AMD Phenom processor operating at a speed of 2.6 MHz with 256 GB of main memory and 16 TB of working disk space.*

Hopefully a match can be found for the user's job with a machine. The ClassAd mechanism is illustrated in Figure 3.14. The best resource for the job is chosen by Condor. For this to work, both the job and all the computing resources (machines) need ClassAds.

The job ClassAd is included as job attributes in the submit description file. (Previously, the ClassAd mechanism was not invoked as there was no job ClassAd given in the job description file.) The resource ClassAd is described using machine attributes and stored in a file with the resource. This ClassAd is set up during the system configuration. The values of some of the attributes can be dynamic and alter during the system operation (for example load).

---

[2] In addition to clock speed, Condor provides for benchmark criteria.

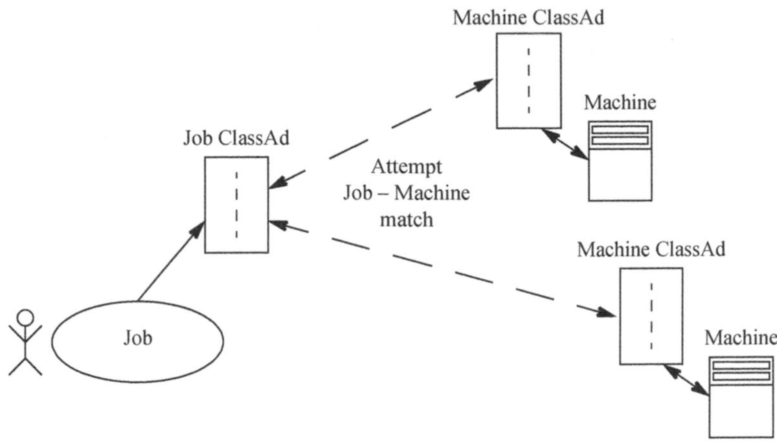

Figure 3.14 Condor's ClassAd matchmaking mechanism.

`MyType` and `TargetType` Commands. As we can see, there are two types of ClassAd, job ClassAds and machine ClassAds. The type of ClassAd is identified by the `MyType` command in the ClassAd, i.e.,

```
MyType = Job
      or
MyType = Machine
```

Another command in the ClassAd, `TargetType`, is used to specify what the ClassAd is to match with. For a job ClassAd, that would be

```
TargetType = "Machine"
```

and for a machine ClassAd, that would be

```
TargetType = "Job"
```

Machine Characteristics. Each machine can be described by a series of ClassAd attributes that specify such things as machine name, architecture, operating system, main memory available for job, disk memory available for job, processor performance, current load, etc. such as listed in Table 3.4. There are many other attributes. The machine characteristics include both static features and dynamically altering parameters. As of Condor version 7, the number of cores is not provided as a default attribute although the number of processors is provided. Additional attributes can be defined.

Job Characteristics. Table 3.5 shows some attributes for a job. There are many other attributes. The job is typically characterized by its resource requirements and preferences.

TABLE 3.4 SOME CONDOR MACHINE CLASSAD ATTRIBUTES (Derived from the Condor Manual (Condor team 2009).)

<table>
<tr><th>Attribute (not all values listed)</th><th>Purpose</th></tr>
<tr><td>Arch = <... INTEL | ... | PPC | ... | X86_64 ></td><td>String identify machine architecture.</td></tr>
<tr><td>Cpu's = <number></td><td>Number of CPUs.</td></tr>
<tr><td>Memory = <number></td><td>RAM in MBytes.</td></tr>
<tr><td>Disk = <number></td><td>Disk space available for job, in Kbytes.</td></tr>
<tr><td>KeyboardIdle = <number></td><td>Number of seconds since keyboard or mouse activity on machine.</td></tr>
<tr><td>KFlops = <number></td><td>Relative floating point performance on Linpack benchmark.</td></tr>
<tr><td>LoadAvg = <number></td><td>Floating point number indicating current load average of machine.</td></tr>
<tr><td>Machine = <string></td><td>Fully qualified hostname of machine.</td></tr>
<tr><td>Mips</td><td>Relative integer performance on a Dhrystone benchmark.</td></tr>
<tr><td>OpSys = < ... | LINUX | ... WINNT51 | ... | WINNT60 ></td><td>Operating system running on machine (WINNT51 = Windows XP, WINNT60 = Windows Visa)</td></tr>
</table>

TABLE 3.5 SOME JOB CONDOR CLASSAD ATTRIBUTES (Derived from the Condor Manual (Condor team 2009).)

Attribute	Purpose
Args	String representing arguments passed to job.
ExecutableSize	Size of executable in Kbytes.
ImageSize = <number>	Job memory image size estimate in Kbytes. If not specified, initially size of executable.
MaxHost = <number>	Maximum number of hosts job would like to claim.
Owner = <string>	String describing user who submitted job.

Matching Commands. There are actually only two matchmaking commands available for both the job ClassAd and the machine ClassAd

- Requirements = <Boolean Expression>
- Rank = <number>

`Requirements` specify the specific machine requirements, whereas `Rank` is used to differentiate between multiple machines that satisfy the requirements and is based upon user preferences. The requirements and preferences are described in C-like expressions.

First let us consider an example of `Requirements`. The Boolean expression in `Requirements` must evaluate to TRUE in both the job and machine ClassAds for job to be allowed to run on the machine. A machine ClassAd might be say

```
MyType = "Machine"
TargetType = "Job"
Machine = coit-grid02.cs.uncc.edu
Arch = "INTEL"
OpSy = "LINUX"
Disk = 1000 * 1024
Memory = 100 * 1024
Requirements = (LoadAvg<=0.2)
```

which includes the requirement that its current load is not too large. Note that the disk and memory attribute values are not the total memory and disk provided on the computer, but that available for a job. The ClassAds of machines in a Condor pool can be displayed with the `condor_status -l` command on the command line, which will display a very long list or attributes, some of which are useful for job scheduling.

The job ClassAd will describe the absolute requirements of the job and also what it prefers. The following is an example of a job requirement:

My job needs a machine with at least 6 GB of main memory and 25 MB of disk space.

A simple job ClassAd would be

```
MyType = "Job"
TargetType = "Machine"
Universe = ...
Executable = ...
Requirements = (memory == 6*1024) && (disk = 25 * 1024)
       ⋮
```

This ClassAd might match with the machine earlier. Afterwards, the available memory of the machine for subsequent jobs might reduce. The Boolean expression in the `Requirements` command can use attributes from both the machine and the job. Requirements can include constraints such as not running a job on a particular machine or one owned by a particular user. To differentiate between the local attribute and an attribute on the matching ClassAd, the attribute is identified with the associated object (e.g., `Target.memory` or `other.memory`, `self.memory`, ...).

***ClassAd Rank Statement*.** Rank can be used in both the job ClassAd and the machine ClassAd and computes to a floating-point number. The job or machine with the highest rank is chosen. An example of a job rank statement is:

```
Rank = Target.Memory * 10000 + Target.Mips
```

Sometimes just `TRUE` (1) or `FALSE` (0) is sufficient for the rank, i.e.,

```
Rank = (Target.Memory > 10000)
```

As an example of rank in a machine ClassAd, one might have

```
Rank = (Department == "Computer Science")
```

where Department defined in the job ClassAd, say as

```
Department="Computer Science"
```

Directed Acyclic Graph Manager (DAGMan) Meta-Scheduler. Condor DAGMan provides the ability to specify dependencies between Condor jobs so that jobs start in a certain order. For example

"Do not run Job B until Job A completed successfully"

This type of activity is especially important to jobs working together (as in Grid computing). In DAGMan, a Directed Acyclic Graph (DAG) is used to represent dependencies, that is, a directed graph with no cycles. Each node in the DAG is associated with one submit description file that describes the job. Each node can have any number of parents and children as long as there are no loops. Figure 3.15 shows a couple of DAGs. In Figure 3.15(a), job *A* must start and complete first, then job *B* can start and must complete before job *C* can start. Figure 3.15(a) implies that job *B* probably requires results produced by job *A* and job *C* requires results produced by job *B*. However, the graph does not show explicitly a data flow, only a control flow. For example, job *C* might use some results of both job *A* and job *B*, or just job *B*. In Figure 3.15(b), job *A* must start and complete first. Then, both jobs *B* and *C* can start. When job *B* completes, job *D* can start but job *E* cannot start until both jobs *C* and *D* have completed. Again, this is only a control flow.

DAGMan handles simple forms of workflow. We will briefly outline Condor's workflow facility, which requires the directed graph to be described in a DAG input file (`.dag` file) and simply schedules jobs according to this input file. (Drag-and-drop graphical workflow editors will be explored in detail in Chapter 8.)

Note that because each node in the DAG has a corresponding submit description file, a node could specify multiple executables within this submit description file. Also each submit description file, and hence node, could specify a different universe.

Defining a DAG. A DAG is defined in a `.dag` file, which lists each of the nodes and their dependencies. There are two basic DAG commands used in this file:

- `JOB` command which maps a node in the graph with its associated submit description file
- `PARENT-CHILD` statement that defines the dependencies between nodes

Each `JOB` statement has a name (say *A*) and a submit description file (say `A.submit`) describing the job or jobs represented by the node. The basic syntax is

```
JOB JobName SubmitDescriptionFileName
```

Originally, each node could only have one job, but this constraint has now been relaxed so that multiple jobs could be described in a single submit description file if for one cluster. Other details can be added to the `JOB` command, see the Condor Manual (Condor team 2009).

The `PARENT-CHILD` statement describes a relationship between two or more jobs and has the syntax

```
PARENT ParentJobName ... CHILD ChildJobName ...
```

where `ParentJobName` refers to a node that is a parent to another node(s) and `ChildJobName` is a child node to another parent node(s), that is, a child node depends upon at least one parent node completing before it can start. There can be more than one `ParentJobName` and more than one `ChildJobName`. Each `ChildJobName` node depends upon each `ParentJobName` node. For example, the DAG of Figure 3.15(a) could be represented with the DAG file

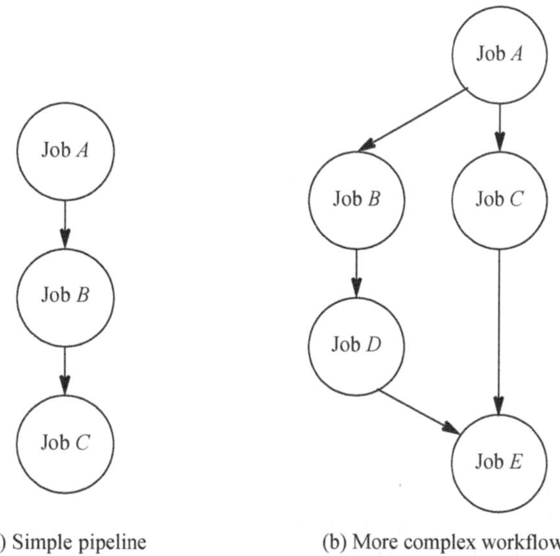

(a) Simple pipeline (b) More complex workflow

Figure 3.15 Directed Acyclic Graph Manager (DAGMan) DAGs.

```
# Figure 3.15(a) DAG
JOB A A.sub
JOB B B.sub
JOB C C.sub
PARENT A CHILD B
PARENT B CHILD C
```

where `A.sub`, `B.sub`, and `C.sub` are the submit description files of jobs *A*, *B*, and *C*, respectively. Multiple entries are needed in the `PARENT-CHILD` statement for Figure 3.15(b), i.e.,

```
# Figure 3.15(b) DAG
JOB A A.sub
JOB B B.sub
JOB C C.sub
JOB D D.sub
JOB E E.sub
PARENT A CHILD B C
PARENT B CHILD D
PARENT C D CHILD E
```

The more general case of `PARENT-CHILD` is the DAG shown in Figure 3.16, described with the DAG file

```
# Figure 3.16 DAG
JOB A A.sub
JOB B B.sub
JOB C C.sub
JOB D D.sub
PARENT A B CHILD C D
```

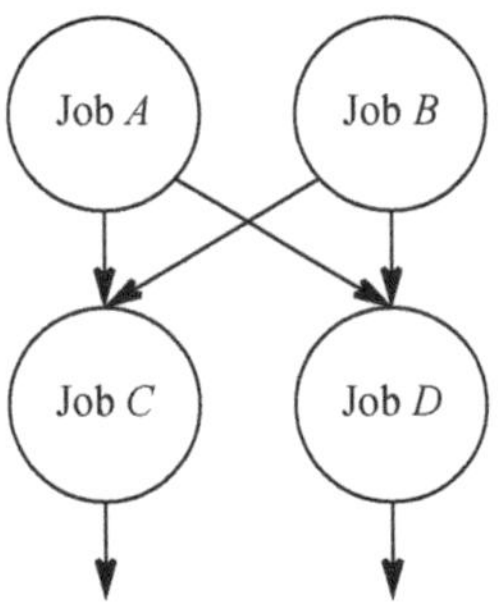

Figure 3.16 DAG for `PARENT A B CHILD C D`.

Running DAG. Once a user has created the submit description file for each node and the DAG file that describes the workflow, jobs are scheduled using the DAGMan scheduler, which is invoked with the `condor_submit_dag` command, e.g.,

```
condor_submit_dag task1.dag
```

where `task1.dag` is the DAG file. DAGMan will submit jobs in the order defined by the DAG file and monitors their progress. It will execute itself as a Condor job in the Scheduler universe.

Job Failures. DAGMan will continue to process jobs in the face of machine failures. If a machine failure occurs, those nodes that can still continue on other machines will do so, but when DAGMan cannot continue any further, it creates a so-called rescue file holding current state of DAG. The rescue file is similar to the original DAG except those nodes that have completed are marked as such with the `DONE` reserved word attached to `JOB` statements. The file allows the jobs to be restarted at a point after where jobs have completed without having to start from the beginning of the DAG. For example in the DAG

```
# Figure 3.15(b) DAG
JOB A A.sub DONE
JOB B B.sub DONE
JOB C C.sub DONE
JOB D D.sub
JOB E E.sub
PARENT A CHILD B C
PARENT B CHILD D
PARENT C D CHILD E
```

nodes *A*, *B*, and *C* have completed before a failure occurred. The DAG can restart with node *D* and then node *E*. The `DONE` keyword can also be inserted by users in their DAGs to restart them manually. This feature is useful during testing and debugging to enable a user not to have to start jobs from the beginning of a DAG on each test, when some jobs have already been tested and produced their desired outputs. Restart is at level of nodes so all jobs within a node have to be performed even though some jobs within the node may have completed without failure.

Summary of Key Condor Features. Condor began in the 1980s as a research project to take advantage of unused cycles of computers on a campus. It has matured into a widely used job scheduler. Apart from its original purpose of using computers in the background for high performance computing and resource sharing, it is also used as a regular scheduler for clusters. Its key features are:

- High throughput reliable computing environment using remote computing resources.
- Job scheduling and submission using matchmaking
- Checkpointing
- Job workflow scheduling using DAGMan, with restart capabilities.

3.3 GRID COMPUTING META-SCHEDULERS

Meta-schedulers schedule jobs across distributed sites and are highly desirable in a Grid computing environment. In a Globus Grid computing environment, the meta-scheduler will interface to the Globus GRAM installation at each site, which in turn interfaces with the local job scheduler as shown in Figure 3.17. Under those circumstances, the meta-scheduler relies on the local scheduler to place a job at each site. The meta-scheduler will also use local Globus components for file transfers and information retrieval. Also when the Grid crosses administrative domains and owners, as is usual, the owner of the local resources can cancel any job on their resources. They have complete control of their own resources. So far, two presentative local job schedulers have been described—SGE and Condor. In the following sections, two Grid meta-schedulers will be described—Condor-G and GridWay.

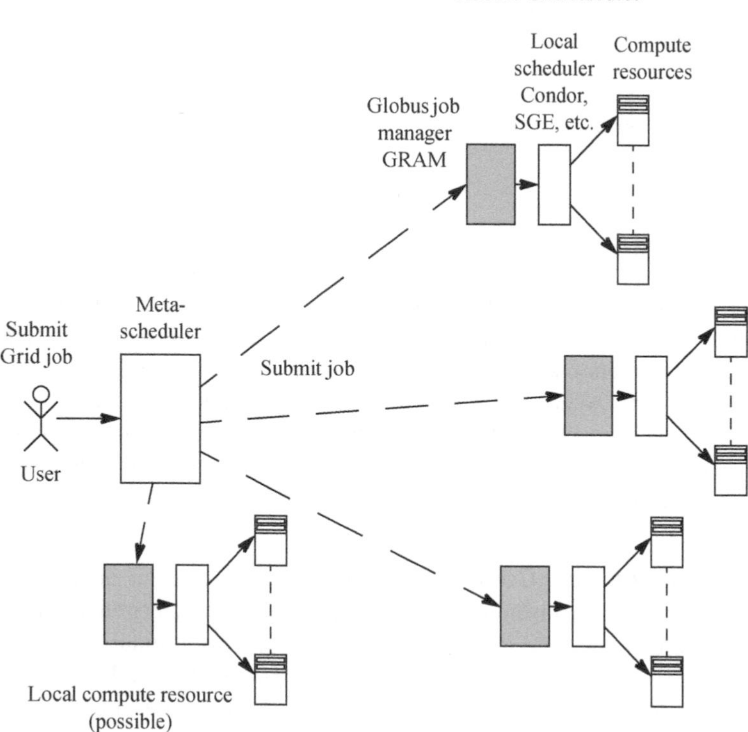

Figure 3.17 Meta-scheduler interfacing to Globus GRAM.

3.3.1 Condor-G

Condor-G is a version of Condor that interfaces to Globus environment. Jobs are submitted to Condor through the Grid universe and directed to the Globus job manager (GRAM), as illustrated in Figure 3.18. Just as for any normal Globus job, it

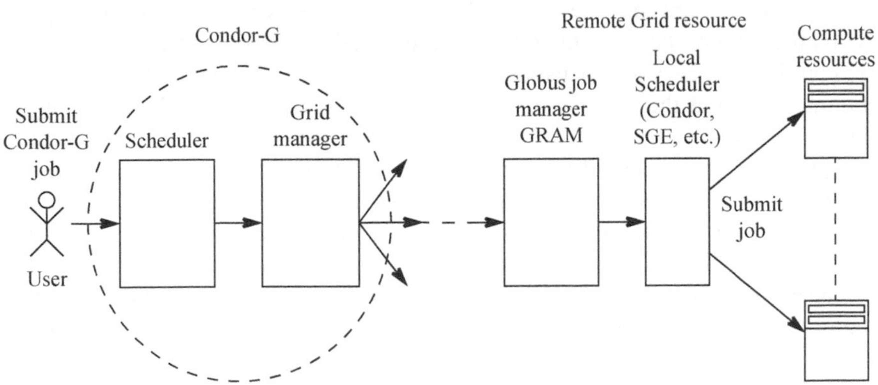

Figure 3.18 Condor-G environment.

is necessary to have a proxy first, typically obtained from the `grid-proxy-init` Globus command or from a MyProxy server.

Suppose for the moment the proxy is obtained through `grid-proxy-init`. Condor-G scheduler was designed for Globus version 2. Version 3 is not supported. Globus version 4 uses a new format for its proxies. If necessary, a Globus version 2 format of the proxy can be created by using `grid-proxy-init` with the `-old` option (`grid-proxy-init -old`). The `condor_status` command will show all the computing resources just as it would if the computing resources were in a local condor pool. A very simple job description file is:

```
universe = grid
grid_resource = gt4 https://coit-grid03.uncc.edu:8440/
wsrf/services/ManagedJobFactoryService Fork
executable = /usr/bin/uptime
log = condor_test1.log
output = condor_test1.out
error = condor_test1.error
should_transfer_files = YES
when_to_transfer_output = ON_EXIT
queue
```

The universe is the Grid universe. The command `grid_resource` has a number of alternative parameters. The first specifies the type of Grid system, in this case Globus version 4 (GT4). Condor-G will work with other Grid environments such as NorduGrid and Unicore, or Condor, or even directly with a scheduler, each case by specifying the environment by name. For GT4, the WS GRAM service to be used can be explicitly stated if necessary, and the scheduler (Condor, PBS, LSF, or SGE). In the above example, WS GRAM service includes its non-standard port. The job (`uptime`) is to be executed locally on `coit-grid03.uncc.edu` using Fork.

Now suppose the MyProxy server is used to generate proxies. This is particularly useful as Condor-G can obtain proxies directly from the MyProxy server and renew them if necessary for long running jobs. Figure 3.19 shows the communication between the user, the MyProxy server, and the Condor-G environment. First, the user stores his/her credentials (user's certificate and private key[3]) on the MyProxy server using the `myProxy_init` command. Then, the user gets a proxy using the command `myproxy-get-delegation` and submits a job submission file to Condor-G to request that the job be executed. The submit description file must identify the myProxy server and the user's name on the certificate (distinguished name) by using the `MyProxyHost` and `MyProxyCredentialName` commands, respectively. Condor will then contact the MyProxy server before the proxy expires if needed to obtain a further proxy, and subsequent ones as they expire. These actions do not require the user to intervene. Condor-G uses the MyProxy `myproxy-get-delegation` command to obtain proxies. However for Condor-G to contact the MyProxy server on the user's behalf, it will need the user's MyProxy server password, which could be included in the submit description file by embedding the `MyProxyPassword` command. If security is a concern, additional measures can be taken, see the Condor manual (Condor team 2009).

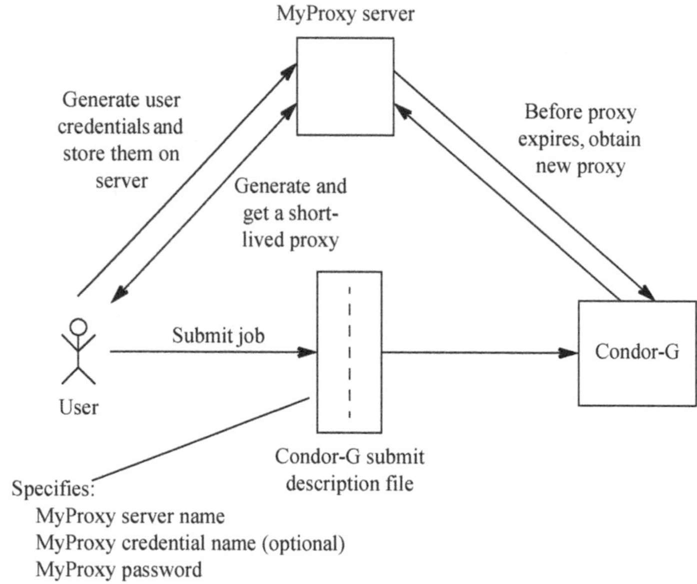

Figure 3.19 Communication between the user, MyProxy server, and Condor-G for long-running jobs.

3.3.2 GridWay

GridWay is a meta-scheduler developed by the Distributed Systems Architecture Research Group at Universidad Complutense de Madrid specifically for a Grid computing environment. The project began in 2002. Its development is open source and became part of the Globus distribution from version 4.0.5 onwards (June 2007).

[3] See Chapters 4 and 5 for an explanation of user's credentials.

As with most modern schedulers, GridWay has the ability to match jobs to resources using both static information about the job and the resources, and dynamic information (resource load). It includes:

- Dynamic scheduling
- Automatic job migration, including controlled by the job during execution
- Checking for both fault tolerance and dynamic job migration
- Reporting and accounting facilities
- Basic job dependencies (workflow)

GridWay can be installed on client machines to interact with a distributed system, or it can be installed on a server where multiple users access it. GridWay uses the file transfer, execution management, and information services of Globus as shown in Figure 3.20.

The GridWay command-line interface has similar commands to that of a local scheduler such as Condor or SGE. For example, `gwsubmit` is used to submit jobs. Common native commands are listed in Table 3.6. If the GridWay interfaces to the Globus Grid computing infrastructure, the user must first obtain a proxy in the usual way before issuing commands.

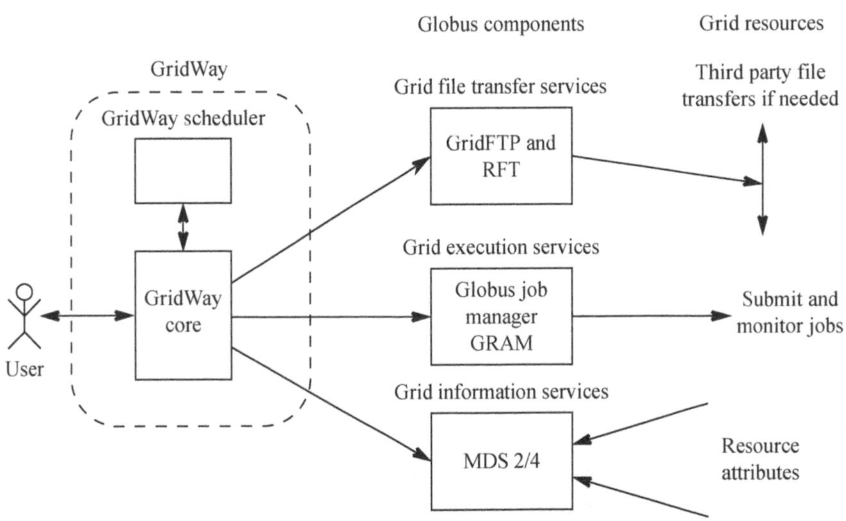

Figure 3.20 Globus components used with GridWay.

Job Description. Jobs are described in a job description file called a *GridWay job template* (GWJT). This file is used as the principal argument for `gwsubmit`. GridWay has its own list of commands for the template. Sample commands are shown in Table 3.7. (This is not a complete list of commands.) The GridWay tool `jsdl2gw` can convert JSDL documents to GWJTs. A simple GWJT template might be:

TABLE 3.6 SOME GRIDWAY COMMANDS (Derived from (GridWay Team 2007).)

Command	Purpose
`gwsubmit [options]`	Submit job. Some options: `-t template` Template file describing job `-n tasks` Submit array job, `n` tasks, same template `-d ID1 ID2...` Job dependencies
`gwhost [options]`	Monitor hosts. Some options: `-m job_id` Prints hosts matching requirements of job host_id Monitor this host_id, also prints queue information
`gwhistory job_id`	Display job history. `job_id` job identification
`gwsp [options]`	Monitor jobs. Some options: `-u user` Monitor only jobs owned by user `-r host` Monitor only jobs executed in host `-c delay` Print job information every `delay` seconds continuously
`gwkill [options] job_id [job_id2 ...]| -A array_id`	Signal job. Some options: `-k` Kill (default, if no signal specified) `-t` Stop job `-r` Resume job `-o` Hold job `-l` Release job `job_id [job_id2 ...]` job identification or `-A array_id` array identification, as given by `gwsp` command

```
EXECUTABLE = /bin/ls
EXECUTABLE = /usr/bin/uptime
STDOUT_FILE = stdout.${JOB_ID}
STDERR_FILE = stderr.${JOB_ID}
```

Variable substitution can be applied using a number of predefined variables such as `JOB_ID` above. If the above GWJT template was saved as the file `myJob.jt`, the command:

```
gwsubmit -t myJob.jt
```

would cause it to be submitted for execution on a Grid resource. Other commands can be used to examine the status of the job. More complex templates can specify multiple jobs, arguments, input and output files, etc. just as previously described local scheduler. The only difference is that the job may be transferred to a remote location on a Grid.

TABLE 3.7 GRIDWAY JOB TEMPLATES (GWJT) ATTRIBUTES (Derived from (GridWay Team 2007).)

Attribute	Purpose
EXECUTABLE	Executable file name
ARGUMENTS	Arguments for executable
ENVIRONMENT	Environment variables, comma separated
STDIN_FILE	Standard input file
STDOUT_FILE	Standard output file
STDERR_FILE	Standard error file
TYPE	"Single," "multiple" and "mpi"
NP	Number of processors in MPI jobs
INPUT_FILES	Local and remote input files, comma separated
OUTPUT_FILES	Local and remote output files, comma separated
RESTART_FILES	Checkpoint files
CHECKPOINT_INTERVAL	Time between checkpoints, in seconds
CHECKPOINT_URL	URL to store checkpoint files (GridFTP)
REQUIREMENTS	If Boolean expression true, host considered for scheduling
RANK	Numerical expression evaluated for each host considered for scheduling
RESCHEDULING_INTERVAL	Time between searches for better resources for the job
RESCHEDULING_THRESHOLD	Threshold below which migration occurs to a better resource
DEADLINE	Job start deadline
CPULOAD_THRESHOLD	Load threshold for CPU assigned to job

Job Matching—REQUIREMENTS and RANK. The arguments in the job template include those needed to describe the job. Also, if desired, arguments that can be used for selecting resources for the job, i.e., job-resource matching. GridWay uses two expressions, REQUIREMENTS and RANK for job matching in a similar fashion to Condor. The REQUIREMENTS expression has to evaluate to TRUE for the execution host to be considered at all for the job. The RANK expression is computed for each host and the hosts with the higher ranks are used first for the job. Variables can be used in REQUIREMENTS and RANK, including those shown in Table 3.8. Each case listed refers to execution hosts.

TABLE 3.8 SOME GRIDWAY VARIABLES FOR REQUIREMENTS AND RANK EXPRESSIONS (Derived from (GridWay Team 2007).)

Variable	Purpose
`HOSTNAME`	Fully qualified domain name
`ARCH`	Architecture (e.g., `"i686"`)
`OS_NAME`	Operating system (e.g., `"Linux"`)
`OS_VERSION`	Operating system version number
`CPU_MHZ`	CPU speed in MHz
`FREE_MEM_MB`	Free memory in MBytes
`FREE_DISK_MB`	Free disk space in MBytes
`LRMS_TYPE`	Type of local DRM system (e.g., `"sge"`)

Array Jobs. An array job is one specifying multiple instances of the same job, each (usually) with different arguments. Array jobs would be specified in the job template using variable substitutions, for example

```
EXECUTABLE = myJob.exe
ARGUMENTS = ${TASK_ID}
STDOUT_FILE = stdout_file.${TASK_ID}
STDERR_FILE = stderr_file.${TASK_ID}
RANK = CPU_MHZ
```

Note here a `RANK` expression is included to show how processors with higher clock frequency are preferred. (Strictly, this should be coupled with a processor type if higher clock frequency is assumed to mean higher performance.) An array of 10 instances of `myJob.exe` could be submitted with:

```
gwsubmit -t myJob.jt -n 10
```

Workflow. Primitive job dependencies can be handled with the `-d` option of the `gwsubmit` command. Job IDs of jobs that must complete first are listed with the `-d` option. For example suppose job *C* described by `C.jt` must wait for jobs *A* (`A.jt`) and *B* (`B.jt`) to complete before starting. Jobs *A* and *B* could be launched with the `gwsubmit` command. Using the `-v` option will cause the job ID to be returned in each case, i.e.:

```
gwsubmit -v -t A.jt
```

which returns the job ID of job *A*, say

```
JOB ID: 25
```

and

```
gwsubmit -v -t B.jt
```

which returns the Job ID of job *B* say

```
JOB ID: 26
```

Then, one could submit job *C* with:

```
gwsubmit -t C.jt -d "25 26"
```

More conveniently than submit dependent jobs individually, the jobs in the workflow could be submitted together in a script. `gwhost` returns the Job ID as an integer.

3.4 DISTRIBUTED RESOURCE MANAGEMENT APPLICATION (DRMAA)

Most schedulers can be controlled by either:

- Command-line commands, or
- APIs in various languages that mirror the command-line tools.

In the Grid computing community, the term *Distributed Resource Management* (DRM) systems is used to cover job schedulers and the like. As we have seen, there are several choices of DRMs for a system, each having different characteristics and modes of operation and different commands and APIs. Consequently, the Grid computing community developed a standard API specification called *Distributed Resource Management Application* API (DRMAA, pronounced "drama") for the submission and control of jobs to DRMs. DRMAA provides a uniform interface irrespective of the actual DRM installed and allows applications to control job submission using standard APIs without needing to tailor the application for a particular DRM. If needed, the specific scheduler can be identified in DRMAA. DRMAA libraries, of course, have to be present for the particular DRM. DRMAA has been developed over several years and now has become a standard through the Global Grid Forum. There are relatively few APIs in DRMAA, making it simple to learn and attractive for adoption. Users can write programs that call DRMAA APIs to communicate with the particular scheduler installed, as illustrated in Figure 3.21.

DRMAA has reference bindings in C/C++ and Java, and bindings in Perl, Python, and Ruby for a range of DSMs including (Sun) Grid Engine, Condor, PBS/Torque, LSF and GridWay. Here, we will concentrate upon the Java binding, which only has a few basic classes and methods. There are only three principal classes:

- `SessionFactory` class
- `Session` class, and
- `JobTemplate` class.

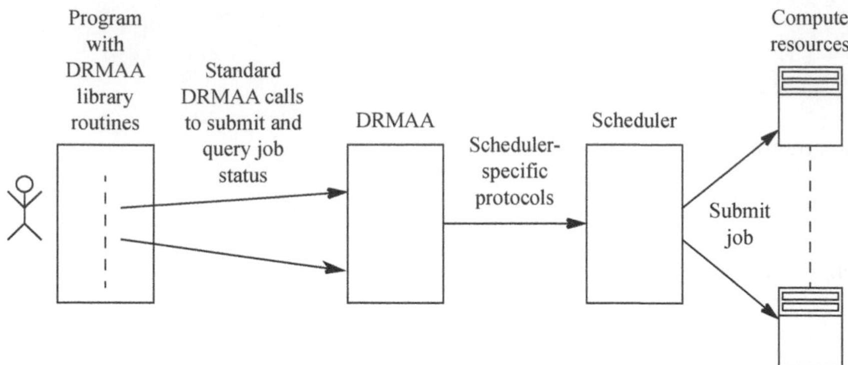

Figure 3.21 Scheduler with DRMAA interface.

that are sufficient of simple job submission requirements. There are a couple of other classes for handling file transfers and informational classes. Table 3.9 lists the two methods for the `SessionFactory` class and Table 3.10 lists the majority of the methods for the `Session` classes. The `JobTemplate` class contains just getters and setters for its parameters.

All DRMAA activities are done within the DRMAA `Session` class. To provide for different implementations and schedulers, the Java binding of DRMAA also specifies the factory class called `SessionFactory`. A factory is obtained from the static `SessionFactory.getFactory()` method. A `session` object is obtained from the factory using the `sesssionFactory getSession()` method.

The job is defined by attributes in a `JobTemplate` class. Attributes that can be set using `JobTemplate` methods include:

- Name of executable
- Directory paths
- Job arguments
- Input and output file streams
- Job status
- Email notifications
- Various timing constraints

TABLE 3.9 SESSIONFACTORY METHODS (Derived from (DRMAA working group).)

Method	Effect
`public static SessionFactory getFactory()`	Returns instance of a `SessionFactory` suitable for DRM in use
`public SessionFactory GetSession()`	Returns instance of a `Session` suitable for DRM in use

TABLE 3.10 SOME SESSION CLASS METHODS (Derived from (DRMAA working group).)

Method	Effect
`public void init(String contactString)`	Initialises DRMAA session. `contactString` specifies DRM. `null` argument would specify default DRM
`public void exit()`	Disengages from DRMAA session
`public JobTemplate createJobTemplate()`	Returns instance of a `jobTemplate`
`public void deleteJobTemplate(JobTemplate jobTemplate)`	Deallocate `jobTemplate` and its memory space. No effect on running jobs
`public String runJob (JobTemplate jobTemplate)`	Submits a job defined by `jobTemplate`. Returns string returned by DRM system
`public List runBulkJob (JobTemplate jobTemplate, int beginIndex, int endIndex, int step)`	Submits a set of parametric jobs defined by `jobTemplate` and loop index. Returns a list of job identifier strings that are returned by DRM system
`public void control(String jobName, int operation)`	Holds releases, suspends, resumes, or kills a job
`public void synchronize(List jobList, long timeout, boolean dispose)`	Waits until all jobs specifed in `jobList` have finished executing. `timeout` is maximum number of seconds to wait. `dispose` specifices how to collect information
`public JobInfo wait(String jobName, long timeout)`	Waits until job specified by `jobName` has finished executing. timeout is max number of seconds to wait
`public int getJobProgramStatus(String jobName)`	Returns status of job. Possible values: UNDERTERMINED, QUEUED_ACTIVE SYSTEM_ON_HOLD, USER_ON_HOLD USER_SYSTEM_ON_HOLD, RUNNING SYSTEM_SUSPENDED, USER_SUSPENDED USER_SYSTEM_SUSPENDED, DONE, FAILED

This is similar to the job description files seen in Chapter 2, except methods are used to set or get values. Sample setter methods of the `JobTemplate` class are given in Table 3.11.

A simple example of the use of DRMAA APIs in a program is shown in Figure 3.22. First, a session is created. Then, a job template is created. Then, the job template can be used to submit the job. In this example, single job is submitted, but one can submit an array of jobs if required.

TABLE 3.11 SAMPLE SETTER METHODS IN JOBTEMPLATE CLASS (Derived from (DRMAA working group).)

Method	Sets
`public void setRemoteCommand(String command)`	Command executed by remote host
`public void setArgs(String[] args)`	Parameters passed as arguments to job
`public void setWorking Directory(String wd)`	Directory where job executed
`public void setEmail(String[] email)`	List of email addresses used to report job completion and status
`public void setInputPath(String inputPath)`	Standard input path of job
`public void setOutputPath(String outputPath)`	Redirection of standard output of job

```
import java.util.*;
import org.ggf.drmaa.*;
public class SampleProg {
   public static void main(String[] args) {
      SessionFactory factory = SessionFactory.getFactory();
      Session session = factory.getSession();
      try {
         session.init("");
         JobTemplate jt = session.createJobTemplate();
         jt.setRemoteCommand("job.sh");
         jt.setArgs(Collections.singletonList("100"));
         String id = session.runJob(jt);
         System.out.println("Job submitted, id = " + id);
         session.deleteJobTemplate(jt);
         session.exit();
      } catch (DrmaaException e) {
         System.out.println("Error occurred " + e.getMessage());
      }
   }
}
```

Figure 3.22 Example use of DRMAA APIs in a program.

3.5 SUMMARY

This chapter introduced the following concepts and terms in job scheduling:

- Static scheduling
- Dynamic scheduling
- Job states

- Checkpointing
- Job migration
- Advanced reservation
- Co-scheduling
- Resource broker
- Job description file sand templates
- Workflows

The following sample schedulers were covered:

- Sun Grid Engine
- Condor and Condor-G
- GridWay

The chapter concludes with the standard job description API called DRMAA.

FURTHER READING

An early Globus paper on advance reservation is by Foster et al. (1999), which introduced the Globus Architecture for Reservation and Allocation (GARA). More information on Sun Grid Engine can be found at (Sun Microsystems 2007). More information on Condor can be found at (Condor team 2009), (Frey et al. 2001) and (Thain, Tannenbaum, and Livny 2003). More information on GridWay can be found at (GridWay team 2007). More information on DRMAA can be found at (DRMAA working group). The Java bindings of DRMAA are described in (Templeton et al. 2007).

BIBLIOGRAPHY

Beckles, B. 2004. Condor: What it is and why you should worry about it. University of Cambridge Computing Service Techlink seminar, June 23rd. http://www-tus.csx.cam.ac.uk/techlink/workshops/2004-06-23_condor.ppt.

Condor team. 2009. Condor version 7.2.1 manual. University of Wisconsin-Madison. http://www.cs.wisc.edu/condor/manual/.

DRMAA working group. http://www.drmaa.org/

Foster, I., C. Kesselman, C. Lee, R. Lindell, K. Nahrstedt, and A. Roy. 1999. A distributed resource management architecture that supports advance reservations and co-allocation. *Proc. 7th Int. Workshop on Quality of Service*, London, UK, May.

Frey, J., T. Tannenbaum, I. Foster, M. Livny, and S. Tuecke. 2001. Condor-G: A computation management agent for multi-institutional Grids. *Proc. 10th Int. Symp. High Performance Distributed Computing* (HPDC-10).

GridWay team. 2007. GridWay 5.2 documentation: User guide. Distributed systems architecture group, Universidad Complutense de Madrid. http://www.gridway.org/documentation/stable/userguide.pdf.

GT 4.0 WS GRAM: Developer's guide. Globus toolkit. http://www.globus.org/toolkit/docs/4.0/execution/wsgram/developer-index.html.

Kosar, T. 2006. A new paradigm in data intensive computing: Stork and the data-aware schedulers. Challenges of Large Applications in Distributed Environments (CLADE 2006) Workshop. *15th IEEE Int. Symp. on High Performance Distributed Computing* (HPDC 2006), Paris, France, June.

Sun Microsystems. 2007. Sun N1 Grid Engine 6.1 user's guide. Part no: 820-0699. http://docs.sun.com/app/docs/doc/820-0699.

Templeton, D., P. Tröger, R. Brobst, A. Haas, H. Rajic, and J. Tollefsrud. 2007. Distributed resource management application API JavaTM language bindings 1.0. Distributed Resource Management Application API (DRMAA) Working Group.

Thain, D., T. Tannenbaum, and M. Livny. 2003. Condor and the Grid. Chap. 11 in *Grid computing: Making the global infrastructure a reality*, ed. F. Berman, A. J. G. Hey, and G. Fox, Chichester England: John Wiley.

Tröger, P., and H. Rajic, A. Haas, and P. Domagalski. 2007. Standardization of an API for distributed resource management systems. *Proc. 7th IEEE Int. Sym. on Cluster Computing and the Grid* (CCGrid 2007), 619–626, Rio de Janeiro, Brazil, May 14–17.

Wilkinson, B., and M. Allen. 2005. *Parallel programming: Techniques and application using networked workstations and parallel computers*. 2nd ed. Upper Saddle River NJ: Prentice Hall.

SELF-ASSESSMENT QUESTIONS

The following questions are multiple choice. Unless otherwise noted, there is only one correct answer for each question.

1. Give one reason why a scheduler or resource broker is used in conjunction with Globus.

(a) Globus does not provide the ability transfer files
(b) Globus does not provide the ability to submit jobs other than local jobs
(c) Globus does not provide the ability to submit local jobs
(d) No reason whatsoever

2. In the context of schedulers, what is meant by the term "Advance Reservation"?

(a) Move onto the next job
(b) Requesting actions to occur at a future time
(c) Submitting more advanced job
(d) Requesting an advance
(e) Reservation that is advanced

3. What is one responsibility of a Resource Broker?

(a) To negotiate for the best price to buy a particular resource outright

(b) To negotiate between individual members of a virtual organization
(c) To optimize the performance of memory on the local resource
(d) To negotiate for the best use of distributed resources

4. What command is used to submit the executable `/bin/hostname` to the SGE scheduler?

(a) `qsub -b y /bin/hostname`
(b) `qsub /bin/hostname`
(c) `sge_submit -Ft SGE -c /bin/hostname`
(d) `qmon /bin/hostname`

5. What Globus command is used to submit the executable `/bin/hostname` to `coit-grid03.uncc.edu` with the SGE local scheduler?

(a) `globusrun-ws -submit -F coit-grid03.uncc.edu -s SGE -c /bin/hostname`
(b) `globusrun-ws -submit -s -F coit-grid03.uncc.edu -Ft SGE -c /bin/hostname`
(c) `globusrun-ws -submit coit-grid03.uncc.edu -Ft SGE -c /bin/hostname`
(d) `qsub coit-grid03.uncc.edu /bin/hostname`

6. What was the original purpose of Condor?

(a) To check the status of computers
(b) To provide the ability to access remote files
(c) To provide the ability to turn off computers remotely
(d) To provide the ability to use idle computers

7. In a Condor environment, what is displayed after issuing the `condor_status` command?

(a) An error message as this is not a valid Condor command
(b) A list of computers in the condor pool and their status
(c) A list of jobs in the Condor queue and their status
(d) The status of the Condor central manager

8. In Condor, where is the job ClassAd?

(a) In the submit description file
(b) In a file that is specified in the `condor_submit` command in addition to the submit description file
(c) In a file that the user transfers to the computer separately
(d) In the local newspaper

9. Identify which of the following are similarities between Condor ClassAd and Globus JDD. (There may be more than one similarity.)

(a) They are both XML languages
(b) They both provide a means of specifying command-line arguments for a job
(c) They both provide a means of specifying machine requirements for a job
(d) There are no similarities

10. What is DAGMan in Condor?

(a) A Data Access Grid Manager
(b) A software tool for providing checkpointing
(c) A scheduler that can schedule jobs in a workflow
(d) A Database Group Manager for Condor

11. What is checkpointing?

(a) Someone pointing to a check
(b) The process of someone checking the grade of an assignment
(c) Saving state information periodically to be able to recover from failures during execution
(d) The process of a compiler finding syntax errors

12. Which of the following is NOT a Condor environment?

(a) Globus/Grid
(b) Vanilla
(c) Chocolate
(d) Standard
(e) Java

PROGRAMMING ASSIGNMENTS

3-1. (No programming) Draw the Condor DAG for the DAG file:

```
# Problem 7.1 DAG
JOB A A.sub
JOB B B.sub
JOB C C.sub
JOB D D.sub
JOB E E.sub
JOB F F.sub
JOB G G.sub
JOB H H.sub
PARENT A B C D CHILD E F G H
```

3-2. (No programming) Examine the documentation of Condor, (Sun) Grid Engine, and GridWay, and compare and contrast the job state diagrams of each scheduler. Discuss why each scheduler has a different state diagram.

The following tasks require access to a system with Condor installed. Log onto this system to perform these assignments.

3-3. *Testing environment with Linux commands*—Create a Condor submit description file to execute the Linux command `hostname` on the Condor pool with output redirected to `hostname.out` and error messages to redirected to `hostname.err`. You will need to provide the full path to `hostname`. Verify that the command completes by using the `condor_q` command and checking the contents of `hostname.out`.

Repeat with the Linux command `sleep 60` but in this case cause the job to be placed in the hold state while running using the `condor_hold` command. Again, you will need to provide the full path. Comment on the effect of this command. Restart the Linux command. Experiment with the following options of `condor_q`: `condor_q -l`, `condor-q -run`, and `condor_q -submitter <username>`. Add lines to the submit description file to cause an email to be sent to you when the job completes and test.

3-4. *Submitting a C program*—Write a C program that computes π by a Monte Carlo method. This method is described in many sources, see for example (Wilkinson and Allen 2005). The method is chosen not because it is fast—it is actually very slow—but this gives an opportunity to examine the condor pool. Set the number of random samples at 10,000,000. Compile the program and test it on one computer in the Condor pool from the Linux command line (`ssh` connection). Then create a Condor submit description file to submit the program as a Condor job to the Condor pool and submit this job. Record the progress of the job and program output. Determine which machine ran the job.

3-5. *Submitting a Java program*—Re-write the program in Assignment 3.4 that computes π by a Monte Carlo method in Java. Compile the program into a class file. Test it locally. Write Condor submit description file for this program to run in the Condor Java universe and test it on the Condor pool.

It is important that the version of Java on the machine running your program is compatible, so first find out the version numbers of Java on each machine for example by using the Linux command `which java`. (This information can also be found from the resource ClassAds, see Assignment 3.6). Then, compile your Java program accordingly. For example, if the lowest version number is 1.5. Compile using the command: `javac -target 1.5 <yourprogram.java>`.

3-6. *Using ClassAds*—Display the resource ClassAds with the command `condor_status -l`. This command will display the resource ClassAds of all resources in the Condor pool, which could be a very long list. Repeat for just one machine using the command `condor-status -l <machine_name>`. Try also using the `-xml` option to display the ClassAd in XML.

Examine the ClassAds and find two machines that have different characteristics (Java version numbers, memory, performance, etc.) and record the differences. Modify the submit description file used in Assignment 3-5 to test that the Condor ClassAd mechanism does in fact work and make jobs run on suitable machines. Include ClassAd statements that will cause the job to run on a machine with the Java version, performance, memory, etc. that matches only one machine in the pool if one machine can be found that is unique, otherwise a subset of machines. Use the command `condor_q -run -submitter <username>` to see the jobs and the machines they are running on.

3-7. *Using DAGMAN*—Write a DAG file to run the Linux command `sleep 60`, and test it with the command `condor_submit_dag <dag_file>`. Record the output that you get. Apart from checking the queue, you can check the corresponding `.log` file to see if the job terminated normally.

Modify the DAG file to execute two programs one after the other. Any two programs can be used. Since DAGMAN uses named log files, choose submit description files with different names for their log files for this test, although this may not be a good approach for a production run with a large number of programs in the DAG.

The following task requires access to a system with Condor-G and Globus installed. Log onto this system to perform this assignment:

3-8. *Using Condor-G*—Create a proxy with the command `grid-proxy-init -old`. Create a Condor submit description file for the Grid universe and your particular Grid resources and the executable `/usr/bin/uptime`. Submit the job using the `condor_submit` command. Check the progress of the job using `condor_q`. It may take several minutes for the job to complete and leave the Condor queue.

CHAPTER 4
Security Concepts

This chapter will look at general methods of making a secure connection. Basic encryption/decryption methods will be examined using symmetric key cryptography and asymmetric key cryptography (public key cryptography) leading to X.509 certificates, certificate authorities, and public key infrastructure. These materials relate to security for networked connections and resources. As such, the materials are also covered in network courses to some extent, and in more detail than here in cryptography and Internet security courses. This chapter presents the materials in the context of connecting users and computing resources, the basis for Grid computing. Chapter 5 will consider the specifics for Grid computing, which has additional security requirements and constraints.

4.1 INTRODUCTION

4.1.1 Secure Connection

Secure connections are needed in many computer-related activities, including e-business and Grid computing. The fundamental purpose of a secure connection is to be able to send confidential information from one point to another in a network without the information being accessible by others not authorized to receive the information along the path or at the end of the connection. The term *data confidentiality* is used to describe an information exchange protected against eavesdroppers. It is also necessary to ensure that the information cannot be altered by others along the path, or if it has been changed, that fact can be recognized by the authorized recipient.

It is an unfortunate reality that there are those that would try to access information not meant for them. The term *data integrity* is used to describe the assurance that the message was not modified in transit (intentionally or by accident). A secure connection requires that the information is only sent by one identifiable by the destination and the destination is identifiable by the source. The term *authentication* is used to describe the process of establishing the identity of a party. Coupled with authentication is allowing access. The term *authorization* is used to describe the process of establishing whether a party is allowed access to a particular resource. Normally the type of access is limited. *Access control* refers to controlling the type of access.

4.1.2 Password Authentication

In dealing with users and computing resources, one first has to establish that both are properly identified. Users must be able to establish that they are connected to the computing resources they believe and also computing resources have to identify themselves to other computing resources to request actions.

For user-computer interaction, the simplest and most well-known way of authenticating a user to the system is password-based authentication. Users enter their username and password that were established for their account. The username and password are sent through network to server. A typical sequence might be

```
login as: abw
password: ********
```

where the user enters his username (`abw` above) and password, shown here as `********`. Typically, the password typed in would not be displayed so that no one can see it. The server validates name and password and responds. If the username and password are valid, access is granted.

As an aside, the password needs to be significantly complex to reduce the possibility of discovering it by simple exhaustive search or from other user information or behavior. For example, all words in a dictionary should not be acceptable by the system as a password, as a computer program could simply go through the dictionary. Exhaustive searches can be counteracted by only allowing a certain number of login attempts, say three attempts before access is permanently denied. Passwords are usually case sensitive and will accept symbols. That feature should be utilized in choosing the password. A sufficient number of characters should be used, at least eight characters.

Coming back to user-system interface, suppose the username is not valid. A possible dialog could be

```
login as: abw
There is no such user
login as:
```

This certainly tells the user that his username is not accepted and authentication failed. A different login prompt behavior

```
login as: abw
password: ********
Access denied
login as:
```

is much better. It is more secure because it reveals less information to a potential intruder. Now it is only known that either the username or the password, or both, are invalid.

4.1.3 Encryption and Decryption

A critical fact for a secure system is how the username and password information is sent through the network to the server. If the username and password are sent in plain text, they are vulnerable to being intercepted along the network. Other computers can be attached directly to the local connection, and as the information moves outside the local network, it passes through intermediate computers as it gets routed to its final destination. Programs exist that can monitor traffic on networks, which could be used to recognize username-password sequences, steal them, and use them. A system is needed in which one can be sure of the integrity of the information being sent. This leads to sending the information in a form that is unintelligible except by the parties that are to read it, in a process called *encryption.* The encrypted message (*ciphertext*) is sent along the network. The information is returned back to its original form at the receiving end in a process called *decryption.* The whole process comes under the term *cryptography.*

Sending secret messages has a very long history, well before the advent of computing. There have been many situations where secret messages are needed especially in wartime. Common encryption methods included simply changing letters in a message with different letters. Such methods are very insecure. With advent of computers, it has become more widespread that information should be sent securely. Since computers send information in a binary representation, the encryption and decryption processes will act upon binary information and produce binary information.

Sending information is actually much more involved than simply sending information in a form that is unintelligible to eavesdroppers, but first let us look a different encryption methods to achieve *data confidentiality* (the aspect of information exchange concerned with protection against eavesdroppers).

Cryptography requires an encryption method, i.e., converting the original message (i.e., plain text or clear text) to an encrypted message (encrypted text or ciphertext), and the reverse process of decryption, i.e., retrieving plain text from encrypted text. An algorithm is needed to encrypt the message and a corresponding algorithm is needed to decrypt the message. One can either make the encryption/decryption algorithms hidden so that an intruder does not know them or better use known algorithms, but to use algorithms that needs a selectable number called a *key.* The algorithm with a key encrypts or decrypts the information. One could view this process as a key to a door lock. The basic lock mechanism is known. However, there are many possible keys but only one will work to unlock and lock a door. As we shall see, and there are significant advantages in doing so, it is possible to develop a method that requires one key to encrypt the information and a different key to decrypt it. Having one key to encrypt a message and another to decrypt it would correspond to one key to lock the door and another to unlock it, if that existed. A key in computer encryption/decryption methods is a number that is sufficiently large to make it impractical to discover it through normal search methods. Also, there should be no way to use mathematics to break the encryption. To reiterate, the encryption/decryption algorithms are not secret. It is the specific key(s) that must be kept secure.

4.2 SYMMETRIC KEY CRYPTOGRAPHY

In *symmetric key cryptography*, the same key is used to encrypt the information and to decrypt it as illustrated in Figure 4.1. To send information, both the sender and receiver have the same secret key in their possession. The sender uses the key to encrypt the information, and the receiver uses the same key to decrypt the information. Only the sender and receiver must know the key, which must be kept secret for the method to be effective. Hence, the method is also called *secret key cryptography*.

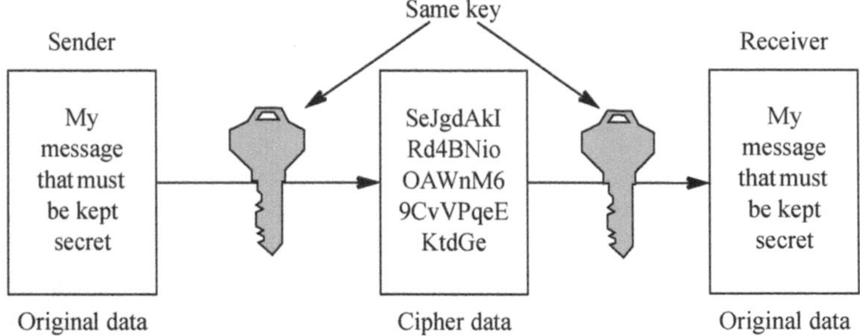

Figure 4.1 Encryption and decryption using symmetric (secret) key cryptography.

Simple Example of Symmetric (Secret Key) Cryptography. A very simple encryption algorithm is to use a Boolean logical operation between the bits of the key and the bits of the data, say exclusive-OR operation. Suppose the data has the binary representation

```
01100001011000100111001001100001
```

Suppose the key is the random string of bits

```
10011101010010001111010101011100
```

(the same number of bits as the data). Performing the exclusive-OR operation to get encrypted message, we get

```
11111100001010101000011100111101
```

This would be the pattern sent. The exclusive-OR operation would result in a 1 only if one bit is a 1 or the other bit is a 1 but not both bits a 1. If both bits are a 1 or a 0, the result bit is a 0.

To get the original message back, the same algorithm and key would be used, i.e., the exclusive-OR operation of the sent data with the same key. Table 4.1 shows that this works. Generally, only by knowing the key would one be able to deduce the original pattern. The random sequence key could be as long as the data, which would provide the strongest encryption but then both the sender and receiver need a very long key in their possession. In most situations, the length of the data will be much longer than of the key, and the data would need to be divided into blocks to perform the encryption process on each block using the same key (a *block* encryption algorithm). Block encryption algorithms are vulnerable to attack, most commonly by a brute-force exhaustive search method. The longer the block and key, the less likely one could deduce data from the encrypted data.

Clearly, the simply exclusive-ORing of patterns is not suitable for a practical encryption and one would never use this algorithm. Secure symmetric key cryptography encodings use much more complicated mathematical operations. An early encryption standard was the *Data Encryption Standard* (DES) developed by IBM in the 1970s originally to encrypt unclassified US government documents. The standard used a 56-bit key plus 8 parity bits. It operates upon 64-bit blocks and employs multiple passes of exclusive-OR operations and scrabbling.

By the late 1990s 56-bit key DES encryption had been broken through exhaustive search within days using either special hardware and by using large numbers of computers collectively. Using a 56-bit key is insufficient nowadays. To improve DES, *triple-DES* was proposed in 1978 by IBM. Triple-DES applies the DES algorithm three times. Three different keys (168 bits plus 24 parity bits) or two different keys, first and last (112 bit key plus 16 parity bits) can be used. It is also possible to use a single key three times.

The *Advanced Encryption Standard* (AES) was proposed in 2001 and essentially replaces DES as a government standard. This encryption is more secure and as yet not readily broken. It uses a different approach to encryption, involving substitutions and permutations. The block size is 128 bit with keys of 128, 192, or 256 bits.

Other encryption algorithms include *RC2* and *RC4* designed by Ron Rivest in 1987. RC2 and RC4 use variable sized keys, often between 40 to 128 bits. RC2 is 64-bit block algorithm. RC4 is stream algorithm that accepts a stream of bits and

Table 4.1 Simple encryption using exclusive-OR operation *(not suitable for real encryption and decryption)*

Original value of bit	0	0	1	1
Possible value of key	0	1	0	1
Encrypted bit	0	1	1	0
Possible value of key	0	1	0	1
Decrypted bit	0	0	1	1

modifies that stream to create the output bits. RC4 has been used in wireless computer connections but has been shown to be vulnerable to attack.

For more information on the algorithms mentioned, see Wikipedia entries on Data Encryption Standard, Triple DES, Advanced Encryption Standard, RC2 and RC4.

Two issues with symmetric (secret) key cryptography are:

- Need a way of both sender and receiver to obtain secret key without anyone else knowing the key.
- Need a different key for each receiver that a sender may communicate with if communicating with different parties is to be secure from each other.

4.3 ASYMMETRIC KEY CRYPTOGRAPHY (PUBLIC KEY CRYPTOGRAPHY)

In *asymmetric key cryptography*, the encryption/decryption algorithm is designed such that two keys are needed, one to encrypt the information and one to decrypt the information. The two keys are associated with an owner. In the most common form, one key is kept secret by its owner. This key is called the *private key*. The other key can be made available to all and is called the *public key*. Although the keys are computed through mathematical processes and are paired, it is not possible, for all practical purposes, to determine the private key from the public key. Because of the terms of public and private keys, the common name for this form of *asymmetric key cryptography* is *public key cryptography*.

Information encrypted with the private key can only be decrypted with the public key and vice versa, information encrypted with the public key can only be decrypted with the private key. In combination, this offers very significant opportunities for establishing secure transmissions. For example, the private and public keys of the receiver of a message can be used. The sender can encrypt a message with the receiver's public key and only the receiver can decrypt it (with its private key), as shown in Figure 4.2. However, anyone can send the message because the public key is known to all, and the receiver has no idea whether the sender is actually who he

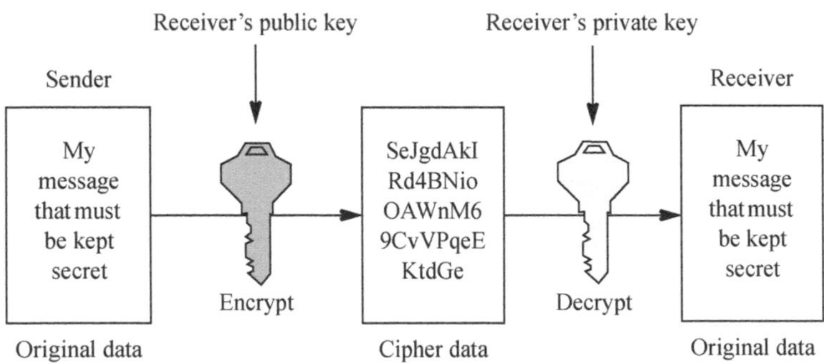

Figure 4.2 Encryption and decryption using asymmetric (public) key cryptography.

says he is, whereas in symmetric (secret) key cryptography if we know for sure only the sender and receiver have the key, then only the sender could have sent the message. So for Figure 4.2 to be acceptable in practice, we also need a way of establishing identity, which we shall look at later.

Public key cryptography uses a clever application of number theoretic concepts of functions and is a very significant advance in cryptography. The first public invention of an asymmetric cryptography method using two keys (although not a public/private pair) was due to Whitfield Diffie, Martin Hellman, and Ralph C. Merkle in 1976. A patent was granted to Hellman, Diffie, and Merkle and assigned to Stanford University in 1980. However, the method was already known in the UK classified community in the early 1970s. It is credited to James H. Ellis, Clifford Cocks, and Malcolm Williamson at the UK Government Communications Headquarters (GCHQ) a British intelligence agency but was not publicly disclosed until 1997 and the Hellman, Diffie, and Merkle patent was awarded without knowledge of the prior classified work (Wikipedia Public key cryptography). The algorithm developed by Diffie, Hellman, and Merkle is called the *Diffie-Hellman key exchange*, and enables two parties each having a private key to compute a common private key in a secure fashion.

The method is elegantly simple. It involves numbers that are agreed upon by the parties, a prime number, p and a base, g, which do not need to be kept secret, and two private keys, one at site A, say a, and one at site B, say b, which do need to be kept secret by their owners. To obtain shared key:

Step 1:

- Site A computes ($g^a \bmod p$) and sends that to B, and
- Site B computes ($g^b \bmod p$) and sends that to A.

Step 2:

- Site A computes $(g^b \bmod p)^a \bmod p$, and
- Site B computes $(g^a \bmod p)^b \bmod p$.

Both A and B have the same result because $(g^a \bmod p)^b \bmod p = (g^b \bmod p)^a \bmod p$. This value is a shared secret key and can then be used by both parties to encrypt and decrypt messages using symmetric cryptography. There are constraints in choosing values to make the method secure.The numbers a, b, and p need to be very large otherwise simple exhaustive search will break the code, and p should be prime or have a large prime factor. g does not need to be large and commonly 2 or 5 is used.

In 1977, shortly after the Diffie-Hellman key exchange was devised, Rivest, Shamir, and Adleman devised a true public key algorithm called *RSA* (for Rivest, Shamir, and Adleman), which has been widely adopted. The RSA algorithm was also known in the UK classified community in the early 1970s but again not publicly disclosed. The RSA concept was described in a classified report in 1970 by James Ellis. It was declassified in 1987. Interestingly, in classified documents, RSA was discovered first, and then Diffie-Hellman, which was the opposite to the order of public discovery.

Keys in RSA are usually between 512 and 2048 bits, the larger the better for security. The RSA algorithm is as follows:

- Find P and Q, two large prime numbers (e.g., 1024-bit).
- Choose E such that:
  1. E is greater than 1,
  2. E is less than PQ, and
  3. E and $(P - 1)(Q - 1)$ have no factors in common other than 1 (relatively prime).

  E does not have to be prime, but it must be odd. $(P - 1)(Q - 1)$ cannot be prime because it is an even number.
- Choose D such that $(DE - 1)$ is evenly divisible by $(P - 1)(Q - 1)$, that is $(DE - 1)/(P - 1)(Q - 1)$ is an integer. Simply find an integer X that causes $D = (X(P - 1)(Q - 1) + 1)/E$ to be an integer, then use that value of D.

Encryption and Decryption Functions. The encryption function is

$$C = (T^E) \bmod PQ$$

where:

C is the encrypted message (ciphertext), a positive integer
T is the message being encrypted (plaintext), a positive integer. T must be less than the modulus, PQ.

The decryption function is

$$T = (C^D) \bmod PQ$$

where:

C is the encrypted message (ciphertext), a positive integer
T is the message being encrypted (plaintext), a positive integer

The public key is the pair (PQ, E). The private key is the number D. PQ is called the modulus. E is the public exponent. D is the secret exponent. There is currently no known easy (computationally tractable) method of calculating D, P, or Q given only (PQ, E) (the public key) if P and Q are very large (1024 bit or more). PQ has to be factored to get P and Q. Once P and Q are in hand (with E), private key D could be obtained. Though it is widely suspected to be true, it is not yet proven that no easy method of factoring exists. It is not yet proven that the only way to crack RSA is to factorize PQ. See (Wikipedia RSA) for interesting attempts to break the code.

Numerical Example of RSA Algorithm. The following is an example of the RSA algorithm using small numbers. (In a real system, one would use very large numbers.)

Choosing P and Q: Choose two prime numbers. Let:

First prime number: $P = 19$
Second prime number: $Q = 13$

Modulus: $PQ = 247$
$(P - 1)(Q - 1) = 216$

Choosing E: Choose E such that $1 < E < 247$ and no factors with 216. Let:

Public exponent: $E = 7$

Choosing D: The simplest choice would be to use $DE - 1 = (P - 1)(Q - 1)$, i.e., $DE = 217$. Then with $E = 7$:

Private exponent: $D = 31$

One must destroy P and Q completely so that no record exists after obtaining PQ, E, and D as P and Q are no longer needed separately and their knowledge provides the way to crack the code. The public key is $(PQ, E) = 247, 7$. The private key is $D = 31$. The encryption function is

$$\text{Encrypt}(T) = (T^E) \bmod PQ = (T^{17}) \bmod 247$$

The decryption function is

$$\text{Decrypt}(C) = (C^D) \bmod PQ = (C^{31}) \bmod 247$$

To encrypt the plaintext value 20, do this:

$$\text{Encrypt}(123) = (20^7) \bmod 247 = 58$$

To decrypt the ciphertext value 58, do this:

$$\text{Decrypt}(58) = (58^{31}) \bmod 247 = 20$$

Padding. Messages to be encrypted can have beginnings and ends that are predictable, for example "Dear Fred, ... Sincerely Tom." In all cases, *padding* is used to both the beginning and end of the message to make it harder to crack the code.

Implementation. Exponentiation with large powers could be done by repeated multiplication. However, the process will be slow as will performing the mod operation on a very large number. Hence, the algorithm is slow and slower than symmetric key cryptography. We shall see shortly how the two methods can be combined to obtain the advantage of public key cryptography and the speed of symmetric key cryptography.

How Secure Is Public Key Cryptography? Like secret key schemes, brute force exhaustive search attack is always theoretically possible with public key cryptography. Also, it is always possible that some theoretical breakthrough leads to a way to crack the code in a practical way but that has not happened as of 2009. Very large numbers are used to make brute force exhaustive search attack infeasible with present-day computers. Of course, computers continue to increase in speed and power and make it more likely the method will be broken eventually, especially by using a collection of computers, as in Grid computing. The number of bits in the keys has to be sufficiently large to stop brute force exhaustive search attacks and one has to look to the future in deciding on the number of bits to use. However, the larger the number of bits, the slower the encryption and decryption in actual use.

As an aside, one of our student Grid computing projects done in 2006 was to break codes by exhaustive search. Exhaustive search can very easily be divided into separate non-communicating parts that can run on separate computers independently. The search space is simply divided into regions and each computer searches through one region. Extreme speed-up can be obtained if the solution is found at the beginning of one region, but of course the time to find the solution is unknown, as where the solution is in the search space also unknown.

Public key cryptography relies on the public keys actually being truly of the respective parties and the private keys are only held by their owners. (How that can be established will be considered later, but for now let us assume this to be the case.)

Non-Repudiation. Public key cryptography can provide for basic *non-repudiation*, that is, a party cannot deny being involved in the message transfer, or in this context, it can be verified that a sender is the source of the message. If a sender encrypts a message with their private key, the message can only be decrypted with their public key as shown in Figure 4.3. Basically, a sender cannot deny they sent the message if they encrypted it with their private key. Everyone has access to the public key and can decrypt it. Note that it is not sufficient simply to encrypt messages with receiver's public key and decrypt with receiver's private key because everyone has access to the public key and could send the message.

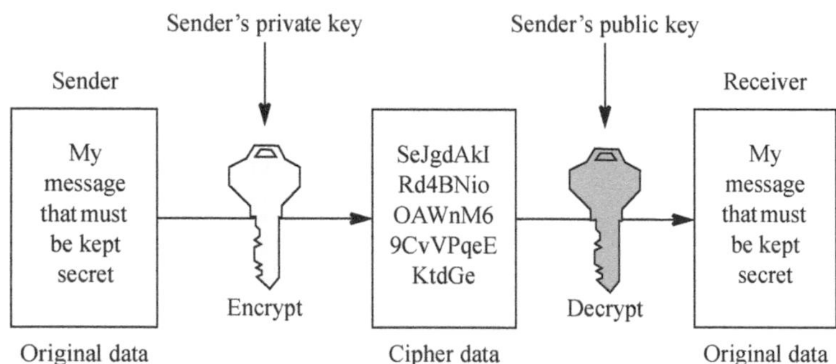

Figure 4.3 Non-repudiation using asymmetric (public) key cryptography.

Public Key Cryptography Double Encryption. One could combine basic public key encryption/decryption with non-repudiation by the sender first encrypting the message with the sender's private key and then encrypting the encrypted message with the receiver's public key. The message can only be decrypted by the receiver, first decrypting with the receiver's private key and then with the receiver's public key, as shown in Figure 4.4. This method guarantees that the message was sent by the sender and only received by the receiver because part of the encryption requires the sender's private key and part of the decryption requires the receiver's private key. However, the method is extremely slow, requiring twice the time of normal public key cryptography because it involves two encryption and decryption processes. Generally, double encryption is not used.

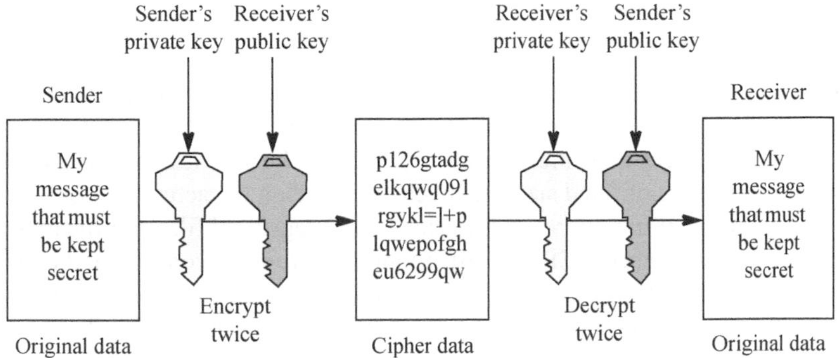

Figure 4.4 Double encryption with asymmetric (public) key cryptography.

Conclusions.

Problems with Symmetric Key Cryptography used alone:

- It requires the key be known by both parties ahead of time or transmitted over a secure channel. (The Diffie-Hellman key exchange algorithm could be used.)
- Each pair of communicating sites needs its own secret key.

Problems with Public Key Cryptography used alone:

- The public key actually must be truly from the owner and the private key must be held securely by the owner.
- If the receiver's public key is used to encrypt the message, anyone can do that as the public key is available to all, so there has to be an additional method to ensure the identity of the sender.
- Public key cryptography is slow.
- Double encryption is even slower.

Public key infrastructure (PKI) provides a solution.

4.4 PUBLIC KEY INFRASTRUCTURE

The term *Public Key Infrastructure (PKI)* is used to describe a structure that binds users to their public keys and leads to a powerful structure for incorporating all the major issues of a secure connection—*data confidentiality, data integrity, authentication*, and indeed *authorization*. Authentication was described earlier as the process of deciding whether a particular identity is who he/she (or it) says he/she (or it) is. The password approach was mentioned. PKI provides a secure scalable way of achieving authentication. Authorization (allowing a certain party to have a certain access) is achieved by separate mechanism but it can use the identity information provided in PKI. First, let us look at the issue of data integrity (making sure the message was not modified in transit).

4.4.1 Data Integrity

Data integrity can be achieved by attaching a binary pattern to the message, computed from the message by some mathematical function. It is expected that this binary pattern changes if the message alters and so can be used to detect an alteration. The binary pattern is of fixed size and much smaller than the actual message.

An example of this approach is to check the integrity of files and file transfers. A cyclic redundancy check (CRC) word is computed from the contents of the file and attached to the file. The CRC word is re-computed when the file is read. A change in the CRC word indicates an error. The function used to compute the CRC word is designed to detect likely errors but it does not guarantee to detect all errors.

Creating a small fixed-sized word from a larger binary pattern also appears where a large binary number is to select a location in a table but there are fewer entries in the table than there are different binary patterns in the number. A *hash function* is used to compute a small binary pattern from the larger binary pattern with a number of different binary values to suit the number of different entries in the table. The small binary pattern is then used to access the table. If the objective is to select entries in the table in a uniform fashion with equal probability of selecting any one location, the hash function is designed carefully to get uniform distribution across all values, that is, with n input numbers and m output numbers where $n < m$ ($m \neq 1$), each output number should be created from n/m numbers. A simple hash function in this case might be $n \bmod m$, if the values of n and m are both uniformly distributed.

A *cryptographic hash function* has two primary purposes:

- Making it not possible to find the original message from the hash value of the message, and
- Making it highly unlikely one can find two messages that have the same hash value (to avoid a message being substituted).

A cryptographic hash function has to be very carefully designed to achieve these objectives. The hash value must be relatively large for security considerations, typically at least 128 bits. Cryptographic hash functions include Message-Digest 5 (MD5) introduced by Rivest in 1991, and Secure Hash Algorithm (SHA-1 and SHA-2) introduced by the National Security Agency. MD5, although widely used, is known to have a security weakness in that it is possible to construct a different message with the same MD5 hash value as the original message.

4.4.2 Digital Signatures

Digital signatures provide a way of achieving authentication and data integrity. A cryptographic hash function is first used to create a *message digest*, a "footprint" of the message, which is then encrypted with the sender's private key to create a digital signature as shown in Figure 4.5. This digital signature could only have been created by the sender as only the sender has the private key. It can be decrypted by anyone with the sender's public key, but it is not possible to obtain original data purely from the digest. Hence, the process is one-way. Changes to the data will usually alter the message digest. It is technically possible for a change in the data to result in no change in the digest, but the probability of that happening is extremely low.

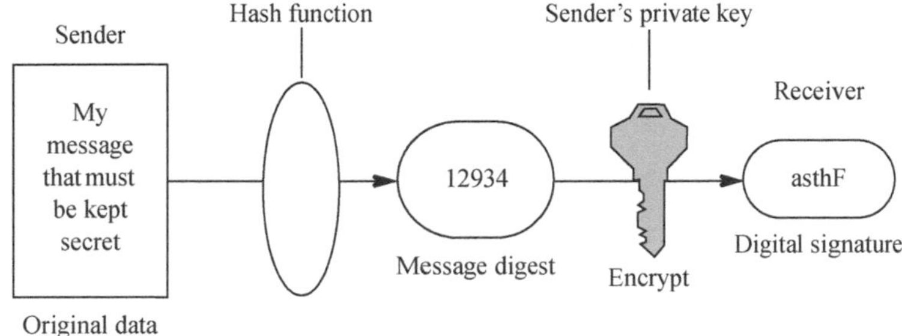

Figure 4.5 Digital signature.

Checking Digital Signature. To check the digital signature, the receiver can do the following:

1. Create a message digest from the message using the same hash function.
2. Decrypt a message digest from the sender with the sender's public key.
3. Compare the two message digests. If they are the same, the message is from the sender and has not been not altered (with very high probability).

The process is shown in Figure 4.6. Note that the use of a digital signature could be considered like a person's signature verifying validity of a check or document but each digital signature will depend upon the message and will in general be different—they cannot be reused even if stolen.

In Figure 4.6, the actual message is not shown encrypted. In such cases, all we are attempting to establish is authentication and data integrity, and not data confidentiality. This may often be sufficient in a Grid environment, but if desired, the data itself can be encrypted. One way is to use the receiver's public key to encrypt the whole message, which is then decrypted by the receiver with their private key, but such public key cryptography is generally too slow for the whole message. A better way is to pass an encrypted secret key to the receiver that way and then encrypt the whole message using secret key cryptography. Specific combined public key/secret key protocols will be described later.

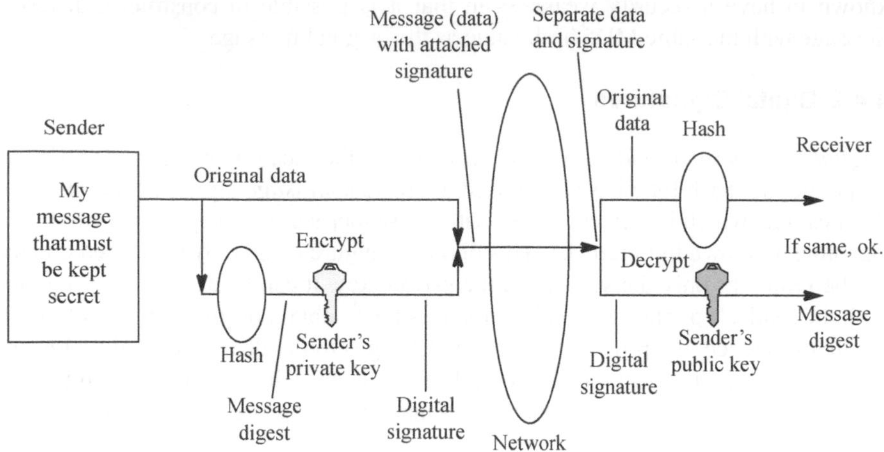

Figure 4.6 Checking digital signature.

4.4.3 Certificates and Certificate Authorities

A digital signature alone is not sufficient to ensure the data is from the sender. It is possible that the public key is a fake and one still could get matching digital signatures. A *certificate* is a digital document belonging to the user, which validates the public key as belonging to the user, or in more generality, validates the public key as belonging to an *end-entity.* The end-entity can be a user or a computing resource but so far, just users are being considered. A certificate contains the owner's name and public key, and other information.

Certificates are comparable to driver's licenses or passports. And just in the cases of licenses or passports, an authority is needed to create the certificates in a recognizable fashion that cannot be forged easily. Such an authority for certificates is called, not unexpectedly, a *certificate authority* (CA) which is a trusted third party that certifies that the information in the certificate is indeed correct, in particular that the public key does in fact belong to the end-entity named on the certificate. The certificate is digitally signed by the CA using its private key, which can be verified using the CA's public key. The principal elements of a certificate—name, public key, and signature—are illustrated in Figure 4.7. The actual certificate is a file, which we shall describe shortly. A certificate authority is comparable to a Department of Motor Vehicles (DMV) for driver's licenses or a passport agency for passports. The signature on a certificate is comparable to a difficult-to-forge seal on a driver's license such as a holographic image.

Driver's licenses and passports use photographs to identify the owner as the name alone is not sufficient. Similarly, the user's given name is not sufficient to identify the person on a certificate. There are many Barry Wilkinsons in the world for example. The way to identify the user in the certificate is to establish a unique name for the user, which we will come back to later.

The certificate authority has to create its own certificate to identify itself and publish its public key (keeping its private key protected). In the event of a single cer-

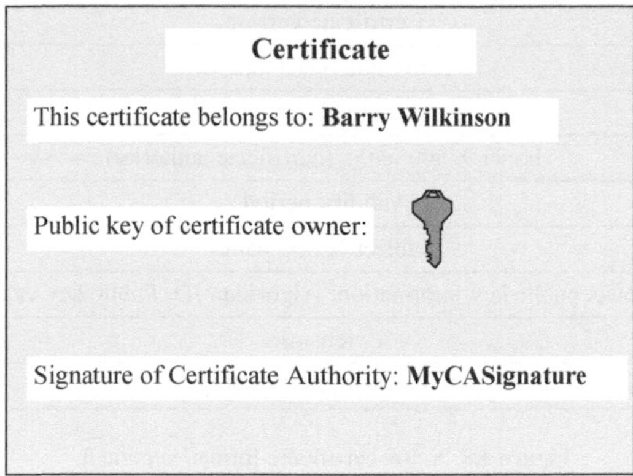

Other information also on certificate, see later.

Figure 4.7 Certificate.

tificate authority (that is, without a certificate authority hierarchy, see later), the certificate authority signs its own certificate, i.e., creates a *self-signed certificate*.

Everybody has to be comfortable with the certificate authority and its published public key. This requires a trusted physical organization that is responsible for the certificate authority. Typically, users create their own private and public key pair rather than the pair being generated by the certificate authority or someone else so that the private key does not need to be transmitted to the user through a network. The user makes a request to the certificate authority for a signed certificate, providing their public key and their unique name to the certificate authority in a form sometimes called an *unsigned certificate*, that is, a certificate that does not have a certificate authority's signature. The certificate authority then issues a signed certificate back to the user. For the certificate authority to accept the request from the user, prior physical communication is usually necessary with "real" humans, that is, between the person who is responsible for the certificate authority and the user. This may take the form of email exchanges, or preferably face-to-face or telephone conversations. A PKI structure needs to specify policies on how to establish a subject's identity. For our Grid computing course, we accepted every student that was registered for the course as a bona fide student needing a certificate.

The certificate format needs to be mutually agreed. Normally, certificates are *X.509 certificates*. The X.509 certificate is a certificate with a format defined by the International Telecommunications Union (ITU). Version 1 was defined in 1988. Version 2 and version 3 added fields. The format for version 3 is shown in Figure 4.8. Names in these certificates follow the naming convention of X.500 namespaces, which provides for an unambiguous hierarchical format, described with a concatenation of attributes to create a *distinguished name* (DN). The attributes are often chosen to match by the user's organizational structure, and the uniqueness relies on that structure being unique. X.500 does not actually specify a string representation for the name, so there are variations in following the X.500 naming conventions. Typically, the hierarchical format is top-down with various levels of attributes finally specifying the user's *common name*. Attributes can be separated with a /. This is not the only

Certificate version
Certificate serial number
Signature algorithm ID
Issuer X.500 name (certificate authority)
Validity period
Subject X.500 name
Subject public key information: Algorithm ID; Public key value
Extensions
Issuer's digital signature

Figure 4.8 X.509 certificate format, version 3.

adopted X.500 format. It is possible to use bottom up with attributes separated with commas.

The following attributes are commonly available:

- Country: `C=`
- State: `ST=`
- Locality: `L=`
- Organization: `O=`
- Organizational unit: `OU=`
- Common name: `CN=`

The organizational unit attribute could appear multiple times to describe a hierarchical organizational structure. As an example, a user might be described with the distinguished name:

```
/O=NCGrid/OU=UNCC/OU=Department of Computer Science/
                CN=Barry Wilkinson
```

where:

Organization: `O=NCGrid`

Organizational unit: `OU=UNCC`

Organizational unit: `OU=Department of Computer Science`

Common name: `CN=Barry Wilkinson`

This distinguished name will work as long as there is not more than one "Barry Wilkinson" in the Department of Computer Science at UNC–Charlotte who is part of NCGrid. The PKI structure needs to specify policies on how to construct the distinguished name for uniqueness.

Figure 4.9 shows an actual user certificate. `Issuer` is the certificate authority. `Subject` is the owner of the certificate. `RSA public key` consists of `Modulus` (*PQ* earlier) and `Exponent` (*E* earlier). As specified in `Signature algorithm`, the signature uses the MD5 hash function with RSA public key encryption algorithm

```
Certificate:
   Data:
      Version: 3 (0x2)
      Serial Number: 38 (0x26)
      Signature Algorithm: md5WithRSAEncryption
      Issuer: O=Grid, OU=GlobusTest,
      OU=simpleCA-coit-grid02.uncc.edu, CN=Globus Simple CA
      Validity
      Not Before: Jan 23 18:12:36 2007 GMT
      Not After : Jan 23 18:12:36 2008 GMT
      Subject: O=Grid, OU=GlobusTest,
      OU=simpleCA-coit-grid02.uncc.edu, OU=uncc.edu, CN=Barry Wilkinson
      Subject Public Key Info:
         Public Key Algorithm: rsaEncryption
         RSA Public Key: (1024 bit)
            Modulus (1024 bit):
               00:d7:a9:cc:42:0f:0d:b4:75:4d:e7:0c:aa:25:11:
               db:b9:fb:e9:e7:e5:76:73:e3:99:3f:07:90:18:41:
               b9:93:5f:16:bc:e0:17:dc:7a:c3:f9:57:ed:b4:4d:
               76:ac:58:91:2d:46:24:5c:ed:06:16:e6:58:11:a2:
               18:19:62:7a:84:d1:09:3b:7f:42:91:e0:38:aa:1c:
               4f:93:15:5a:ba:76:8e:6e:a3:4c:5e:85:42:c8:2a:
               9c:52:b7:29:20:eb:c1:bb:b2:6f:b7:d3:35:a1:e0:
               98:69:6c:2a:1a:d1:e8:6c:68:b8:7b:16:19:33:22:
               eb:31:7c:18:fa:dc:5b:93:db
            Exponent: 65537 (0x10001)
            X509v3 extensions:
               Netscape Cert Type:
               SSL Client, SSL Server, S/MIME, Object Signing
   Signature Algorithm: md5WithRSAEncryption
      8c:f0:b6:df:81:41:bf:cf:34:c3:47:96:38:4d:0e:ac:0e:10:
      f7:e6:b2:21:c0:c0:47:95:9f:3f:48:42:6c:e9:9a:8e:78:20:
      ce:e9:7a:8f:e8:b3:e4:a3:87:20:c9:74:6a:dd:dc:c5:b0:a7:
      72:29:59:82:93:0d:7f:35:e4:01:2f:68:77:d1:65:14:dd:28:
      e3:1d:97:db:d1:85:1f:3f:89:da:d5:fb:e0:a5:c9:bc:b2:59:
      f7:8d:a1:89:4e:04:5f:d2:a8:53:f9:9f:2e:6f:e4:4d:c2:f8:
      4e:b0:16:69:48:5a:36:2c:03:e8:08:3c:2a:ac:29:eb:69:26:
      97:c1
-----BEGIN CERTIFICATE-----
MIICazCCAdSgAwIBAgIBJjANBgkqhkiG9w0BAQQFADBnMQ0wCwYDVQQKEwRHcmlk
MRMwEQYDVQQLEwpHbG9idXNUZXN0MSYwJAYDVQQLEx1zaW1wbGVDQS1jb2l0LWdy
aWQwMi51bmNjLmVkdTEZMBcGA1UEAxMQR2xvYnVzIFNpbXBsZSBDQTAeFw0wNzAx
MjMxODEyMzZaFw0wODAxMjMxODEyMzZaMHkxDTALBgNVBAoTBEdyaWQxEzARBgNV
BAsTCkdsb2J1c1Rlc3QxJjAkBgNVBAsTHXNpbXBsZUNBLWNvaXQtZ3JpZDAyLnVu
Y2MuZWR1MREwDwYDVQQLEwh1bmNjLmVkdTEYMBYGA1UEAxMPQmFycnkgV2lsa2lu
c29uMIGfMA0GCSqGSIb3DQEBAQUAA4GNADCBiQKBgQDXqcxCDw20dU3nDKolEdu5
++nn5XZz45k/B5AYQbmTXxa84BfcesP5V+20TXasWJEtRiRc7QYW5lgRohgZYnqE
0Qk7f0KR4DiqHE+TFVq6do5uo0xehULIKpxStykg68G7sm+30zWh4JhpbCoa0ehs
aLh7FhkzIusxfBj63FuT2wIDAQABoxUwEzARBglghkgBhvhCAQEEBAMCBPAwDQYJ
KoZIhvcNAQEEBQADgYEAjPC234FBv880w0eWOE0OrA4Q9+ayIcDAR5WfP0hCbOma
jnggzul6j+iz5KOHIMl0at3cxbCncilZgpMNfzXkAS9od9FlFN0o4x2X29GFHz+J
2tX74KXJvLJZ942hiU4EX9KoU/mfLm/kTcL4TrAWaUhaNiwD6Ag8Kqwp62kml8E=
-----END CERTIFICATE-----
```

Figure 4.9 Sample user certificate.

described earlier. Notice that certificates have a period when they are valid. The period is usually set by their owners or by the system. Figure 4.9 shows the user certificate valid for one year from a given date. The certificate of the certificate authority is likely to be valid for much longer, for example 15 years. Limiting the time that a user certificate is valid provides some level of protection, and user certificates especially could have quite short time periods, but mechanisms need to be in place to handle certificates being compromised.

Using Signed Certificate to Send a Secure Message. Figure 4.10 shows the actions taken between a sender, certificate authority, and receiver. The sender has to obtain a signed certificate from the certificate authority as described earlier. Then the sender can send their message to the receiver with their certificate if the receiver does not already have the sender's certificate in their possession. The receiver has to trust the certificate authority that signs the sender's certificate and have in its possession the public key of the certificate authority, as this key is needed to check the signature of the certificate authority on the sender's certificate. In summary, *if* you trust the certificate authority *and* you are confident that the key that you have is really the public key of the certificate authority, *then* you can obtain the sender's public key from the sender's certificate with confidence. Note that the certificate authority's public key can be obtained from its self-signed certificate and this can be sent from the certificate authority.

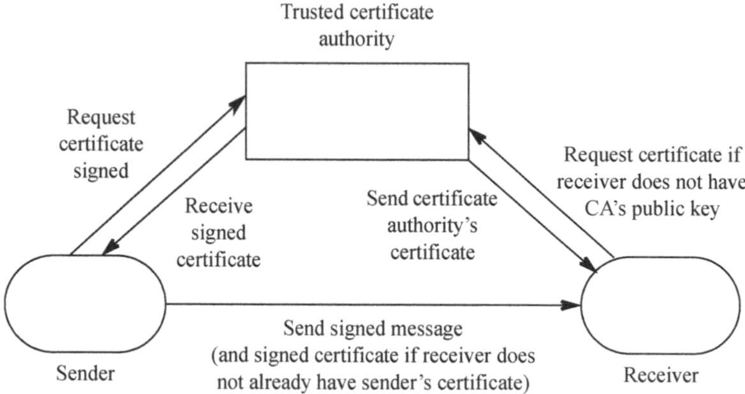

Figure 4.10 Actions to send a message between sender and receiver who both trust CA.

Combining Public Key and Secret Key Cryptography. Using public key cryptography to encrypt all message transactions would be very slow because of the encryption and decryption process. Therefore, public key and secret key cryptography is generally used together. The basic concept is to use public key cryptography with certificates and a certificate authority to establish a secure authenticated connection between parties and pass a secret key between parties in secure fashion. Then, secret key cryptography can be used with the secret key to encrypt data, which is much faster than public key cryptography.

Several network protocols have embedded public key and secret key cryptographic algorithms. Most notable is the SSL (*Secure Sockets Layer*) protocol, which is described later. SSL can be added to insecure protocols such as HTTP to get a secure protocol. SSL with HTTP creates HTTPS. Other secure protocols include S/MIME (*Secure Multipurpose Internet Mail Extensions*) for secure email, developed by RSA Data Security Inc., and SET (*Secure Electronic Transaction*) for secure e-commerce, developed jointly by Visa, Mastercard, IBM, and other companies, for secure credit card transactions over the Internet. For all of these protocols, certificates are used and need to be signed by a certificate authority. For SET, the certificate would be signed by the bank's certificate authority. In other cases, a commercial certificate authority might be used.

Commercial Certificate Authorities. Many commercial certificate authorities now exist. Notable ones include VeriSign Inc. and Entrust Technologies Inc. Web browsers have built-in recognition for such trusted certificate authorities, allowing SSL and other secure connections to use them. One can see the certificate authorities that a browser will recognize by selecting `tools` → `internet options` → `content` → `certificates` on Internet Explorer (version 7.0), which displays the window such as shown in Figure 4.11 with a long list of recognized certificate authorities for various purposes. Certificates may include expired certificates as shown in Figure 4.11. One can view the contents of each certificate if one wishes by going to `view` → `details`. The term *thumbprint* and *thumbprint algorithm* are used for *signature* and *signature algorithm*. If the certificate authority is trusted, more information is available. Its primary purposes might be listed, such as:

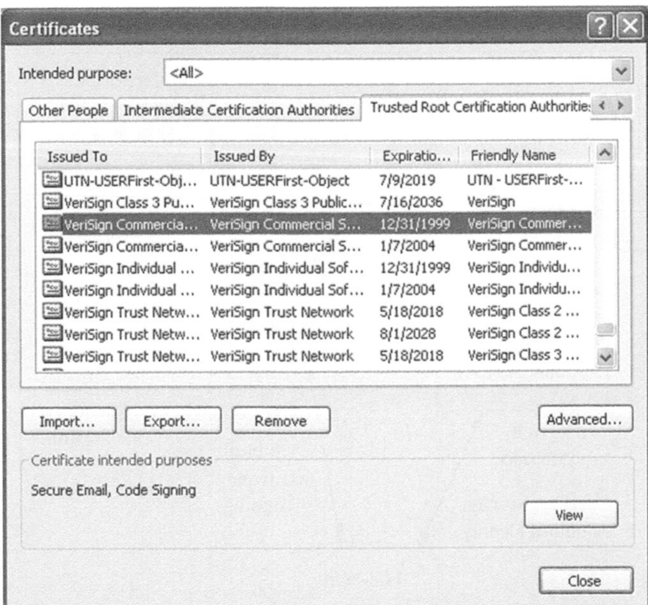

Figure 4.11 Sample trusted certificate authority list in Internet Explorer.

```
This certificate is intended for the following purpose(s)
    •  Protects email messages
    •  Ensures the identify of a remote computer
    •  Ensures software came from software publisher
    •  Protects software from alteration after publication
    •  All issuance policies
    •  All application policies
```

One can dig deeper by going to the on-line Issuer Statement (from `General →` `Issuer Statement`).

Components of Public Key Infrastructure. Public key infrastructure (PKI), the structure that binds users to their public keys, has a number of activities and can take on different organizations' structures. In Figure 4.12, the functions are divided into three components: a *certificate authority*, a *registration authority*, and a *certificate repository*. The certificate repository stores a list of issued certificates and might be accessed through LDAP (Lightweight Directory Access Protocol). It might be possible for external entities and applications to query this database directly. In addition, it is necessary to maintain a list of those certificates that are known to have been compromised. These certificates are revoked and held in a *Certificate Revocation List* (CRL). The *registration authority* (RA) acts for a certificate authority for some management functions, including processing user requests, confirming their identity and entering their information into the certificate repository database. The

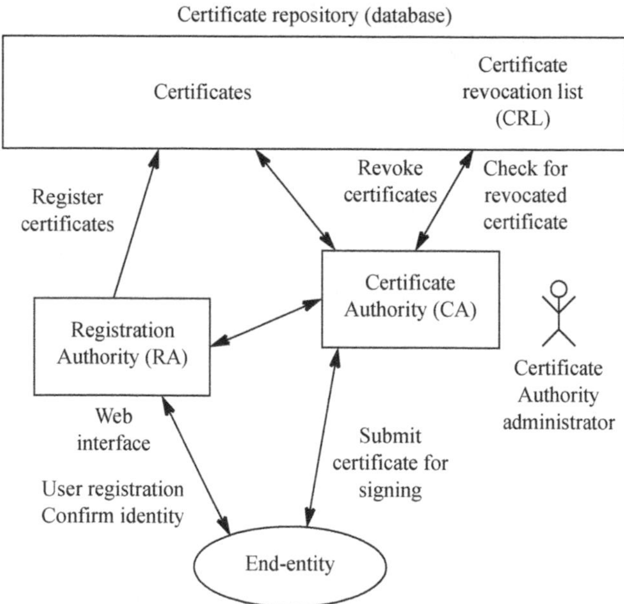

Figure 4.12 Public key infrastructure with a certificate repository and registration authority.

registration authority is not strictly necessary as certificate authorities could do all these functions, but if present, could be run on a separate localized computing resource. It typically would have a Web interface for user interaction. The certificate authority is responsible for signing user certificates, and other functions such as revoking certificates. Notice that the PKI components are managed by an administrator (a person), and generally require the administrator to respond manually to signing requests.

Public key infrastructures could take on larger hierarchical organizational structure, such as shown in Figure 4.13. This figure shows multiple *local registration authorities* that service end-entities (users) and report to a registration authority.

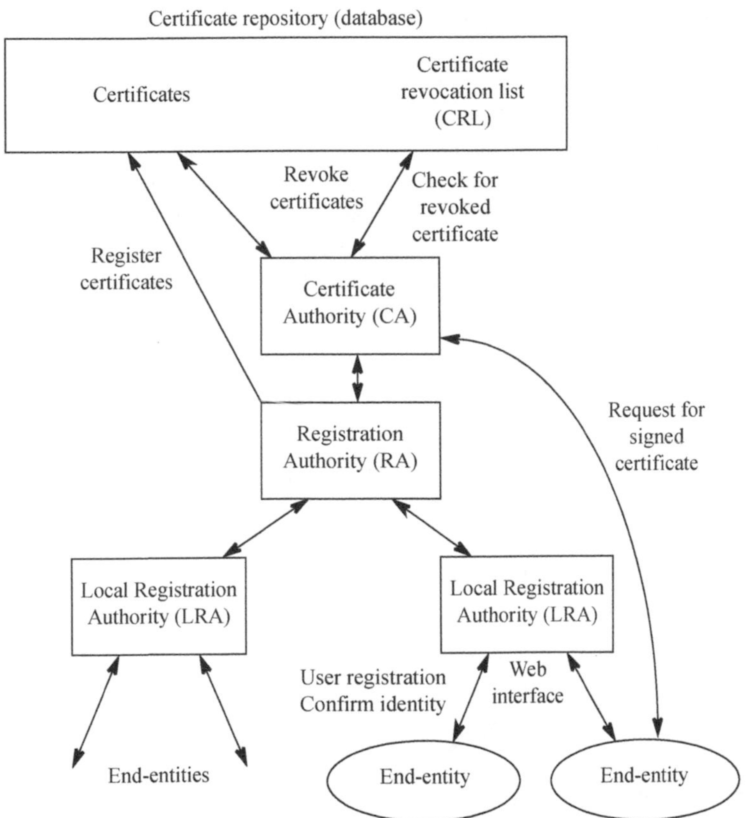

Figure 4.13 General public key infrastructure (with a single certificate authority).

Setting Up and Using a Certificate Authority. In a Grid computing project, a single certificate authority might be set up (as described in Chapter 5). In fact, it is quite easy to set up a certificate authority for a small project using the OpenSSL toolkit, which has commands for creating certificate authorities (`CA -newca`) and for creating a request to create an unsigned certificate (`CA -newreq`) and for the certificate authority to sign the unsigned certificate (`CA -sign`). Assignment 4-3 explores using OpenSSL. OpenSSL has version for all major operating systems (Linux, Mac, Windows). More details when the actions are done in a Grid computing environment can be found in Chapter 5.

Multiple Certificate Authorities. To accept certificates signed by different certificate authorities, the system can simply hold certificates from each certificate authority it trusts (and a file describing the format of the distinguished names of the certificates, see Chapter 5 for more details). As we have seen, browsers maintain the certificates of many certificate authorities that it personally will accept self-signed.

Certificate authorities can also be organized in a hierarchy, starting from a root certificate authority. Figure 4.14 shows a two-level hierarchy. Suppose Alice wishes to send a message to Bob.[1] Alice sends her certificate and that of her certificate authority, *A*, to Bob. Alice's certificate is signed by certificate authority *A*, which is not directly recognized by Bob. However, Bob recognizes the root certificate authority (and his own certificate authority *B*). The root certificate authority has signed the certificate of certificate authority *A*, which Bob can verify using the public key of the root certificate authority. Once he verifies that, he can use the public key of *A* to verify Alice's certificate. The process can be extended to multiple levels but requires one to transverse the tree accessing certificate authority certificates. Also, if the root certificate authority is compromised, the whole infrastructure is compromised.

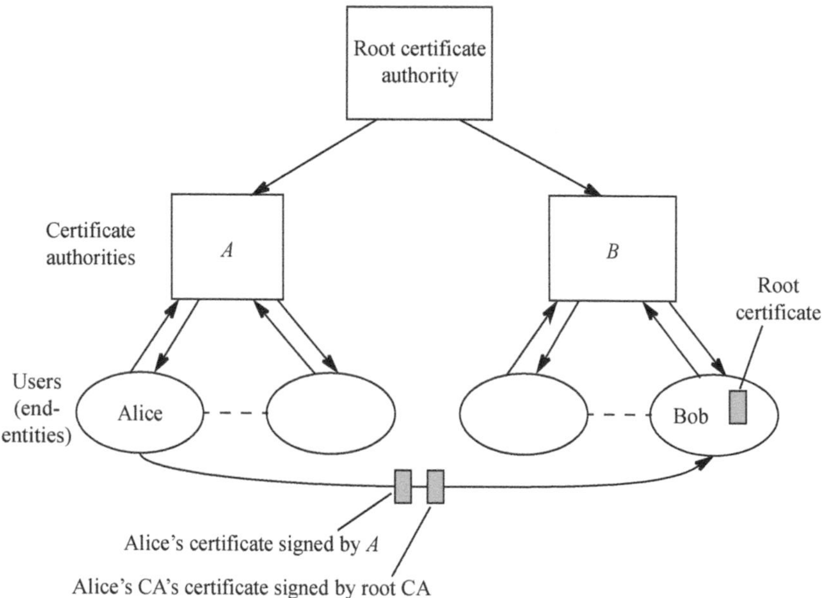

Figure 4.14 Two levels of certificate authorities, with Alice sending a message to Bob.

---

[1] In this domain, it is very common to use the names starting with different letters of the alphabet to identify users sending messages, such as Alice, Bob, Carol, Daniel, Eric, ...

Cross-Certification. Another way of handling multiple certificate authorities is to *cross-certify* the certificate authority with another certificate authority. This process operates with pairs of certificate authorities as illustrated in Figure 4.15. It requires each certificate authority to sign the certificate of the other certificate authority in the pair. Suppose there are two certificate authorities *A* and *B*, and users (end-entities) Alice and Bob. Alice has her certificate signed by certificate authority *A*. Bob has his certificate signed by certificate authority *B*. If Alice wants to communicate with Bob, she sends her certificate to Bob together with the certificate of her certificate authority *A*. Bob will accept *A*'s certificate because it is signed by his certificate authority *B* and can extract the public key of certificate authority *A*, from which he can accept Alice's certificate.

Cross-certifying certificate authorities is convenient for a pair of certificate authorities and can be extended to more certificate authorities, but the cross-certification process has to be done between each pair of certificate authorities for all certificates to be accepted by all participants. Hence with n certificate authorities, there would be $n \times (n - 1)$ cross-certified certificates in total. Each user would need the certificates of all n certificate authorities. A sender would need to select the appropriate certificate authority certificate signed by the receiver's certificate authority.

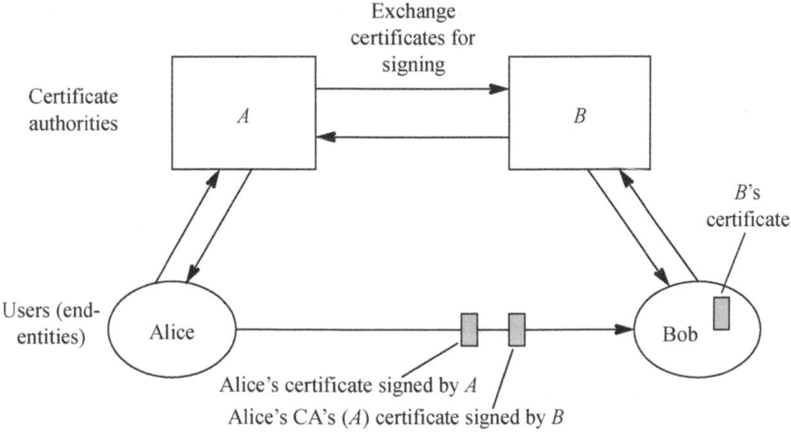

Figure 4.15 Cross-certified certificate authorities.

Bridge Certificate Authorities. Cross-certification can be applied between a single certificate authority called a *bridge certificate authority* and local certificate authorities. Typically, one local certificate authority exists for one administrative domain or institution. The certificate of each local certificate authority is signed by the bridge certificate authority, which is done by local certificate authority making a certificate signing request to the bridge certificate authority. A certificate of the bridge certificate authority is signed by the local certificate authority, which is done by bridge certificate authority making a certificate signing request to the local certificate authority. This will result in the bridge certificate authority having a certificate signed by each local certificate authority. However, each local certificate authority only has one certificate, which is signed by the bridge certificate authority.

The bridge certificate authority approach enables a trust model to be constructed in which all certificates signed by authorities cross-certified with the bridge can be trusted. Examples of production Grids that use a bridge certificate authority include SURAGrid and the Higher Education Bridge Certification Authority (HEBCA). More details of bridge certificate authorities can be found in (Humphrey, Basney, and Jokl 2005).

4.5 SYSTEMS/PROTOCOLS USING SECURITY MECHANISMS

4.5.1 Mutual Authentication and Single-Sided Authentication

Mutual authentication is central to security. Mutual authentication between Alice and Bob means that Alice satisfies herself that Bob is who he says he is, and Bob satisfies himself that Alice is who she says she is. This can simply be achieved by Alice and Bob exchanging their certificates. Alice sends her certificate to Bob who checks that the certificate is signed by a certificate authority that Bob recognizes, and Bob sends his certificate to Alice who checks that the certificate is signed by a certificate authority that Alice recognizes, as illustrated in Figure 4.16. Note that to check the signature on the certificate, Alice and Bob will need the public key of the respective certificate authority, which is obtained from the certificate authority's own certificate. So one needs the certificate of the certificate authority in their possession. It is not encrypted in any fashion and available to all, but the certificate authority's certificate itself has to be trusted if it is self-signed.

Mutual authentication is point-to-point. Mutual authentication could be used between communicating computing resources or between a user and a computing resource and typically would be in Grid computing (see Chapter 5). Mutual authentication could be used for confidential email exchanges to authenticate the source of emails, and also to encrypt messages for confidentiality although this is not done much for personal emails. In the Microsoft Outlook email tool, user certificates are

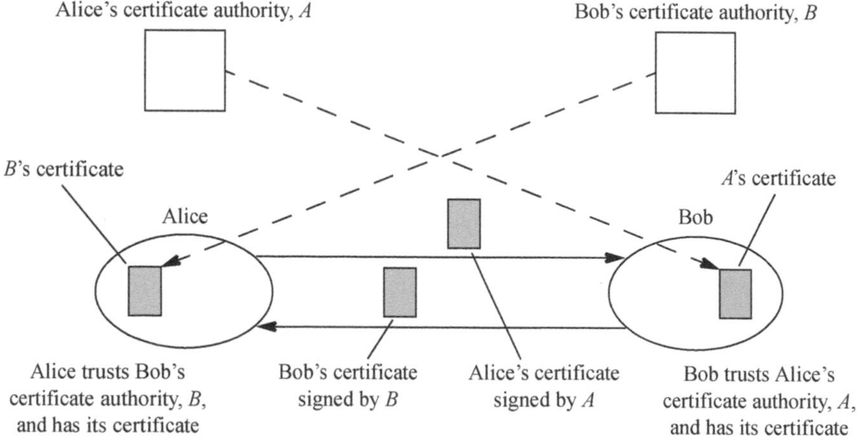

Figure 4.16 Mutual authentication.

called digital IDs and messages can be digitally signed by first obtaining a digital ID (certificate) and selecting "digitally sign a message." Messages need not be encrypted, just signed to confirm that you are the sender. They can also be encrypted for confidentiality, using the basic public key encryption process previously shown in Figure 4.2, i.e., the message is encrypted with the recipient's public key and decrypted by the recipient using his private key. This requires the sender to have the recipient's public key, which generally means the recipient's certificate is published somewhere for access, for example in the Active Directory directory service for Microsoft Outlook.

In some instances, only one-way digital signature authentication is needed or is possible. For example, a user accessing a server might need to be assured that the server is who it is supposed to be by requiring the server's certificate, but the user does not have a certificate and relies on password-authentication for access as illustrated in Figure 4.17. This is very common for users making access to remote servers. The client-server connection software (next section) will check the digital signature.

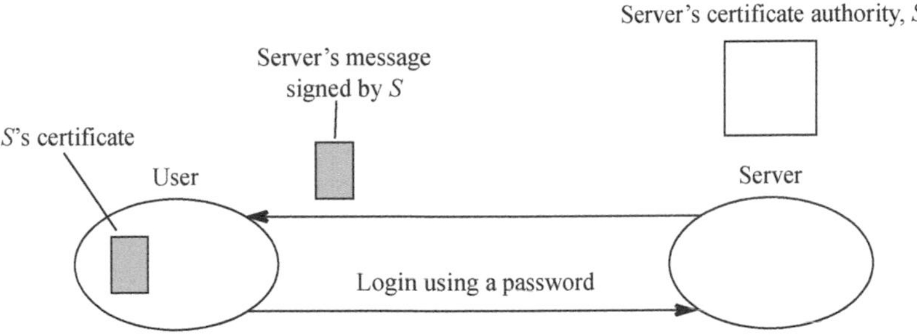

Figure 4.17 One-way digital signature authentication with user password authentication.

4.5.2 Secure Shell (SSH)

The *Secure Shell* (SSH) program was originally developed in 1995 by Tatu Ylönen, a researcher at Helsinki University of Technology, Finland, to make a secure connection between two computers and to exchange data in a secure fashion (Wikipedia Secure Shell). It has since been refined by Ylönen and others and the SSH protocol has become the de facto standard for users to log into remote computers securely. SSH uses encryption for data confidentiality and integrity, and public key cryptography for authentication. Both public key authentication of the remote computer and authentication of the user is available, although often only authentication of the remote computer is carried out, and basic password authentication is done for authenticating the user, as mentioned in the last section. All Computer Science students will be familiar with SSH login to remote computer systems using a client such as Putty. SSH uses port 22.

SSH provides additional features such as *tunneling*. In *SSH tunneling*, a secure connection is forged through an insecure network by encrypting and encasing the messages using the SSH protocol with the SSH port (port 22) as illustrated in Figure 4.18. The client will connect to a local port that will be forwarded through the tunnel to a port on the remote server. Tunneling also enables higher-level protocols and packets to be encapsulated. For example in *X11 tunneling,* Unix X Window system (X11) graphics output is redirected from a remote system to the local system for display. The X Window system is a standard graphical interface for Unix/Linux systems and is very widely used for such systems. Version 11, introduced in 1987, is still the current version and de facto standard—hence the term X11—which is synonymous with the X Window system. The full history of X11 can be found at (Wikipedia X Window System). More details on tunneling can be found at (Wikipedia Tunneling protocol).

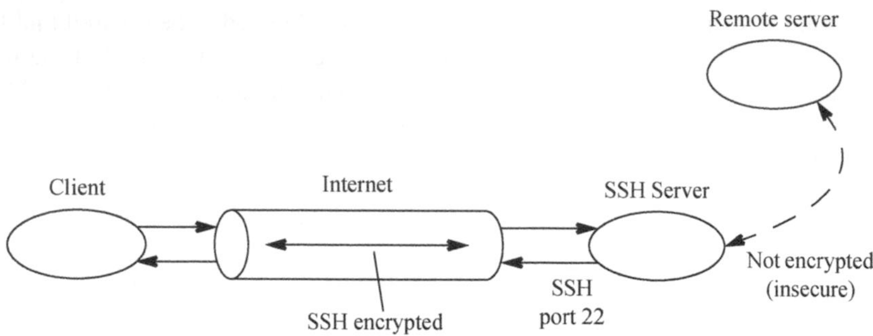

Figure 4.18 SSH Tunneling.

4.5.3 Secure Sockets Layer (SSL) Protocol

The *Secure Sockets Layer* (SSL) protocol provides for secure connection and uses PKI certificates. The protocol was first introduced by Netscape in 1994. Version 3 was introduced in 1996. It has been widely adopted for secure client-server connections and is supported by Netscape and Microsoft Internet Explorer browsers. The *Transport Layer Security* (TLS) protocol is newer and based upon SSL version 3.

The SSL protocol requires the generation of a random number at each end of the connection and exchanged together with certificates. Several messages are exchanged between the client and the server. The protocol can be described in four phases as illustrated in Figure 4.19. Briefly:

Phase I

Client starts handshake and sends:

- A random number, *X*.
- A list of supported ciphers and compression algorithms.

Phase II

Server selects cipher and compression algorithm, and notifies the client. Then it sends:

- Another random number, *Y*.
- A server certificate, which includes the public key.

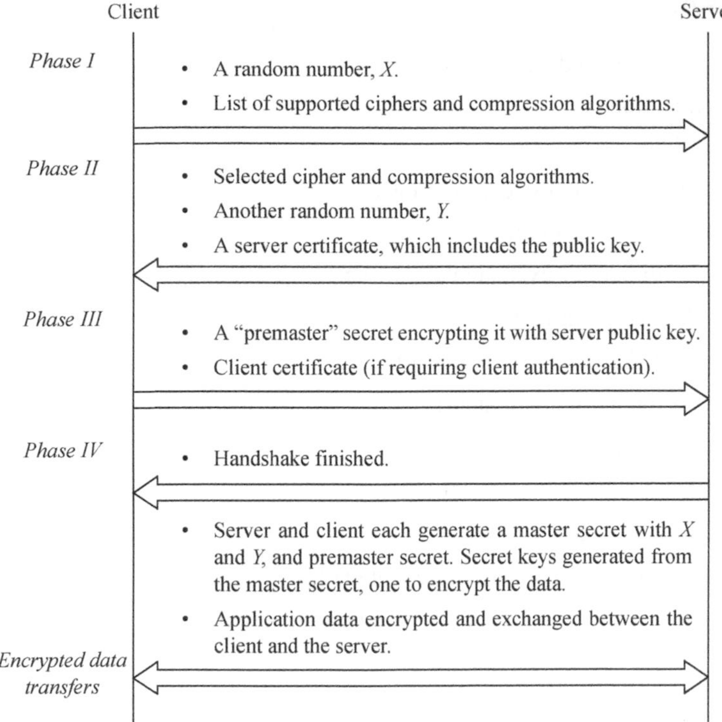

Figure 4.19 Phases of SSL (Secure Sockets Layer).

Phase III

Client sends:

- A "pre-master" secret encrypting it with server public key.
- Possibly a client certificate which includes the client's public key (see below).

Phase IV

Handshake finished. Message is sent to inform client.

Then, the server and client each generate a master secret by combining the random numbers *X* and *Y*, and the pre-master secret. Several secret keys are generated from the master secret, one to encrypt the data. Then application data can be encrypted and compressed and exchanged between the client and the server.

Typically in phase II, the client does not provide its certificate, which would provide two-way authentication. Instead, only the server is authenticated. SSL ensures:

- Authentication (by verifying certificates)
- Integrity (by including a digital signature)
- Confidentiality (by encrypting data with a secret key)

Non-repudiation is not ensured.

4.6 SUMMARY

This chapter introduced the following:

- Authentication
- Encryption
- Symmetric (secret key) cryptography
- Asymmetric cryptography
- Diffie-Hellman key exchange algorithm
- Public key cryptography
- RSA public key encryption algorithm
- Data confidentiality and integrity
- Digital signatures
- Certificates and certificate authorities
- Public key infrastructure (PKI)
- Structures for multiple certificate authorities
- Protocols for secure connections using public key infrastructure

Security issues not covered in this chapter include the written policies enacted for each certificate authority. It is necessary for all parties to agree or accept pre-agreed formal administration policies. Sometimes these policies are complex and can take many pages of description for a particular PKI.

FURTHER READING

The mathematics behind creating a secure encryption method is beyond the scope of this book, and the reader is referred to Ferguson and Schneier (2003) and Schneier (1996) for more information on security of encryption methods. A clear concise description of digital signatures can be found in Youd (1996). A course textbook on cryptography and network security is by Stallings (Stallings 2006).

BIBLIOGRAPHY

Ferguson, N., and B. Schneier. 2003. *Practical cryptography.* Indianapolis, Indiana: John Wiley & Sons.

Humphrey, M., J. Basney, and J. Jokl. 2005. The case for using bridge certificate authorities for Grid computing. *Software Practice and Experience* 35:817–826.

OpenSSL: Documents CA.pl(1) http://www.openssl.org/docs/apps/CA.pl.html

Schneier, B. 1996. *Applied cryptography*. 2nd ed. New York: John Wiley and Sons.

Stallings, W. 2006. *Cryptography and network security*. 4th ed. Upper Saddle River, NJ: Prentice Hall.

Wikipedia Advanced Encryption Standard. http://en.wikipedia.org/wiki/Advanced_Encryption_Standard

Wikipedia Data Encryption Standard. http://en.wikipedia.org/wiki/Data_Encryption_Standard

Wikipedia Public-key cryptography. http://en.wikipedia.org/wiki/Public-key_cryptography

Wikipedia Public Key Infrastructure. http://en.wikipedia.org/wiki/Public_key_infrastructure

Wikipedia RC2. http://en.wikipedia.org/wiki/RC2

Wikipedia RC4. http://en.wikipedia.org/wiki/RC4_(cipher)

Wikipedia RSA. http://en.wikipedia.org/wiki/RSA

Wikipedia Secure Shell. http://en.wikipedia.org/wiki/Secure_Shell

Wikipedia Triple DES. http://en.wikipedia.org/wiki/Triple_DES

Wikipedia Tunneling protocol. http://en.wikipedia.org/wiki/Tunneling_protocol

Wikipedia X Window System. http://en.wikipedia.org/wiki/X_Window_System

Youd. D. 1996. What is a digital signature? An introduction to digital signatures. http://www.youdzone.com/signature.html

SELF-ASSESSMENT QUESTIONS

The following questions are multiple choice; unless otherwise noted, there is only one correct answer for each question.

1. In security, what is meant by the term *authentication*?

(a) The process of deciding whether a particular identity can access a particular resource
(b) The process of giving authority to another identity
(c) The process of deciding whether a particular identity is who he says he is
(d) None of the other answers

2. In security, what is meant by the term *non-repudiation*?

(a) Means that one cannot recognize errors in a message
(b) Means that one can deny that they sent or received a message
(c) Means that one can deny they provided the data for service
(d) Means that one cannot deny that they sent a message
(e) None of the other answers

3. Which of the following is an advantage of secret key cryptography? (Select all that apply.)

(a) Parties do not need to establish previously a secure channel to exchange keys
(b) It is faster than public key cryptography
(c) It allows for non-repudiation
(d) It allows for the creation of a digital signature

4. What is public key cryptography?

   (a) Cryptography that uses a single key called a public key
   (b) Cryptography that must use double encryption
   (c) Cryptography that uses a pair of keys
   (d) Cryptography that uses keys held in a public library

5. Which of the following is an advantage of public key cryptography? (Select all that apply.)

   (a) It allows for non-repudiation
   (b) It allows for the creation of a digital signature
   (c) Parties do not need to establish previously a secure channel to exchange keys
   (d) It is faster than secret key cryptography

6. In the RSA algorithm, P and Q must be: (Select all that apply.)

   (a) Different
   (b) Prime
   (c) 1024 bits
   (d) Even

7. Why is $(P - 1)(Q - 1)$ an even number in the RSA algorithm?

   (a) P and Q are odd numbers
   (b) P and Q are large numbers
   (c) $(P - 1)(Q - 1)$ is not necessarily even
   (d) Two numbers multiplied together always result in an even number

8. What statements are true about digital signatures? (Select all that apply.)

   (a) Each user has an unchanging digital signature
   (b) Each message has the same digital signature
   (c) Ideally, each message has a different digital signature
   (d) Digital signatures cannot be decoded except by the intended recipient
   (e) Only the sender of the message can create the digital signature for the message

9. Why do users create their own private and public key pair rather than be given them from the certificate authority?

   (a) It is quicker
   (b) For security
   (c) They do not create their own private and public key pair

10. The certificate authority's public key is needed to check a certificate signed by it. How does one obtain this public key?

   (a) Ask the owner of the certificate you wish to check
   (b) Obtain it from the certificate authority's own certificate
   (c) You do not need the public key
   (d) It is in your certificate

11. What is a hash function?

   (a) A function that encrypts data

(b) When applied to data, a function that creates small fixed-sized data pattern
(c) Used in cooking breakfast
(d) When applied to data, a function that creates a digital signature

12. In public key cryptography, which key does one use to create a digital signature?

(a) The sender's private key
(b) The destination's pubic key
(c) The destination's private key
(d) The sender's public key

13. In public key cryptography using single encryption, which key or keys could one use to encrypt information to be sent a destination such that the destination could decrypt the message? There may be more than one key. Select all suitable keys.

(a) The destination's private key
(b) The destination's public key
(c) The sender's public key
(d) The sender's private key

14. Identify those items included in a owner's X.509 certificate. (Select all that apply.)

(a) The owner's private key
(b) The public key of a certificate authority
(c) The owner's public key
(d) The private key of the certificate authority
(e) The distinguished name of the user
(f) The distinguished name of the certificate authority
(g) The names of the browsers supported

15. What is the purpose of a digital signature? (Select all that apply.)

(a) To make it difficult for someone to modify the message without detection
(b) To give authority to a user to access a resource
(c) To verify that a Certificate Authority authenticates a user's public key
(d) To verify that a user is who they claim to be

16. What is the purpose of a certificate? (Select all that apply.)

(a) To verify the public key really belongs to the owner
(b) To provide the owner's public key
(c) To provide the CA's public key
(d) To provide the owner's private key

17. What must be true before you should accept a client's certificate?

(a) You must have the client's private key
(b) You must sign the certificate
(c) You must trust the CA that signed the certificate
(d) The client must have signed the certificate

18. Suppose a hierarchical binary tree structure of certificate authorities consists of a root certificate and two levels of certificate authorities below the root. Suppose Alice and

Bob do not share the same certificate authority. What certificates must each hold for them to exchange messages?

(a) None
(b) Certificate of each other's certificate authority
(c) Certificate of the root certificate authority only
(d) Certificate of the root certificate authority and each other's certificate authority
(e) Certificates of all the certificate authorities
(f) Other

19. What is SSL?

(a) Secure Session Layer
(b) Secure Sockets Layer
(c) Secret Session Layer
(d) Simple Security Logic
(e) None of the above

PROGRAMMING ASSIGNMENTS

4-1. Write a program that will perform a brute-force exhaustive search method of cracking the DES algorithm on a single 64-bit block. Demonstrate your program.

4-2. Install the OpenSSL package on your computer system. Check where the OpenSSL commands have been installed. Create a certificate authority. Make a certificate request and sign the request. Display the contents of the request and the signed certificate.

CHAPTER 5
Grid Security

In Chapter 4, the fundamental aspects of creating a secure infrastructure were described. It began with encrypting communications, which has a very long history, although the specific encryption methods now actually used are quite recent. Encryption itself is not sufficient. It is critical to establish *authentication* and *authorization* mechanisms for users and resources. Public key infrastructure was described, with digital certificates and certificate authorities, and communication protocols using certificates. Collectively, this is the principal way of obtaining secure authenticated communication in the Internet. Grid computing uses Internet protocols whether using the actual Internet, which is most likely, or dedicated high performance networks. Therefore, the Internet security mechanisms described in Chapter 4 are directly applicable and are used. However, there are aspects in Grid computing that do not exist in general Internet communications, and need to be addressed for Grid computing security infrastructure. This chapter will explore these aspects and how security is specifically implemented in a Grid computing environment.

5.1 INTRODUCTION

5.1.1 Grid Environment

Let us first review the Grid environment and types of user activities. In a Grid computing environment, we would like to run jobs on various resources within the Grid, possibly run jobs that use multiple resources, transfer files from one resource to another to be located where they are needed, and not necessarily to or from the computer system that we are logged onto (so-called third-party transfers between two remote computer systems). We might have very large data files that have to be replicated or divided among the remote resources. We would like to develop programs that can run on different types of computers as Grid resources are often heterogeneous. Ideally, we would like to develop programs that can automatically discover resources and distribute the work accordingly. A key aspect of much of what we want to do is the ability to have remote resource act on our behalf, so-called *delegation*, which has to be incorporated into the overall authentication/authorization mechanisms.

The data used or created in a Grid computing project may be provided under certain policies of use. The data may be confidential and may lead to publications and patents. Hence, it may be necessary to maintain the data in protected storage areas and to encrypt any data transfers. Data encryption will not be universally done on all projects as it incurs a performance overhead. Typically, it will not be the default for Grid computing data transmissions and would be selected when needed. Programs being executed on a Grid resource may be subject to licensing agreements. Licensing agreements are rarely handled at the Grid infrastructure level.

There are a number of significant socio-political issues as well as technical issues regarding the existing infrastructure at each site. In most Grid projects, resources are provided by different administrative domains and each administrative domain controls their own resources. Each administration is responsible for adding users to their resources and controlling access policies. Such local administration has to cooperate with other local administrations to build an overall Grid computing environment. It has to be agreed which resources are to be used by the Grid collective, how much they are to be used by others, and when. The members and resources of a virtual organization for a Grid computing project can vary. It might start small and grow. It is necessary to provide for changing membership and for growth. Ideally, setting user account and access privileges at each site should be automated. Possibly a centralized or distributed database is created for such information, which then can be accessed locally. Many Grid projects do not involve financial aspects—resources are provided for the common good of all in the Grid collective (virtual organization) under certain agreed rules. Local administrators are heavily involved. It can be difficult to orchestrate shared use of local resources and networks, which may be heavily used by their local users already.

There are issues regarding the existing security mechanisms at different sites, which could be different. Grid security is based upon PKI. A separate PKI security structure might be chosen to be used for the Grid platform to that existing at the local site rather than attempt an integration. Low level details such as how to navigate across firewalls need to be addressed. Grid computing middleware software, as will be described in subsequent chapters, has a set of components and services each with default network ports for access by remote clients. Clearly, the ports cannot be used by other pre-existing system components and must be accessible through firewalls. There could also be IP address restrictions. Local administrators will need to allow Grid traffic on certain agreed ports, and be given the confidence that security is not compromised.

5.1.2 Authentication and Authorization Aspects for a Grid

As described in Chapter 4, central security aspects are:

- *Authentication*—The process of deciding whether a particular identity is who he says he is (applies to humans and systems)
- *Authorization*—The process of deciding whether a particular identity can access a particular resource

which carry over to Grid computing environment, but are more complicated as they must also address delegation and distributed users and resources.

Grid computing resources need to be made accessible only to bona fide members of the virtual organization within agreed virtual organization policy agreements. There is much more to this than simply making secure access to remote computing resources through remote login. Secure access to remote computing resources is well known and typically uses `ssh` as described in Chapter 4. All Computer Science students will be familiar with remote `ssh` login to a remote computer system using an `ssh` client such as Putty. Often one gets the question from one new to the Grid concept. "What is new? I can log into remote machines using `ssh`. I do not need anything else."

Figure 5.1 compares the scenario using `ssh` with the ideal Grid computing approach of an integrated mechanism. Using `ssh` (Figure 5.1(a)), users need an account on every computing resource they wish to access. Then, a user could simply `ssh` into each computing resource with PKI-based server authentication. It is up to the user to accept the public key of each server. The user will be prompted upon first contact with the particular server. Usually, the user will establish its own identity using password user authentication in the subsequent response with the server. If PKI-based user authentication is needed in addition to PKI-based server authentication, the user would need to generate a public/private key pair typically using the `ssh-keygen` command to generate the key pair locally and then transfer the public key to *each* remote resource storing it in `~/.ssh/authorized_keys`. The Grid resources are geographically distributed and do not share a common file system. Once logged into one computer, a user could submit jobs to that computer or transfer files between that computer and the computer they are using for the login session. File transfers from one server to another server can be made once a connection is made to one server and then the other server using Linux commands or an `sftp` client. Running a distributed job on multiple computers simultaneously requires much more effort.

In a Grid computing environment, we would like most of the previous operations eliminated so that the user can sign onto the Grid environment once with a single application of a password (*single sign-on*) and get access to all computing resources. It should be unnecessary for the user to provide additional authentication during the activities and when programs are executing. Figure 5.1(b) shows the PKI-based Grid environment. The user interacts with resources either through a command-line interface or (preferably) through a Web-based portal as described in

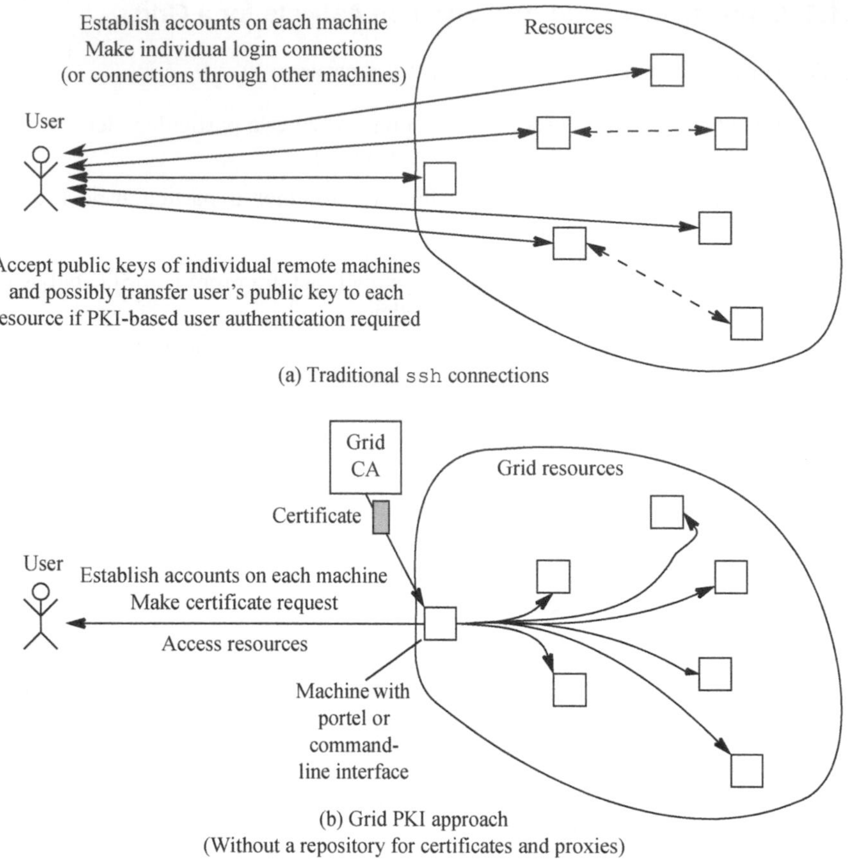

Figure 5.1 `ssh` login to remote machines compared to a Grid computing environment.

Chapter 1. The user first obtains a signed certificate from the designated certificate authority. This certificate will be good for any Grid resource, whether local or remote. Delegation is achieved by using *proxy certificates* (*proxies*). A proxy certificate is similar to a user certificate except that proxy certificate is signed by the entity that wishes to pass its authority to another to act upon their behalf. The user will create a proxy from their existing user certificate. The proxy will be set with short validity period. A proxy is handed to resources to be used to act on the user's behalf in the short term. Such delegation through proxy certificates can have a chain of trust whereby resources themselves can create self-signed proxies. Once this process is complete, the user can cause remote actions to take place such as third-party transfers and remote job execution without any need to log into separate resources. As we shall see, typically a repository holds the user and proxy certificates to alleviate the user from managing them.

A Grid environment still needs accounts at each site, either individual user accounts or group accounts. Users have a unique distinguished name as given in their certificate. This name identifies them and can be used in place of local account names. The only additional requirement is to map each distinguished name to the corresponding local account name, which is done in a file called a *gridmap* file. This method is not scalable although widely used. (This method and more powerful ways are described in detail later in the chapter.)

Once the identity of a user has been established (*authentication*) and it has been decided that a particular user is allowed a particular access to a computing resource, i.e., accounts have been set up and mapped (the overall *authorization* step), one has to decide what type of access is allowed (*access control*), which is a finer level of authorization rather than blanket ability to make any type of access. Users may only have access to their own files or they may be allowed to read files of other users in certain circumstances such as in collaborative projects. Controlling the type of access to files and directories to those who need access is a well-known situation that is applied to all computer systems, distributed or not. The most common approach is with the use of *access control lists* (ACLs). Access control lists have been around for many years and existed in the 1970s. They are used in Unix/Linux and in Windows systems, including Windows Vista, even though the method is simplistic and can be broken into. In the ACL method, a list is maintained that defines what accesses are allowed for each object, usually files or directories. There will be individual entries in this list for each file or directory. In Linux system, the type of access can be defined as read access (`"r"`), write access (`"w"`), and execute access (`"x"` for programs or executables). Those that can make the access can be defined as individual user, groups of users, or others. This leads to nine file/directory permission bits (see Appendix B). Windows XP has a much finer access control list with 15 different types of objects and 30 different privileges (Govindavajhala and Appel 2006). Unfortunately, that does not necessarily make access control better as it is important to set these privileges correctly so as not to allow sneak entry through another way. It has long been recognized that access control lists are vulnerable to security breaches and in the 1970s an alternative was devised called *capabilities*, see (Wikipedia Capability-based security). However, this approach has not been widely adopted in production operating systems. As for Grid computing environments, the simplest approach is to rely on the existing file system permissions for local access control, but this requires an access control list maintained at each site and is less than ideal.

With that outline, let us now look at actual Grid computing security infrastructure. This book concentrates upon the Globus toolkit, so let us look at the specific details of how that toolkit implements the security concepts.

5.2 GRID SECURITY INFRASTRUCTURE (GSI)

5.2.1 Component Parts

The Globus project addressed security from the beginning. The toolkit provides a set of tools, libraries, and protocols to allow users and applications to access resources securely in a Grid computing environment. An early Globus security paper by Foster et al. (1998) describes the "Grid security problem" and the Globus solutions. That paper describes the problem in somewhat abstract terms without dictating the precise security technology, but the Globus project, as virtually all other Grid projects, settled upon public key infrastructure (PKI) with certificate authorities and X.509 certificates.

The security part of the Globus toolkit is called the Globus *Grid Security Infrastructure* (GSI) and is composed of a number of components to implement PKI for Grid computing. It began with basic pre-Web services authentication and authorization component. Globus versions 3 and 4 (GT3 and GT4) embodied additional components (Web services where appropriate). The components of the Globus toolkit version 4 currently provided for security are:

Web service components:

- WS Authentication and Authorization
- Community Authorization Service
- Delegation Service

Non-Web service components:

- Pre-WS Authentication and Authorization (GT2 legacy component)
- Credential Management

The pre-WS Authentication and Authorization GT2 legacy component does not conflict with the Web service version and can be used at the same time. It was provided because of the widespread use of GT2, and can be found in many production Grids including Open Science Grid (OSG) and SURAGrid, although over time this component will be fully depreciated and not be seen. We do not include it in the Globus installation described in Appendix D.

5.2.2 GSI Communication Protocols

The security aspects of the communications in GT4 (and GT3) are based upon recent Web services security protocols. First let us differentiate between

- Transport-level protocols, and
- Message-level protocols.

We have already seen transport-level protocols in SSL (Secure Sockets Layer)/TLS (Transport Layer Security) in Chapter 4 although not by this transport-level classification.[1] In transport-level protocols, the whole message is encrypted before being sent and decrypted when received. When SSL/TLS is used with HTTP (called HTTPS), the whole HTTP data transmission is encrypted. This works fine for messages that are sent point-to-point. However, in multi-hop communication where the message is processed at intermediate sites, each site would need to decrypt the whole message and then re-encrypt it in its entirety before sending it onwards. Under

[1] In the TCP/IP layered network communication model, SSL/TLS sits at the "application" level, above TCP/IP's so-called "transport" layer.

those circumstances, the intermediate sites need the ability to decrypt and encrypt messages using a process that allows the destination to still decrypt the message. In addition, the information while decrypted at the intermediate sites is not secure, and sensitive data may be available at these intermediate sites that they may not need or should not see. This is particularly relevant in a service-oriented architecture using Web services. Intermediate services may need certain data or alter certain data but not all of the data to process a request. The transport-level approach exposes the whole message at intermediate sites. The problem also can occur with the use of proxy servers that act on behalf of a client. The client sends its requests to the proxy server, which then makes requests to other servers on behalf of the client and expects direct reply from these servers. "Proxy" situations are particularly relevant in Grid computing. The transport-level approach however may be the fastest solution for point-to-point communication as no processing is needed to select that part to decrypt.

In message-level protocols, only the message content, or specific portions of the message contents are encrypted. This provides greater flexibility than with transport-level protocols. Various authentication schemes and intermediate message processing can be employed. Conceptually, message-level protocols are at a higher level than transport-level protocols and can use various transport-level protocols. However, message-level protocols are slower than transport-level protocols. The concepts of both protocols are shown in Figure 5.2.

Web service security with message-level protocols has been addressed by OASIS (Organization for the Advancement of Structured Information Standards) and W3C (World Wide Web Consortium) with the introduction of a number of related

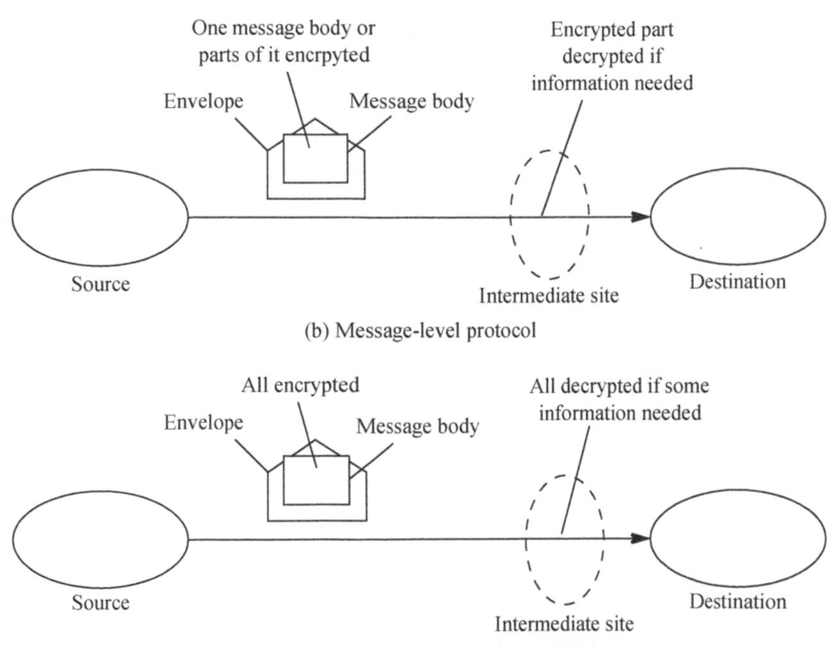

Figure 5.2 Transport-level and message-level protocols.

standards and specifications. The W3C specification *XML Encryption* allows parts of a SOAP message to be encrypted and made confidential and to be acted upon by different parties. The W3C specification *XML Signature* is a W3C standard for signing parts of a document, typically an XML document. The OASIS framework *WS-Security* provides mechanisms for secure SOAP message exchanges including using XML signature for message confidentiality and XML encryption for message integrity. The OASIS protocol called *WS-SecureCommunication* provides additional support for establishing and sharing security contexts. The details of the WS-Security framework and related specifications are outside the scope of this book. It is sufficient to say that these specifications address secure multi-hop message-level communications for Web service and other applications, including Grid computing that leverages upon Web services.

GT2 only used the SSL protocol described in Chapter 4 for mutual authentication and message protection. Both GT3 and GT4 provided additional support for message-passing protocols as they became available. GT4 provides for three different communication methods, namely:

- GSI Transport
- GSI Secure Message
- GSI Secure Conversation

GSI Transport is a transport-level protocol and uses TLS (Transport layer Security, the updated version of the original SSL protocol). *GSI Secure Message* is a message-level protocol and uses WS-Security. *GSI Secure Conversation* is also a message-level protocol but uses WS-SecureConversation. GSI Secure Conversation is the only one that provides support for credential delegation. The default protocol is GSI Transport (TLS), which has the best system performance to date in terms of communication speed.

Users can choose to use GSI encryption security mechanisms. GSI-Transport sends Globus (Web) service messages using HTTP over SSL/TSL, i.e., by HTTPS. A secure channel is opened. The Globus container that hosts the Globus services is started with the command `globus-start-container` (see Chapter 7). It can be started as a non-secure HTTP container with the command `globus-start-container -nosec`. Starting a non-secure (HTTP) container only disables GSI transport-level security. Message-level security still can be used.

5.2.3 GSI Authentication

Globus authentication, *GSI Authentication,* is basically the same as regular PKI authentication. Users have credentials they use to prove their identity on the Grid. These credentials consist of:

- X.509 certificate and
- Private key

The private key is kept secret by the owner (or on the owner's behalf at a secure repository) and encrypted with a *passphrase*. In this context, a passphrase is similar to a password but implies that it can be very long and incorporate complete sentences with spaces. A properly chosen long passphrase makes it more secure. This is good for security, but inconvenient for repeated usage and mechanisms are in place to reduce repeated access to the private key (see later). The X.509 certificate is available to all, as described in Chapter 4.

Certificate Authorities for Grid Computing. A Grid computing group (virtual organization) requires one or more certificate authorities to sign the certificates. Generally, one cannot leverage upon existing commercial certificate authorities that one would find recognized by a browser (such as listed in Internet Explorer, see Chapter 4, Figure 4.11) because the virtual organization wants to control who becomes a member of the organization and this is done by issuing certificates signed by a certificate authority of the virtual organization. Globus does provide an on-line certificate signing service called the *Globus certificate service* that issues "low-quality" GSI certificates to users who want to experiment with Grid software without having to set up or have access to a certificate authority. This certificate authority can be used for testing and can be used in Grid service training tutorials but is insufficient for any serious Grid computing work. The Globus certificate service has no means of verifying users nor does it ensure that distinguished names provided are in fact unique.

Globus also provides a simple implementation of a certificate authority called *simpleCA*, which is part of the Globus toolkit and can be installed easily under the direct control of a system administrator. One can install it on a personal computer running Linux for experimentation. SimpleCA commands creates the OpenSSL certificate authority configured to work with Globus and in fact the OpenSSL could be used directly to create a certificate authority that will function with Globus. The installation of SimpleCA is covered in Appendix D. SimpleCA is used in our Grid computing course. In our course, we use multiple certificate authorities, one at each institution for signing the certificates of students at that institution and then arrange for Globus to accept certificates signed by multiple certificate authorities. With only a few sites, we do not employ any hierarchy structure or a bridge certificate authority—rather we simply configure each Globus installation to accept the certificates signed by each certificate authority, which corresponds to a browser set to accept multiple certificate authorities.

Large Grid computing projects might have multiple certificate authorities with bridge or hierarchical certificate authorities as described in Chapter 4. Some projects keep to one certificate authority, which certainly simplifies management and creates centralized point for signing. Some national Grids have established a single certificate authority for the virtual organization, for example the UK e-Science national Grid has a centralized certificate authority but uses a multitude of registration authorities spread around the country at universities IT services departments and research laboratories for identity management and accepting the initial request for a certificate. These registrations are manned by named individuals who will require positive proof of identity (photo ID), as illustrated in Figure 5.3. This process is analogous to Department or Division of Motor Vehicles in the US and other countries that process registration of motor vehicles and driving licenses on behalf of the region. Vehicle registration paperwork is sent back to a centralized data center.

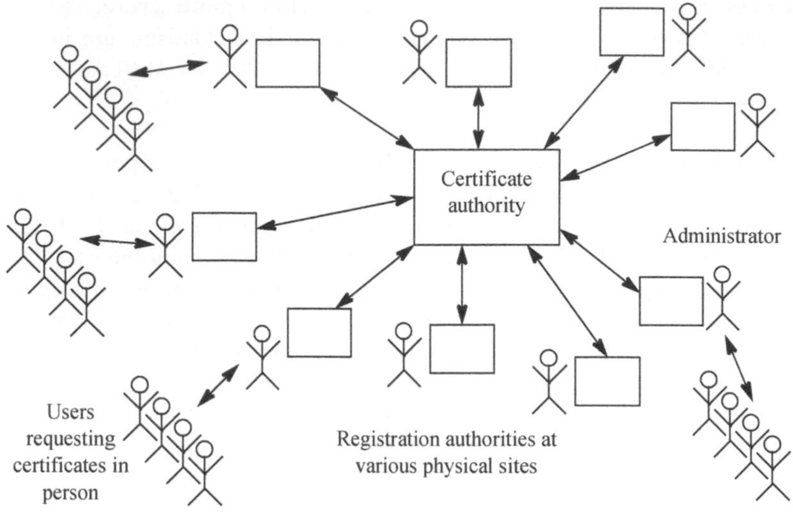

Figure 5.3 Single national certificate authority with multiple registration authorities.

First, let us consider a single certificate authority for a Grid, which combines the activity of user registration with signing certificates. After a certificate authority is established for the Grid, users have to get their signed certificates from the certificate authority. Users joining a Grid from geographically dispersed locations must communicate with the certificate authority system administrator to verify their identity and to get a signed certificate. The communication is often done by email using non-secured channels! Ideally, users should present themselves to the certificate authority system administrator just as one might to get one's first passport or driver's license, but this is rarely done. Sometimes, a secure Web page is used for users to enter confidential information and get their signed certificate.

For our Grid course, we used the PURSe (Portal-based Registration Service) software for students to get their signed certificate as described in Chapter 1, together with the MyProxy credential repository. This software structure hides the details of what goes on behind the scenes. Without such software, users and the system administrator have to co-operate with placing the signed certificate in the correct directory. It is a useful learning process to go through such actions (see below and Assignment 5-1).

Getting a Certificate using Globus Commands. Before you can ask a certificate authority to create a signed certificate, you need to generate your own public/private key pair. You do not actually need the public key separate from your certificate as your certificate will hold your public key, so in the process described, you will be left with your private key and your signed certificate.

The Globus command

```
grid-cert-request
```

will create a private key pair and request for a signed certificate, that is, an unsigned certificate containing the subject name and public key. A default distinguished name (certificate subject) will be displayed for the user as part of the message. The command requires that you create a passphrase, which will be used to encrypt the private key and must be remembered. Three files are created by the command in the user's `.GLOBUS` directory, namely

- User request: `usercert_request.pem`
- User's private key: `userkey.pem`
- Empty file: `usercert.pem`

The user request file, `usercert_request.pem`, can be considered as an unsigned certificate containing the subject name and public key. The `grid-cert-request` command corresponds to the OpenSSL command `CA.sh -newreq`, which produces similar files.

It is critical that your private key is not compromised. The empty file, `user-cert.pem`, is a placeholder for where the signed certificate will be put later, which will have the same name. The file `usercert_request.pem` is the request to be sent to the certificate authority. Typically, this file is sent by email to the administrator of the certificate authority. The `grid-cert-request` command creates a message telling the user what to do.

The administrator will run the command

```
grid-ca-sign -in usercert_request.pem -out signedcert.pem
```

(Again, this corresponds to an OpenSSL command.) It will need the passphrase used to encrypt the certificate authority's private key. The command generates the signed certificate called `signedcert.pem` (in the command as shown). The certificate authority administrator will return the signed certificate to the user, typically by email. In that case, the user then has to replace the empty `usercert.pem` with this file (rename it to be `usercert.pem`). The complete set of actions using Globus are shown in Figure 5.4. There are other ways of getting the signed certificate back to the user, including letting the administrator access the user's account using root privileges. This was the way we chose for the Fall 2005 Grid course before we switched to using PURSe.

Finally, we have the complete set of *user credentials*:

- User's private key: `userkey.pem`
- User's signed certificate: `usercert.pem`.

Certificates for Resources. Computing resources used in a Grid also need their identity verified in a formalized manner when added to the Grid infrastructure and will need their own *host certificate* signed by a certificate authority trusted by the Grid. Only such machines will be allowed to participate in the Grid activities. They might be used under certain access rights and at certain times or with certain users, especially if the computers are shared with non-Grid activities.

Host credentials consist of the following two files:

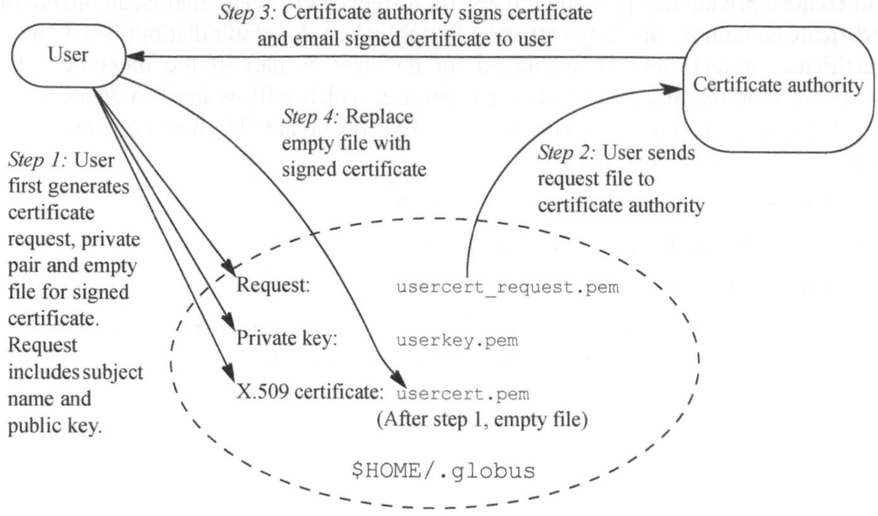

Figure 5.4 Getting a signed certificate using Globus commands.

- Host's private key: `hostkey.pem`
- Host's signed certificate: `hostcert.pem`

typically located in `/etc/grid-security/`.

The `grid-cert-request` command to create the host credentials is

```
grid-cert-request -host hostname
```

for example

```
grid-cert-request -host coit-grid01.uncc.edu
```

Signing follows the same procedure as for user certificates with submission to the certificate authority. The private key, `hostkey.pem`, will initially only be accessed by root, but users running a container for services (Chapter 7) will need access to it and so the privileges have to be altered to allow this, typically by creating a `globus` user that has the privileged access to the key.

A certificate can be created for a specific Web service with the `grid-cert-request` command with two options, `-host` and `-service`

```
grid-cert-request -host hostname -service service_name
```

In this case, a password will be not be used.

Certificate of Certificate Authority. When a certificate authority is created, it will self-sign its own certificate. The certificate authority has the two files:

- <cert_hash>.0
- <cert_hash>.signing_policy

where <cert_hash> is the hash code of the identity of certificate authority (a 32-bit number, given in hexadecimal). cert_hash.0 is the actual certificate of certificate authority and cert_hash.signing_policy is a file that defines the format of the distinguished names of certificates signed by the certificate authority.

Configuring GT4 to Trust a Particular Certificate Authority. Just as a browser can recognize many certificate authority signatures, Globus can be configured to accept certificates from multiple certificate authorities. It is just a matter of loading the two files of each certificate authority (<cert_hash>.0 and <cert_hash>.signing_policy). One can see the list of certificate authorities recognized and choose one to sign your certificate request by issuing the command

```
grid-cert-request -ca
```

A sample output is shown below:

```
nondefaultca=true

The available CA configurations installed on this host are:

1)  61de2736 - /O=Grid/OU=GlobusTest/OU=simpleCA-coit-
grid02.uncc.edu/CN=Globus Simple CA
2)  76cc56e4 - /O=Grid/OU=GlobusTest/OU=simpleCA-coit-
grid03.uncc.edu/CN=Globus Simple CA
3)  f3117196 - /O=Grid/OU=UNCW Dept of Computer
Science/OU=torvalds.cis.uncw.edu/CN=Globus Simple CA

Enter the index number of the CA you want to sign your cert
request:
```

GSI Authentication Protocol. The GSI Authentication protocol for establishing trust was originally based on the SSL protocol described in Chapter 4. For *A* authenticating your certificate on *B*, the basic process is:

- Host *B* send your certificate to host *A*.
- Host *A* gets your public key and name from the certificate using the public key of the certificate authority.
- Host *A* creates a random number and sends it to host *B*.
- Host *B* encrypts random number with your private key and sends it to host *A*.
- Host *A* decrypts number and checks number. If correct, host *A* authenticates the certificate.

Mutual authentication (two parties proving to each other that they are who they say they are) involves the previous process done both ways. Both parties need to trust the certificate authorities that signed each other's certificates.

5.2.4 GSI Authorization

To recap, authorization is the process of deciding whether a particular identity can access a particular resource and in what fashion. Apart from users and computing resources having valid signed certificates, which provides proof of identity (authentication), users need authorization to access the resources. Currently in Globus toolkit proper, this is a separate mechanism and only basic access control facilities are provided. There are other software packages to aid the process, and there are also wide-ranging proposals to merge authentication and authorization in a single certificate. Authorization to access a resource implies that the user needs an account to access.

Accounts. Accounts have to exist on each computer system that users wish to access with permissions that allow the user to access the accounts. Each user might have an individual account within the virtual organization structure or group accounts. Setting up individual accounts is time-consuming as these computers will be geographically distributed and the computer systems have different owners so many system administrators could be involved. A mechanism for creating and managing these accounts is very desirable, possibly using a LDAP (Light Weight Directory Access Protocol) database, a network accessible database. The database lists the users using their unique X.500 distinguished names format found in their X.509 certificates and their access privileges.

In a Grid computing environment, sometimes it is convenient to have a group account for the virtual organization and all users in the virtual organization have access or share this account. In a training environment, it might be convenient to establish generic accounts prior to users being identified, e.g. `student1`, `student2`, `student3`, ... This can be helpful as then the user certificates can be created at the same time and everything is set up ready for a training course before participants join. We have done this in our Grid computing course when we set up certificates for students, but have now moved on to having students more involved in the process with individual accounts set up after students make requests through PURSe on the first day of the course. But this does require very short deadlines to make account requests so as not to delay subsequent activities.

Mapping Distinguished Names to Accounts. Globus provides a very basic way of mapping each user's distinguished name to account names and so gives a mechanism to access to the accounts via their distinguished names. The mapping is done in a file called a *gridmap* file, which contains a list of distinguished names and their local username mappings. Each user entry takes the form

```
Distinguished_name     local_user_account_name
```

For example

```
"/O=Grid/OU=GlobusTest/OU=simpleCA-coit grid02.uncc.edu/
    OU=uncc.edu/CN=abw" abw
```

Each distinguished name is given in quotation marks to allow spaces and must exactly match the way it appears in the user's certificate. The gridmap file is typically kept in `/etc/grid-security/grid-mapfile`. For security reasons, this file should not be assessible by users. The command `grid-map-add_user -dn distinguished_name` is used to add entries, which would only be done by administrators. The distinguished name can be obtained from the user's certificate with the command `grid-cert-info --subject --file certificate_file`.

Each computing resource that supports an account needs a gridmap in this mechanism. Each gridmap file must contain entries for all users allowed to access particular accounts on the resource as illustrated in Figure 5.5. Users might have the same account name on each machine, although that is not necessary. A user with the common name `bob` for example is shown with different account names on different systems (`bob`, `bob1`, `bob3`). If all account names are the same, a single gridmap file can be created and copied to all sites.

The gridmap file approach is a very primitive mechanism, and if used ideally the gridmap file needs to be populated with information in an automated process. Sometimes, a script is used to create the gridmap file from an LDAP directory holding user account information. Even that is somewhat primitive. A tool similar to gridmap files is GUMS (Grid User Management System), which will do the same

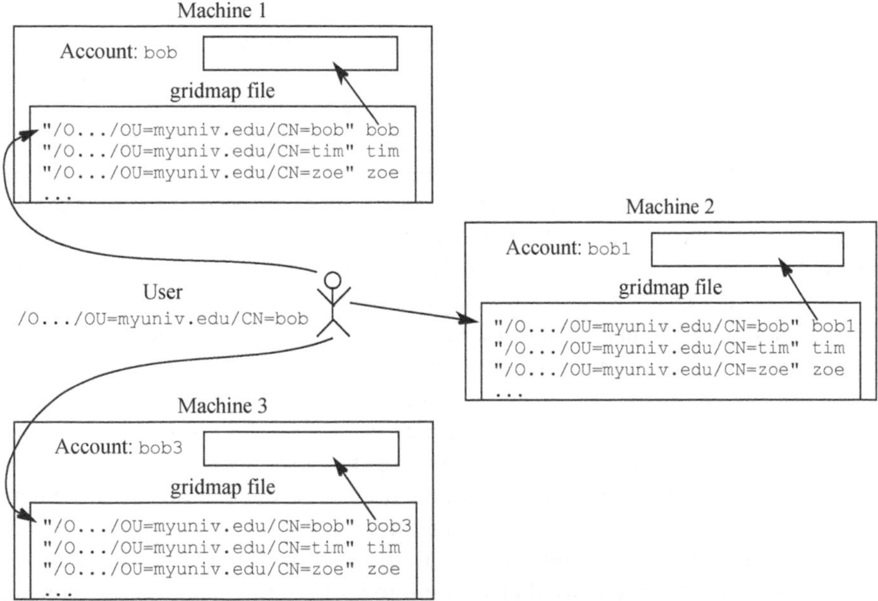

Figure 5.5 Mapping accounts using gridmap files on distributed computers.

mapping as gridmap files. In Section 5.4, higher-level authorization tools will be described.

Account Privileges with Gridmap Files. Gridmap files are often compared to access control lists, but as described they only provide blanket access and they do not provide specific types of access (levels of permissions, read/write/execute, group memberships, etc.). User access privileges, without any further mechanisms, will derive from the local system access control lists. Generally, we need a more powerful mechanism to control the type of access. Computers might be used under certain access rights. There may be times of the day even that they cannot be used by Grid users. There may be other limitations. The system may be dynamically changing. Higher level tools that are better suited to complex Grid computing environments will be outlined in Section 5.4. First let us discuss delegation as delegation is a critical feature of a Grid computing environment and some higher-level authorization tools use delegation.

5.3 DELEGATION

5.3.1 The Need for Delegation

Delegation is the process of giving authority to another identity (in this context, a computer) to act on your behalf. This is a critical requirement for Grid computing. Coupled with delegation is *single sign-on*, which enables a user and its agents to acquire additional resources without repeated physical authentication by the user (that is, submitting passwords/passphrases). Delegation is desirable in most tasks that a Grid user requests.

A common example for delegation occurs in third-party file transfers (Tuecke et al. 2004). A user might need to arrange that data files are located at specific places in the Grid. The files may have to be moved from one remote site say site *A* to another remote site say site *B*, initiated from where the user is located. The user interacts with a local file transfer service and mutually authenticates with that service. With the user's delegated authority, the local file transfer service then contacts file transfer services at site *A* and *B* that are to do the actual file transfer. Both site *A* and site *B* receive delegated authority from the local file transfer service to perform the transfer, and the local file transfer service has delegated authority from the user. The remote file transfer services mutually authenticate each other and perform the transfer with delegated authority as illustrated in Figure 5.6.

Clearly a user could log into site *A* and from there make a connection to site *B* manually and perform the file transfers, which has always been possible for many years (using Linux commands), but that requires two user authentications, one for site *A* and one from site *A* to site *B*. Ideally, the user wants to do these remote actions without having to submit passwords multiple times. Even running a job on a remote site requires login onto the remote site in a traditional distributed system. In a Grid system, it is also possible one would like site *B* to contact site *C* and site *C* to contact site *D* and so on in the process of doing complex tasks. Delegation provides a way of

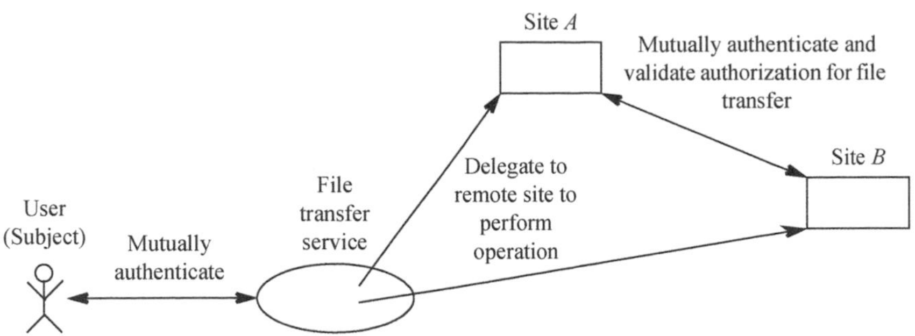

Figure 5.6 Delegation in third-party file transfers.

achieving remote actions without direct user intervention, and allows the user to leave and do something else while the remote actions are taking place, confident in the notation that delegated authority allows it to continue without the user's intervention. The next question is how should delegation be implemented in a Grid system.

5.3.2 Proxy Certificates

Using a *proxy certificate* is a way of implementing delegation, which was introduced by the Globus project and used extensively in GSI. In fact, a proxy certificate must be created to run even a simply Globus job. A *proxy certificate*, usually simply called a *proxy*, gives the resource possessing the proxy the authority to act on your behalf, just as proxy vote can be used for someone to place a vote on your behalf. The proxy credentials consist of a proxy certificate (with its public key) and proxy private key. The subject of the proxy certificate is the identity of the entity giving the proxy authority (the user in this case) with `/CN=proxy` added to the name to show that the certificate is a proxy certificate. The proxy certificate signed by the subject, not the certificate authority. This process can be compared to a proxy vote being signed by you but placed by someone on your behalf. Your signature should be checked. With proxy certificates, that is done by checking the signature on the proxy certificate, which requires the public key of the user. The public key of the user requires checking the signature on the user's certificate, which requires the public key of the certificate authority. So the proxy certificate alone is not sufficient. The user's certificate and certificate of the certificate authority (or at least their public keys) are needed also by the resource to act on your behalf—three certificates in all—as illustrated in Figure 5.7.

User certificates usually have long lifetimes, maybe a year, and the user's private key is kept very secure in an encrypted form based upon a passphrase established by the user. Each time the user performs a PKI authentication protocol (e.g., SSL/TSL), the user's private key must be decrypted with the passphrase. At that point, there is a brief opening for a breach of security, but as soon as possible the decrypted private key must be destroyed. If the proxy's private key was also encrypted, it would need a passphrase to decrypt it each time, which would defeat the purpose of just having a single sign-on with one application of a password/pass-

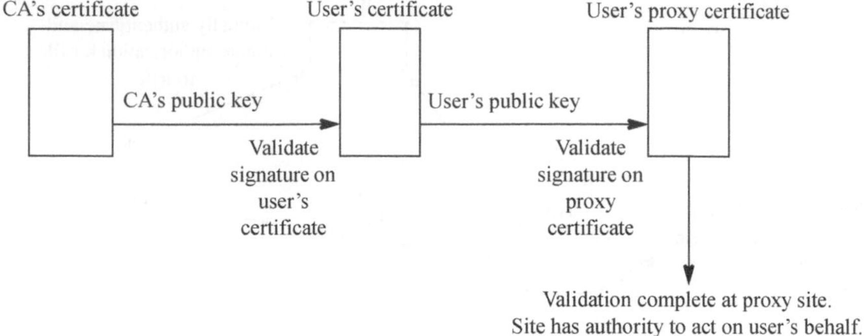

Figure 5.7 Validation at proxy site.

phrase. Consequently, the proxy's private key is simply protected by operating system file permissions and does not need a decrypting passphrase to access. The proxy is given a limited lifetime, say 2 hours, so that any breach of security is limited in time and the potential damage is limited.

Because the proxy has its own subject name (user's name plus being a proxy), it can be used in authentication and authorization mechanisms as a separate entity. For example, it would be possible, and a good idea, for a proxy of a user not to carry the full access rights of the user. It could be limited to the type of actions contemplated. For example in file transfer operations, it is not necessary nor desirable for the proxy to have execution rights on the files being transferred, in case the proxy is compromised. Delegation rights can be encoded into the certificate in a "ProxyCertInfo" X.509 extension field. The actual language used in the delegation has not currently been standardised and all parties would need to agree on the language.

In a Globus environment, one immediately provides delegated authority. The user must create a proxy which can be done manually with the `grid-proxy-init` command. The proxy can be destroyed with `grid-proxy-destroy`, and its contents can be examined with `grid-proxy-info`.

Now let us outline the process in a little more detail in terms of hosts. Suppose host *A* wishes to delegate authority to host *B* to act on its behalf with host *C*. First *A* requests B to create a proxy certificate request. *B* makes the request to *A* for a signed proxy certificate. This request is similar to a user making a request for a signed user certificate from a certificate authority—*B* generates a public and private key pair locally with the public key contained as part of an unsigned proxy certificate (`usercert_request.pem` file). The request (`usercert_request.pem`) is sent to *A*. *A* signs the request acting as a certificate authority for the proxy and sends both its own certificate and the self-signed proxy certificate back to host *B*. Host *B* then sends both certificates to host *C*. Host *C* needs the public key of the certificate authority that signed *A*'s certificate and trust it to validate *A*'s certificate and trust *A*'s public key. *C* then uses *A*'s public key to validate the proxy certificate and trust the proxy public key. It checks that the subject on *A*'s certificate and *A*'s proxy match. (The proxy will have additional `/CN=proxy` as mentioned earlier.) Now an encrypted message can be sent from *B* and *C* encrypted with the proxy's private key. Host *C* can decrypt the message with the proxy's public key with trust. The process is illustrated in Figure 5.8. The description of this process is derived from Ferreira et al. (2003).

After host *A* has delegated authority to host *B*, the process can be continued with host *B* delegating authority to host *C*, host *C* to host *D* and so on in a *chain of trust*. Each host will sign the proxy for the next host as illustrated in Figure 5.9. The validation path is traced back to the certificate authority. Notice that the authorization at each level is limited by the previous level and cannot be extended, and could become more restrictive.

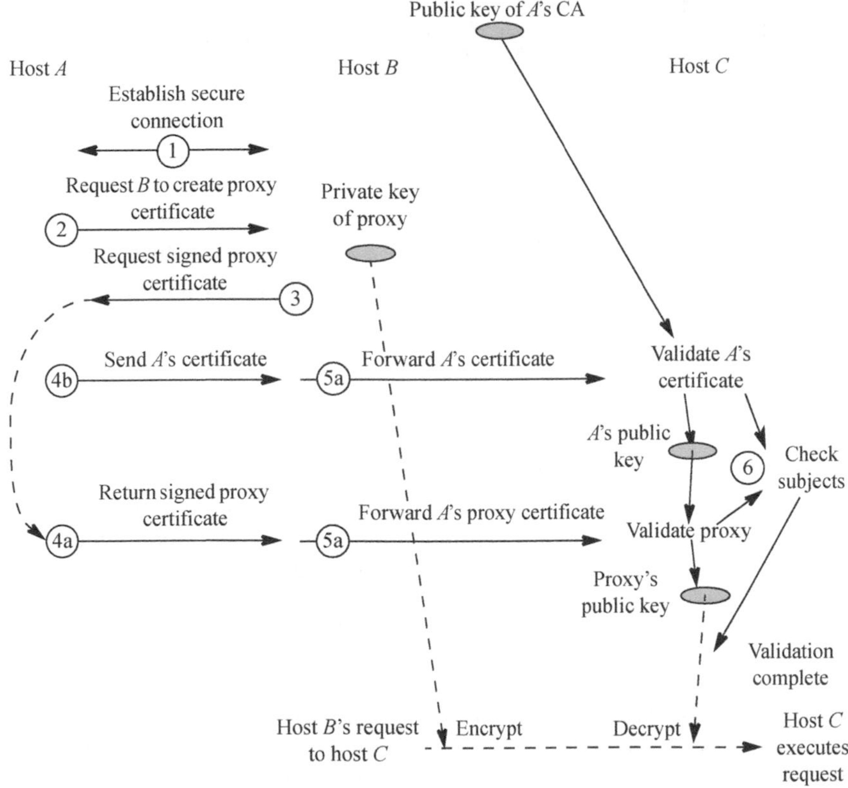

Figure 5.8 Delegation of authority to another host prior to sending an encrypted message.

5.3.3 MyProxy Grid Credential Repository

MyProxy (NCSA 2007) is a repository for certificates, and widely used in Grid computing. As its name suggests, MyProxy is used specifically to provide short lifetime proxies upon request. MyProxy was originally a separate service to those provided with Globus, but was incorporated into Globus 4.0, and it has become an integrated part of the Globus environment especially with the GridSphere portal. There are two ways that MyProxy might be used—as a repository for user's credentials or as a repository for proxy credentials.

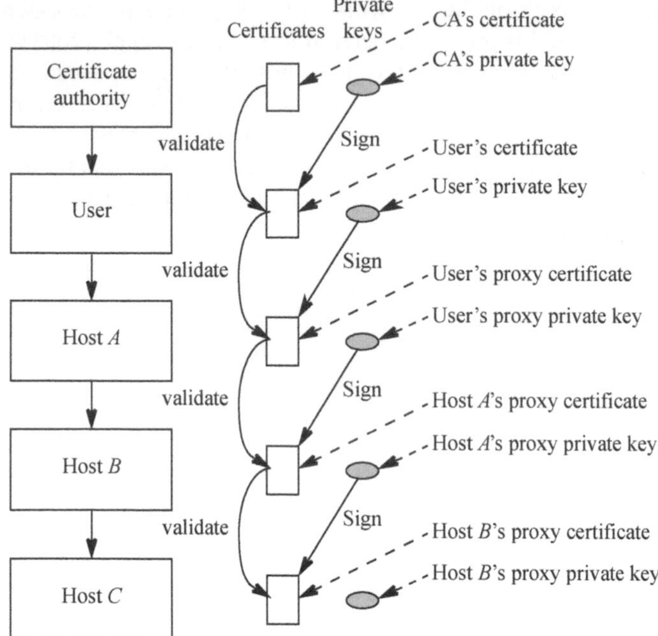

Figure 5.9 Chain of trust.

Repository for User's Credentials. MyProxy can be used as a repository to store the user's credentials (user's certificate and private keys), from which proxy certificates are created by MyProxy upon request. Storing the user's credentials relieves the users of the responsibility for keeping them locally. The private key is encrypted with the same user's passphrase as used for encrypting the original private key.

Repository for Proxy Credentials. MyProxy can generate proxies for a user if it is given delegated authority to do so by having the user's proxy in its possession. (It does not need the actual user's certificate.) So MyProxy need only store the user's proxy. It could be given a proxy certificate with a short lifetime, say a 7-day lifetime, from which even shorter proxy certificates are provided by MyProxy, say proxies with 12 hours or whatever the user wishes within the limits of the stored proxy.

Once users have their user or proxy credentials stored in the MyProxy service as illustrated in Figure 5.10, local and remote services can retrieve short life-time proxies as required. For example, a user might submit a job to GRAM (Grid Resource Allocation and Management), the Globus job submission service. GRAM will need to obtain a proxy from MyProxy to initiate the request. A long-running job might need to request subsequent proxies over time to keep the job running.

MyProxy can be used in conjunction with a registration service such as PURSe (Portal based User Registration Service). Users can retrieve and see their proxy through the PURSe portlet within the Grid portal such as Gridsphere, as was illus-

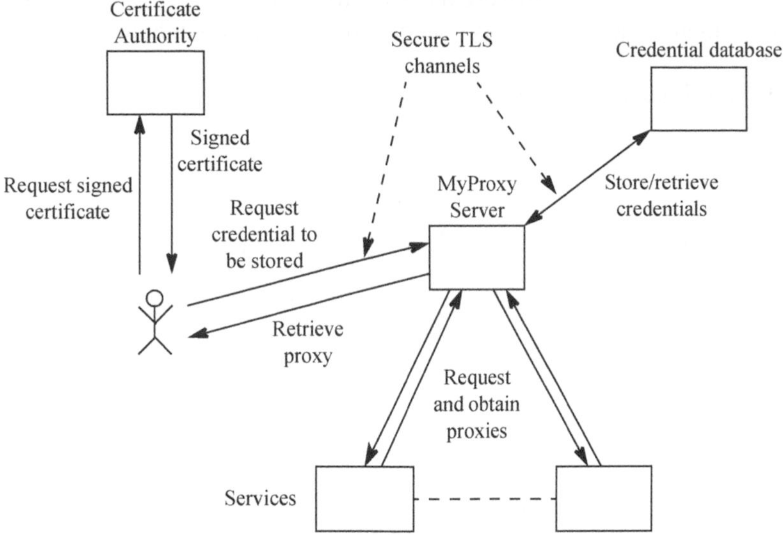

Figure 5.10 MyProxy server with a separate certificate authority.

trated in Chapter 1 (Figure 1.11), but now the proxy should be more clearly understood. The user simply selects the Proxy management tab from the PURSe portlet and can see available proxies. Proxies are loaded automatically when the user logs in. There could be more than one valid proxy including from previous sessions. Once a proxy expires, it will disappear. If necessary, the user uses the "Get New Proxy" button to request a new proxy, which will be created and displayed. The email address at the end of the distinguished name in Figure 1.11 is added by PURSe.

MyProxy assists in the use of proxies and reduces risk of management of many copies of credentials. On a Grid, every party identifies itself with credentials and these credentials can be stored in a MyProxy server. MyProxy has a number of commands that a user can issue on a command line (or through the MyProxy portlet), notably:

- `myproxy-store` Store user credentials found in `~/.globus/usercert.pem` and `~/.globus/userkey.pem` in MyProxy server
- `myproxy-init -t hours` Store a proxy where `hours` is the lifetime (default 12 hours), as an alternative to storing the user's credentials
- `myproxy-logon` (formerly `myproxy-get-delegation`) Retrieve a proxy
- `myproxy-info` Query stored credentials
- `myproxy-destroy` Remove credential

A typical command line sequence would be as follows. The user first gets valid long-term credentials by making a request for a signed certificate from a certificate authority, as described earlier. The credentials are typically stored in `$HOME/.globus/` as `$HOME/.globus/usercert.pem` (signed certificate) and `$HOME/.globus/userkey.pem` (private key). The user then issues the `myproxy` command `myproxy-store` to store the user credentials or `myproxy_init` to store proxy credentials. The command is issued on the computer that holds the user's credentials.

A typical sequence using `myproxy-init` is

```
myproxy-init -s myproxy.coit_grid02.uncc.edu

Your identity: /C=US/O=UNCC/CN=Barry Wilkinson
Enter GRID pass phrase for this identity:
Creating proxy ......................................... Done
Your proxy is valid until Fri Sep 13 13:52:56 2008
Enter MyProxy Pass Phrase:
Verifying password - Enter MyProxy Pass Phrase:
A proxy valid for 168 hours (7.0 days) for user abw now exists
on myproxy.coit_grid02.uncc.edu.
```

`GRID pass phase` above is the passphrase used when the user credentials were created and is needed to validate the request. It will not be needed again as your proxy credentials will be used subsequently. The MyProxy passphrase is used to retrieve the proxy credential from the MyProxy server. The default lifetime is 12 hours. One could select a different lifetime for the proxy using the `-c` option. One could even select the same lifetime as the user's long-term credentials by using the `-c 0` option (maximum lifetime).

Then, the user might wish to retrieve the proxy, which can be done with `myproxy-logon` command, i.e.,

```
myproxy-logon -s myproxy.coit_grid02.uncc.edu

Enter MyProxy Pass Phrase:
A proxy has been received for user abw in /tmp/x509up_u500
```

Using MyProxy with PURse registration within a portal would typically store the long-term user credentials in the MyProxy server using MyProxy administrator commands. A certificate authority has been integrated into the MyProxy software (from version 3.0 onwards) specifically for issuing short lifetime credentials. When enabled, the `myproxy-login` command will return a signed user certificate from this certificate authority.

5.4 HIGHER-LEVEL AUTHORIZATION TOOLS

Gridmap files described in Section 5.2.4 are the basic way that Globus provides for mapping distinguished names to local machine accounts but it is a very primitive way that does not scale well. Gridmap files also only provide a basic mapping of distinguished names to machine accounts and do not include any finer access control or any higher-level control of authorization for a Grid environment. Several tools have been developed to provide higher-level authorization. Here, some of these tools will be briefly identified.

5.4.1 Security Assertion Markup Language (SAML)

Security Assertion Markup Language (SAML) is a framework that embodies both an XML language for making assertions for authentication and authorization decisions and a request-response protocol for such assertions. SAML was actually developed for business Web sites but has been applied to Grid computing in concert with other tools. It was developed by OASIS for facilitating the exchange of security information between business partners, in particular to obtain single sign-on for Web users (Cantor et al. 2005). It addresses the situation where a Web user accesses a Web site that might require the user's request to be redirected to another affiliated site after being authenticated, for example travel bookings and automobile reservations. It has applicability to Grid computing because of the similar nature of accessing a chain of resources. SAML does not use certificates and certificate authorities. The identity of a user is asserted at any point in the network by an *identity provider* and it is up to the application to trust this assertion. Hence, the approach is completely decentralized. Information being transmitted can be encrypted using SSL/TLS protocols.

There are three principal components—the *user agent* (Web browser), the *identity provider*, and the *service provider*, as illustrated in Figure 5.11. The user agent might first interact with the initial Web site. In the process of conducting the transactions, the user is redirected to another Web site that will process the user's request (service provider). The first Web site acts as the identity provider and will make an SAML assertion to the service provider that the user is authenticated for the requested action. SAML enables multiple servers to share authentication information. The SAML framework does not actually perform the authentication service. It just passes the information in an XML format, and usually by HTTP/SOAP[2]. It is still necessary to have authentication software/server.

SAML provides for the communication of user authentication, authorization and attribute information and covers

- Assertions — the information being communicated

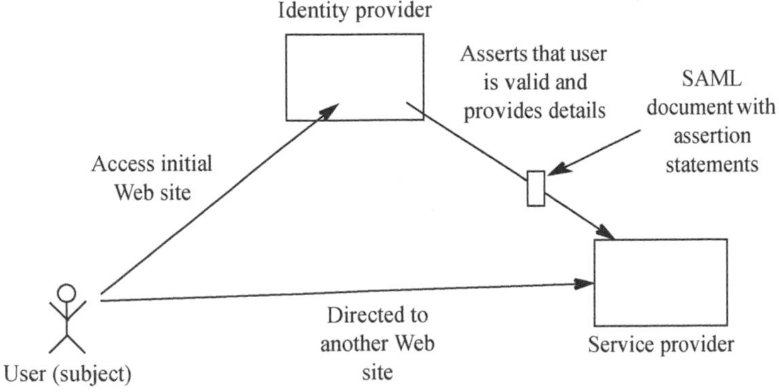

Figure 5.11 SAML single sign-on.

[2] SOAP is a communication protocol for passing XML documents, see Chapter 6 for more details.

- Protocol — the way that the message exchanges are done
- Binding — the mapping to concrete SOAP exchanges and specific protocols (usually HTTP)

There are three forms of assertions:

- Authentication statements
- Attribute statements
- Authorization decision statements

Authentication assertion statements confirm the user's identity to the service provider. *Attribute assertion statements* detail specific information about the user that can be used to establish access decisions. This information can then be passed via the *authorization assertion statements*. Attributes might for example include that a user is an administrator (root privileges) or has limited user privileges. A SAML authorization decision might state that the subject (user) is allowed to perform the specified operation on the specified resource. Apart from the pattern shown in Figure 5.11, other patterns are accommodated, for example when the server provider requests confirmation of authorization from the identity provider in a scenario that the user first contacts the service provider.

5.4.2 Using Certificates for Authorization

Using Non-Critical X.509 Extensions. The original purpose of X.509 certificates is to prove identity (*authentication*). Since these certificates are passed to resources, and both authentication and authorization decisions are often made at the same time, it is obvious that one could add additional information in the certificates for authorization. The X.509 format allows additional information in so-called *non-critical X.509 extensions* in the certificate, enabling existing software to at least pass on such certificates for authorization decisions. One of the early software components using this approach is the *Community Authorization Service* (CAS), developed by the Globus community to provide an authorization service in a Globus environment using proxy certificates (Pearlman et al. 2002). CAS is part of Globus 4.0. According to Pearlman et al. (2003), CAS addresses the need for:

- Scalability
- Expressiveness that is not provided in the authorization mechanisms of native operating systems
- Consistency across operating systems
- Distribution of any changes in policy

The approach taken is to have a centralized server, a so-called *CAS server* that issues a proxy to a user to include authorization assertions inserted as non-critical X.509 extensions in the certificate. This approach enables the proxy certificates to be processed by existing software. Originally, these proxies were CAS proxies, that is, having the subject as CAS, but quite quickly they were changed to be user proxies, i.e., having the subject of the user. The latter enables the identity of the user to be established directly for the proxy. The basic process is for a user to request the proxy from the CAS server that is established for the virtual organization. This server will have access to the issuer's privileges through a virtual organization database, and will add authorization assertions to the user's proxy. Although not initially in CAS proxies, these assertions are now using SAML. The user presents the access request with the proxy to the resource as it would in normal GSI. The resource acts as it would with a normal proxy, checking the identity of the user and any local authorization policies. In addition, it will read the CAS assertions and restrict access further according to these assertions. The mechanism is shown in Figure 5.12. The approach requires the resource to be "CAS-enabled," i.e., be able to recognize CAS-SAML assertions and take appropriate actions.

In a Globus installation where CAS is installed, a user can obtain a proxy modified by CAS with assertions by using the command `cas-proxy-init -t tag`. The CAS modified proxy and `tag` is used in the subsequent `cas-wrap` command to run programs. To run a program `prog1`, the `cas-wrap` command would be `cas-wrap -t tag prog1` (GT4 CAS User's Guide). CAS can combine local course-grained access control policies with more fine-grained access control policy management of the community.

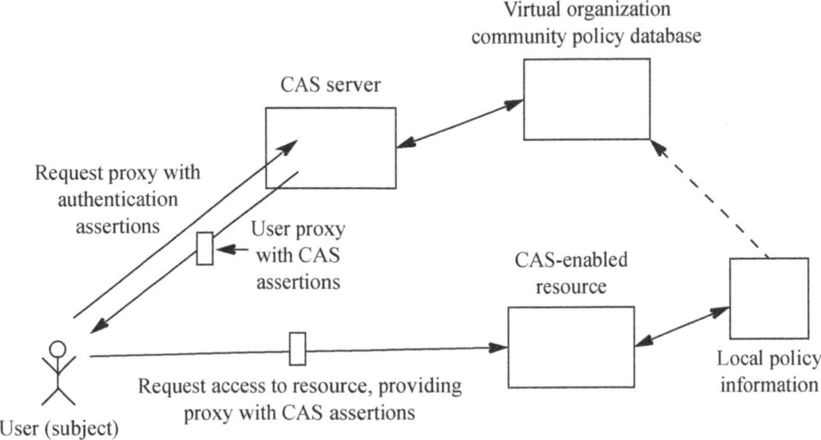

Figure 5.12 CAS structure.

Attribute (Authorization) Certificates. Another approach rather than place authorization assertions in the user's certificate as an extension is to create a completely separate so-called *attribute certificate* (also called an *authorization certificate*). A X.509 attribute certificate has been proposed as a standard for broad applicability. An attribute certificate is bound to a particular identity and digitally signed to validate it. The full treatment of attribute certificates is beyond the scope of this book but the basic idea is shown in Figure 5.13. In this figure, a regular proxy certificate is derived from the user's certificate issued from a certificate authority and is used to delegate authority. A separate attribute certificate issued by an attribute authority is used to define specific authorization policies attached to the particular

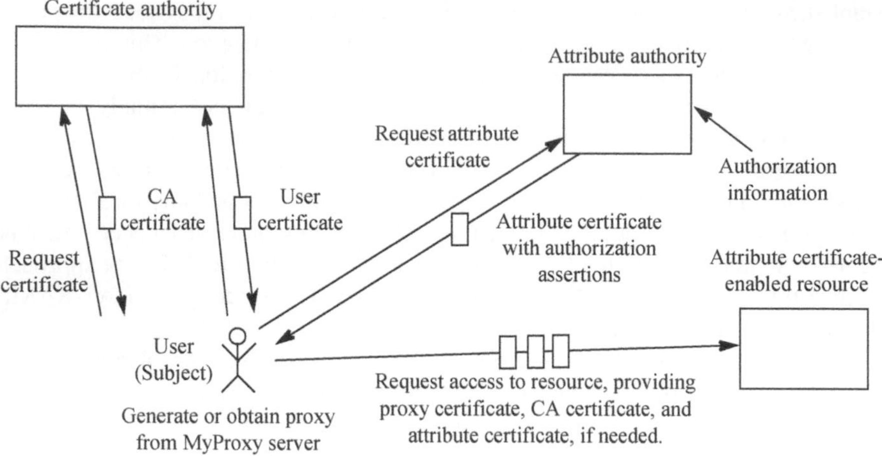

Figure 5.13 Attribute certificate mechanism—concept.

user accessing the resource. Using a separate certificate for authorization enables different issuing bodies to be used. It also enables the lifetime of the authentication to be different to that of authorization. Lifetime of authorization is not normally the same as for authentication. Combining authorization and authentication in one certificate would mean both would have to be given the same lifetime.

An example of using a separate authorization document is in the VOMS (Virtual Organization Membership Service). There are several other approaches being developed to address authentication and authorization including Shibboleth and GridShib, which are outside the scope of this book.

5.5 SUMMARY

This chapter follows on from Chapter 4 on general Internet security concepts and introduces how security concepts are applied to a Grid computing environment and the new aspects that Grid computing introduces. The chapter introduces:

- Grid Security Infrastructure (GSI)
- GSI Authentication mechanisms
- Transport and message-level protocols
- Certificate authority structures for Grid computing
- GSI Authorization mechanisms
- Delegation
- Proxy certificates
- Authentication mechanism such as using attribute (authorization) certificates.

FURTHER READING

The Globus home page, http://www.globus.org/, offers material on Grid security as used in Globus and in fact in Grid computing in general. Although based upon the now obsolete Globus version 3, the IBM Redbook "Introduction to Grid Computing with Globus, Fundamentals and Concepts" (Ferreira et al. 2003) is an excellent reference book describing security aspects. The concepts have not changed significantly through each version of Globus.

BIBLIOGRAPHY

Cantor, S., J. Kemp, R. Philpott, and E. Maler, eds. 2005. Assertions and protocols for the OASIS Security Assertion Markup Language (SAML) V2.0 OASIS Standard. 15 March. http://docs.oasis-open.org/security/saml/v2.0/

Ferreira, L., V. Berstis, J. Armstrong, M. Kendzierski, A. Neukoetter, M. Takagi, R. Bing-Wo, A. Amir, R. Murakawa, O. Hernandez, J. Magowan, and N. Bieberstein. 2003. *Introduction to Grid computing with Globus, fundamentals and concepts.* IBM Redbooks. http://www.redbooks.ibm.com/redbooks/pdfs/sg246895.pdf.

Foster, I., C. Kesselman, G. Tsudik, and S. Tuecke. 1998. A security architecture for computational Grids. *5th Conference on Computer & Communication Security*, San Francisco.

GT4 CAS User's Guide. http://www.globus.org/toolkit/docs/4.0/security/cas/user-index.html

Globus Certificate Service. http://gcs.globus.org:8080/gcs/index.html

Globus Simple CA package. http://www.globus.org/toolkit/docs/2.4/gsi/simple-ca.html

Govindavajhala, S., and A. W. Appel. 2006. Windows access control demystified. Princeton University, Jan. 31. http://www.cs.princeton.edu/~sudhakar/papers/winval.pdf

NCSA (National Center for Supercomputing Applications). 2007. MyProxy credential management service. http://grid.ncsa.uiuc.edu/myproxy/

Pearlman, L., V. Welch, I. Foster, C. Kesselman, and S. Tuecke. 2002. A community authorization service for group collaboration. *IEEE 3rd Int. Workshop on Policies for Distributed Systems and Networks*.

Pearlman, L., C. Kesselman, V. Welch, I. Foster, and S. Tueck. 2003. The Community Authorization Service: Status and future. *Conference for Computing in High Energy and Nuclear Physics (CHEP03)*, March 24-28, La Jolla, California.

Tuecke, S., V. Welch, D. Engert, L. Pearlman, and M. Thompson. 2004. Internet X.509 Public-Key Infrastructure (PKI) proxy certificate profile, June. http://www.ietf.org/rfc/rfc3820.txt

Wikipedia Capability-based security. http://en.wikipedia.org/wiki/Capability-based_security

SELF-ASSESSMENT QUESTIONS

The following questions are multiple choice. Unless otherwise noted, there is only one correct answer for each question.

1. If a third party intercepts a communication using (encrypted) transport-level protocol, what can it discover?

(a) Nothing
(b) Everything about the message and its contents, i.e., it can read the whole message
(c) It can discover who sent the message and who will receive the message
(d) None of the above

2. If a third party intercepts a communication using (encrypted) message-level protocol, what can it discover?

(a) Nothing
(b) Everything about the message and its contents, i.e., it can read the whole message
(c) It can discover who sent the message and who will receive the message
(d) None of the above

3. What is the difference between the user's credentials and the user's certificate?

(a) Nothing
(b) User credentials includes both the user's certificate and the user's private key
(c) User's certificate includes the user's private key
(d) User's credentials includes the user's public key

4. What is the difference between a *password* and a *passphrase*?

(a) Nothing
(b) A password has to have digits
(c) A passphrase has to be a sentence
(d) A passphrase allows spaces

5. In Grid security, what is delegation?

(a) Process of making use of many processors to speed up the computation
(b) Process of giving authority to another identity to act on your behalf
(c) Process of assigning work to another processor
(d) Process of creating a certificate request and getting it signed by a certificate authority

6. What is a disadvantage of using gridmap files for access control? (May be more than one.)

(a) It is difficult to maintain for large Grids
(b) It does not apply fine-grain access control
(c) It is difficult to verify user credentials
(d) It is difficult to map distinguished names to local accounts
(e) It is difficult to maintain in a dynamically changing virtual organization

7. What type of security mechanism does Globus 4.0 use?

(a) Password-based protection
(b) Public key cryptography
(c) Secret key cryptography
(d) None

8. What is a proxy?

(a) A certificate provided to enable resources to be acquired on the user's behalf
(b) Secret key
(c) A third party given authority to acquire resources on the user's behalf
(d) A computer given authority to acquire resources on the user's behalf

9. Which of the following are encrypted? (May be more than one.)

(a) User's certificate
(b) User's private key
(c) Proxy certificate
(d) Proxy's private key

10. Suppose there is a chain of trust with hosts *A*, *B*, *C*, and *D*. How many certificates must host *D* receive to establish trust?

(a) 0
(b) 1
(c) 2
(d) 3
(e) 4
(f) More than 4

11. Suggest one advantage of using attribute (authorization) certificates rather than using assertions held into non-critical X.509 extensions of users certificates.

(a) There are no advantages
(b) Allows authorization policies to have a different period of validity
(c) Allows authorization policies to have the same period of validity
(d) It is faster

PROGRAMMING ASSIGNMENTS

The following assignments assume access to a Grid platform with a certificate authority and a Globus installation. Individual installations will have different operating arrangements so additional operating procedures will need to be provided.

5-1 Connect to a server with a Globus installation. Make a request for a signed certificate using the Globus command

```
grid-cert-request
```

You will be prompted for a passphrase. Devise a passphrase, which should be different from your login password. It is important to remember this passphrase. As described in the chapter, three files are generated in your `.globus` directory

```
usercert_request.pem
usercert.pem (an empty file)
userkey.pem
```

The file `usercert_request.pem` will be sent to the certificate authority administrator. Depending upon your system, arrange that this file is sent to the certificate authority administrator for signing. (That may be done by email or by the system administrator directly accessing your `.globus` directory.)

Once your certificate is signed, test that it works by creating a proxy with the command

```
grid-proxy-init
```

You will need your passphrase that you selected when creating the certificate request. Check that the proxy was created using the command

```
grid-proxy-info
```

CHAPTER 6
System Infrastructure I: Web Services

This chapter will discuss the background for how the Grid infrastructure can be implemented. It begins with a review of early distributed computing concepts for obtaining remote actions, as such actions form the basis of recent methods. Then, the basic technology of Web services is described. Web service technology overcomes certain disadvantages of earlier methods. It has been adopted for many Grid middleware components and is a candidate for incorporating into distributed Grid applications. It uses standard Internet technologies for communication and information representation. In particular, XML (eXtensible Markup Language) is adopted as the standard way of representing information and is key to the success of Web services. XML is also used in job description languages (Chapters 2 and 3) and elsewhere in Grid computing. Appendix C provides an outline of XML. Those readers that do not already know XML should read Appendix C. Chapter 7 continues the topic of Web services in a Grid computing environment.

6.1 SERVICE-ORIENTED ARCHITECTURE

Distributed computing such as Grid computing relies on causing actions to occur on remote computers. Taking advantage of remote computers was recognized many years ago well before Grid computing. One of the underlying concepts is the *client-server model*, as shown in Figure 6.1. The client in this context is a software component on one computer that makes an access to the server for a particular oper-

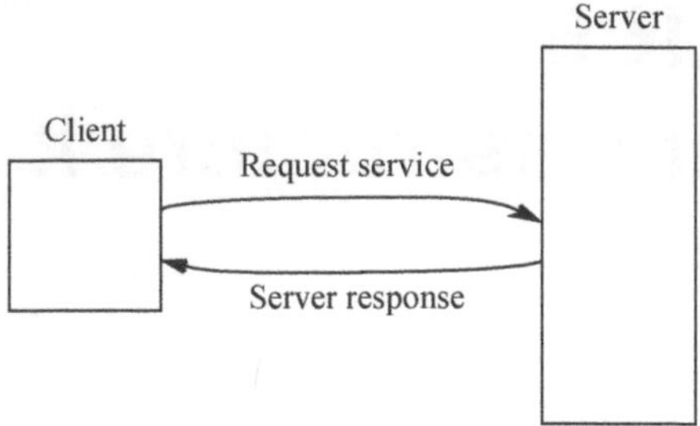

Figure 6.1 Client-server model.

ation. The server responds accordingly. The request and response are transmitted through the network from the client to the server.

An early form of client-server arrangement was the *remote procedure call* (RPC) introduced in the 1980s. This mechanism allows a local program to execute a procedure on a remote computer and get back results from that procedure. It is now the basis of certain network facilities such as mounting remote files in a shared file system. For the remote procedure call to work, the client needs to:

- Identify the location of the required procedure.
- Know how to communicate with the procedure to get it to provide the actions required.

The remote procedure call introduced the concept of a *service registry* to provide a means of locating the service (procedure). Using a service registry is now part of what is called a *service-oriented architecture* (SOA) as illustrated in Figure 6.2. The sequence of events is as follows:

- First, the server (service provider) publishes its services in a service registry.
- Then, the client (service requestor) can ask the service registry to locate a service.
- Then, the client (service requestor) binds with service provider to invoke a service.

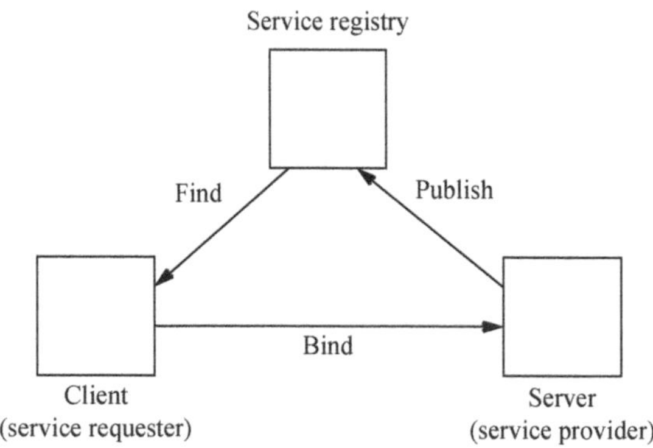

Figure 6.2 Service-oriented architecture.

Later forms of remote procedure calls in 1990s introduced distributed objects, most notably, CORBA (Common Request Broker Architecture) and Java RMI (Remote Method Invocation).

A fundamental disadvantage of remote procedure calls so far described is the need for the calling programs to know implementation-dependent details of the remote procedural call. A procedural call has a list of parameters with specific meanings and types and the return value(s) have specific meaning and type. All these details need to be known by the calling program, and each remote procedure provided by different programmers could have different and incompatible arrangements. This led to improvements including the introduction of *interface definition* (or *description*) *languages* (IDLs) that enabled the interface to be described in a language-independent manner and would allow clients and servers to interact in different languages (e.g., between C and Java). However, even with IDLs, these systems were not always completely platform/language independent.

Some aspects for a better system include:

- Universally agreed-upon standardized interfaces
- Inter-operability between different systems and languages
- Flexibility to enable different programming models and message patterns
- Agreed network protocols (Internet standards)

Web services with an XML interface definition language offer the solution.

6.2 WEB SERVICES

6.2.1 Concept

Web services were introduced in 2000 as software components designed to provide specific operations ("services") that are accessible using standard Internet technology. Web services have similarities with RMI and other distributed object technologies (CORBA, etc.) but they use XML and standardized network protocols and are platform independent. A standardized XML language is used for the interface definition language called *Web Services Description Language* (WSDL).

Web services are designed for machine interaction over a network and invoked by other programs. They are usually addressed by a URL (Uniform Resource Locator), for example, the URL:[1]

```
http://www.cs.uncc.edu/webservices/math1
```

This particular URL does not exist, and if it did, it would only be meaningful to Web service software. The Web service itself may be located anywhere that can be reached on the Internet. One can visualize a situation where the user interacts with an appli-

[1] Web services can use more powerful addressing mechanisms, especially for Grid computing, which is discussed in Chapter 7.

cation, which itself may use Web services located elsewhere to perform the actions requested by the user, as illustrated in Figure 6.3. The Web services could perform generally useful functions for the application. Later, we shall see a simple Web service that performs mathematical operations. A Web service could perform operations related to a specific application. For example, a retail business might use Web services to access an inventory of items they hold.

An alternative scenario for using Web services is to use them as front-ends to applications, making the applications accessible on the Internet, as shown in Figure 6.4. This particular arrangement finds applicability for Grid-enabling applications (Chapter 9).

For Grid infrastructure implementation, Web services are attractive for the various distributed intercommunicating Grid components, and also as an interface for users to access the Grid resources and perform other Grid related activities as illustrated in Figure 6.5. Using Web services in Grid computing infrastructure are explored in Chapter 7.

Generally, Web services alone are regarded as *stateless*, that is, they do not remember or store information themselves from one invocation to another. This is reasonable since a Web service might be accessed by many requestors in no specific order. The same characteristic can be found when accessing Web pages. One can

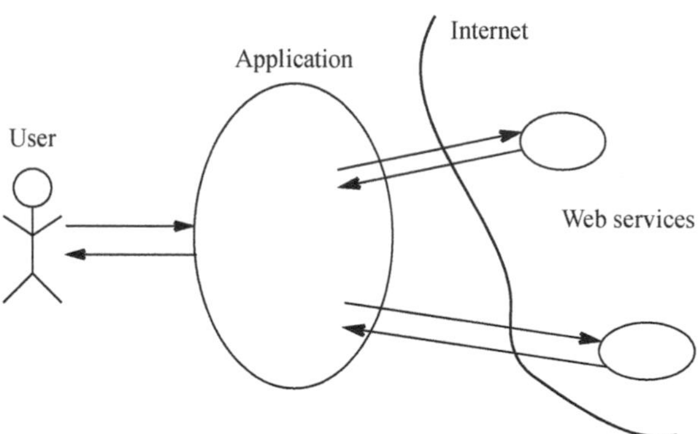

Figure 6.3 Application employing Web services.

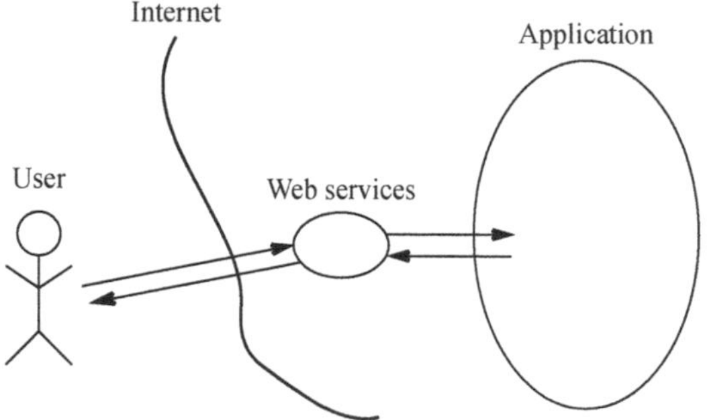

Figure 6.4 Web service front-end to an application.

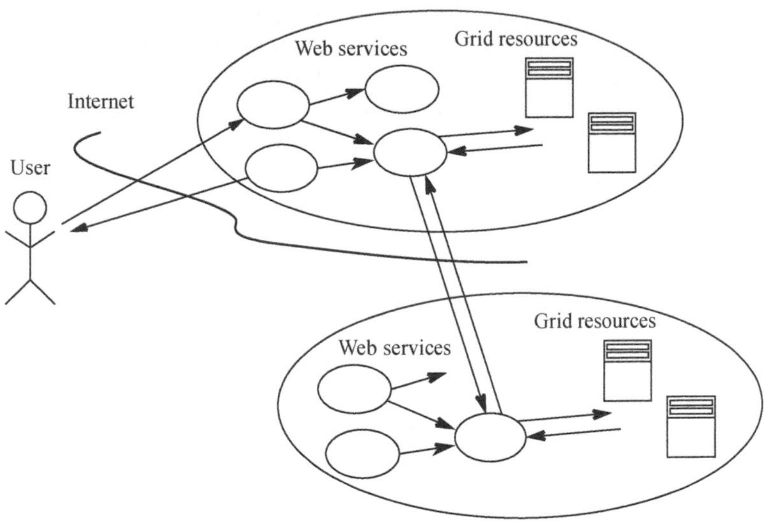

Figure 6.5 Web services for distributed Grid components.

move from one Web page to another, so can other users without interference. (With Web pages, user-specific information could be stored separately in cookies on the user's computer.) Web services can be the front-end to a stateful resource. For example, a retail business inventory might be accessed through a Web service and the Web service can return information to a requestor about say the level of inventory of a product, which can change over time. A Web service can incorporate state in a standard called *Web Service Resource Framework* (WSRF), which we consider in Chapter 7. Grid computing needs state represented in its services. First though, let us look at the design of a traditional stateless Web service.

6.2.2 Communication Protocols for Web Services

Web services use XML documents. Hence in Web services, one needs a communication protocol for passing XML documents. SOAP is a communication protocol for passing XML documents, standardized by W3C organization (World Wide Web Consortium). SOAP was originally an abbreviation of Simple Object Access Protocol but now is simply SOAP and is not object oriented. It provides mechanisms for:

- Defining communication unit — a SOAP message
- Error handling
- Extensions
- Data representation
- Remote procedure calls (RPCs)
- Document-centric approach for business transactions
- Binding to HTTP

SOAP messages are transported using standard Internet protocols, most likely the HTTP protocol. The message that "goes down the wire" consists of an HTTP packet containing details about routing, reliability, security, the SOAP message and possible attachments. One can picture the collection as shown in Figure 6.6. The SOAP envelope format is shown in Figure 6.7. Contained in the body is the transported XML document. The XML/SOAP standardization body, World Wide Web Consortium (W3C) covers SOAP and its attachments.

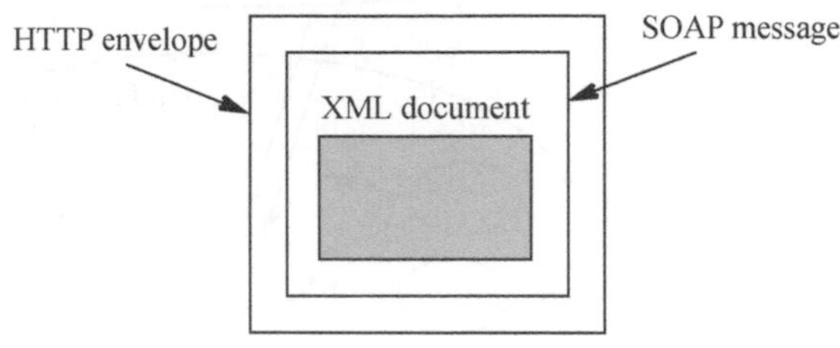

Figure 6.6 Message package.

Figure 6.7 SOAP envelope.

6.2.3 Defining a Web Service Interface — WSDL

We need a way of formally describing a service, what it does, how it is accessed, etc. i.e., an interface description language (IDL). In keeping with the overall philosophy of using XML, this interface description language should be an XML language. The World Wide Web Consortium (W3C) has published an XML standard for description of Web services called *Web Service Description Language* (WSDL). A WSDL document describes three fundamental properties of a service:

- What it is — operations (methods) it provides
- How it is accessed — data format, protocols
- Where it is located — protocol specific network address

WSDL Version 1.1 was introduced in 2001. Version 1.2 was initially proposed in 2003 as a working draft and renamed as WSDL version 2 in 2004. It finally became an official W3C recommendation after four years of development in June 2007. WSDL 2.0 was intended to develop upon on WSDL 1.1 but has significant differences to WSDL 1.1. Here, WSDL 1.1 are outlined as it became widely used and adopted in Grid computing software such as the Globus tool (together with WSRF to make the service stateful as discussed in Chapter 7). WSDL 2.0 is beyond the scope of this book. WSDL 1.1 and WSDL 2.0 are not compatible.

To describe a WSDL document in some detail but keeping it generic, let us consider a service that has one function (operation) called `funct1`. Suppose this function has one argument supplied, `arg1`, and returns one result, `result`, based only upon the supplied argument, as shown in Figure 6.8. In the following for simplicity, both `arg1` and `result` are integers. The service is stateless and does not store any values from one invocation to the next. The WSDL document does not actually define the specific operations of a service, just message interface, so the function could do anything. Typically, the names used in the WSDL document are related to the actual messages, operations, and arguments, but here they are given generic names.

A WSDL Version 1.1 document has the following structure:

- Root definitions and namespaces
- Abstract definition of service
- Message definitions (parameters in method signature)
- Type definitions – data types
- Binding definitions to network protocols
- Service definitions (where service is, ports)

The basic structure and interrelationships are shown in Figure 6.9. What the service provides (operations) is defined by an *abstract definition* of the service, which specifies the messages that must be sent to the service and the messages that must be returned, in abstract terms, that is, as inputs or outputs. The abstract definition of the

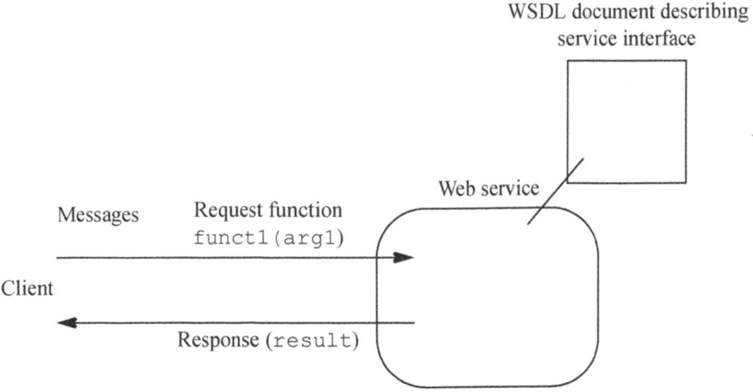

Figure 6.8 Generic Web service with a WSDL document.

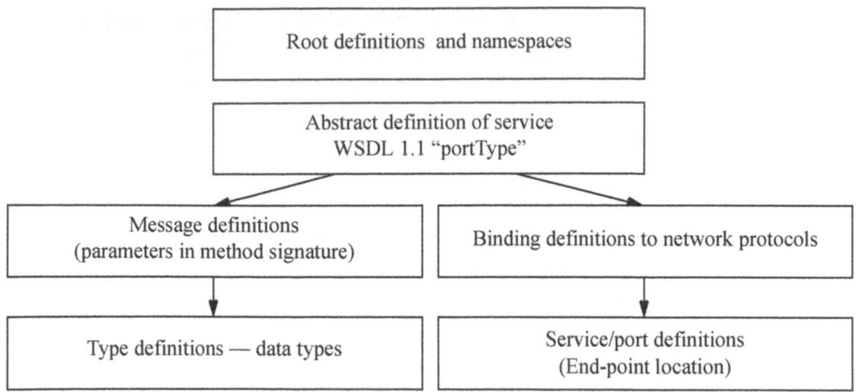

Figure 6.9 Basic parts of a Web service WSDL 1.1 document.

service is called a *portType* in WSDL version 1.1, a not very obvious name. WSDL version 2 changed the name of the abstract service definition to the name *interface.* The portType (or interface) identifies and names the messages to and from the service. The composition of the messages is then defined in the *message definitions*, which will refer to datatypes used in the messages. If these datatypes are not predefined primitive datatypes in the underlying Schema Definition Language (XSD) such as integers or string, then separate *type* definitions are needed to define the complex datatypes in terms of primitive datatypes, and referred by names in the message definition. The messages also need to be bound to specific network protocols, that is, the way that the messages are carried. This is achieved in the *binding* element. Messages are usually "SOAP over HTTP." The *port* element is the endpoint for the service and describes the location (URL) for specific binding. The *service* component is a named collection of ports.

A simple WSDL document for our generic service is shown in Figure 6.10 and is structured conceptually in a top-down fashion. As we shall see, the programmer will not necessarily have to write this file, as it can be generated automatically from Java class of the Web service with Web service tools, and in fact one of the assignments at the end of the book calls for using such tools.

Root Definitions. The root definitions define the namespaces[2]. The namespace `xmlns:xsd="http://www.w3.org/2001/XMLSchema"` is the namespace for the XML schema. The namespace `xmlns:wsdl="http://schemas.xmlsoap.org/wsdl/"` is the WSDL schema. There are namespaces for SOAP.

`<portType>` **(Interface) Element.** The `portType` (interface) element corresponds to the Java interface of the Web service, that is, its input/output functionality but not its implementation. Our service is simply called `GenericService`. The

[2] See Appendix C for more details on namespaces.

```
<?xml version= "1.0" encoding="UTF-8">
<!-- NAMESPACES -->
<wsdl:definitions targetNamespace="http://DefaultNamespace"
xmlns:apachesoap="http://xml.apache.org/xml-soap"
xmlns:impl="http://DefaultNamespace"
xmlns:intf="http://DefaultNamespace"
xmlns:soapenc="http://schemas.xmlsoap.org/soap/encoding/"
xmlns:wsdl="http://schemas.xmlsoap.org/wsdl/"
xmlns:wsdlsoap="http://schemas.xmlsoap.org/wsdl/soap/"
xmlns:xsd="http://www.w3.org/2001/XMLSchema">
<!-- P O R T T Y P E -->
<wsdl:portType name="GenericService">
   <wsdl:operation name="funct1">
      <wsdl:input message="impl:Funct1In"name=Funct1In"/>
      <wsdl:output message="impl:Funct1Out"name=Funct1Out"/>
   </wsdl:operation>
</wsdl:portType>
<!- M E S S A G E S -->
   <wsdl:message name="Funct1In">
      <wsdl:part name="in0" type="xsd:int"/>
   </wsdl:message>
   <wsdl:message name="Funct1Out">
      <wsdl:part name=""out0" type="xsd:int"/>
   </wsdl:message>
<!- BINDINGS -->
   <wsdl:binding name="GenericServiceSoapBinding"
                                    type="impl:genericService">
      <wsdlsoap:binding style="rpc"
             transport="http://schemas.xmlsoap.org/soap/http"/>
      <wsdl:operation name="funct1">
      <wsdlsoap:operation soapAction=""/>
         <wsdl:input name="Funct1In">
            <wsdlsoap:body
               encodingStyle="http://schemas.xmlsoap.org/soap/encoding/"
               namespace="http://DefaultNamespace" use="encoded"/>
         </wsdl:input>
         <wsdl:output name="Funct1Out"">
            <wsdlsoap:body
               encodingStyle="http://schemas.xmlsoap.org/soap/encoding/"
               namespace="http://DefaultNamespace" use="encoded"/>
         </wsdl:output>
      </wsdl:operation>
   </wsdl:binding>
<!- SERVICE DEFINITIONS -->
   <wsdl:service name="GenericServiceDef">
      <wsdl:port binding=
             "impl:GenericServiceSoapBinding" name="GenericService">
         <wsdlsoap:address location=
             "http://localhost:8080/axis/testaccount/GenericService"/>
      </wsdl:port>
   </wsdl:service>
</wsdl:definitions>
```

Figure 6.10 WSDL document for generic Web service.

body of the portType element will define operations. This corresponds to the methods of the Web service. Our generic service has a single operation called funct1. The service has one input message and one output message. The output message is given as a response to the input message. The input message corresponds to the parameters of the Web service method. The output message returns the Web service result. The final WSDL structure for the portType element is

```
<!-- P O R T T Y P E -->
<wsdl:portType name="GenericService">
  <wsdl:operation name="funct1">
    <wsdl:input message="impl:Funct1In"/>
    <wsdl:output message="impl:Funct1Out"/>
  </wsdl:operation>
</wsdl:portType
```

The message attributes in the input and output elements refer to named message elements given later in the WSDL document. The input and output message elements can also be given names using a name attribute but this is optional. Also, the attribute called parameterOrder can be added to the portType element to clarify the order of the attributes but this is also optional. Here, we are presenting a simple design. Next, the input and output messages need to be defined in more detail in terms of their datatypes. This is done in the message element.

<message> **Element.** A message element defines the name of a message that is referred to in the portType element and consists of one or more part elements, which constitute the message

```
<!- M E S S A G E S -->
  <wsdl:message name="Funct1In">
    <wsdl:part name="in0" type="xsd:int"/>
  </wsdl:message>
  <wsdl:message name="Funct1Out">
    <wsdl:part name=""out0" type="xsd:int"/>
  </wsdl:message>
```

In the example above, each message has a single part and carries a single integer value but a message could be decomposed into multiple parts and complex data structures, and then the actual data structure of each part would need to be specified in a types element. In our case, we are using a single primitive datatype (xsd:int), which is predefined in XMD and so types elements are not needed.

Binding. Binding provides how the abstract interface maps to a concrete protocol. In our example, it is SOAP over HTTP:

```
<!- BINDING -->
<wsdl:binding name="funct1SoapBinding"
              type="impl:genericService">
```

```
    <wsdlsoap:binding style="rpc"
        transport="http://schemas.xmlsoap.org/soap/http"/>
    <wsdl:operation name="funct1">
    <wsdlsoap:operation soapAction=""/>

      <wsdl:input name="Funct1In">
        <wsdlsoap:body encodingStyle=
            "http://schemas.xmlsoap.org/soap/encoding/"
          namespace="http://DefaultNamespace" use="encoded"/>
      </wsdl:input>

      <wsdl:output name="Funct1Out"">
        <wsdlsoap:body encodingStyle=
            "http://schemas.xmlsoap.org/soap/encoding/"
          namespace="http://DefaultNamespace" use="encoded"/>
      </wsdl:output>

    </wsdl:operation>
  </wsdl:binding>
```

In the above, the binding element specifies the transport in the `wsdlsoap:binding` element and then specifies SOAP envelope for each message. Note that the names of the messages are the names given in the `portType` element. The binding style is `rpc` (remote procedure call).

Service Definition. The service definition will collect ports together. In our example, there is a single port:

```
  <!- SERVICE DEFINITIONS -->
  <wsdl:service name="GenericServiceDef">
    <wsdl:port binding="impl:GenericServiceSoapBinding"
        name="GenericService">
      <wsdlsoap:address
      location=
    "http://localhost:8080/axis/testaccount/GenericService"/>
    </wsdl:port>
  </wsdl:service>
```

In our example, the service is located on the `localhost` at: `http://localhost:8080/axis/testaccount/genericService`. The binding attribute refers to the name of the binding.

Message Patterns. In the previous example, a request is made to a Web service and the Web service replies with the result. There are actually four message patterns:

- Request-response — a client makes a request and gets a response from the service. This pattern requires both the input and output messages. The previous example uses this pattern.

- One-way — a client makes a request and the service acts upon it without a response. This pattern only requires an input message. There is no output message.
- Notification — the service initiates the message transfer and sends a message to the client, which does not respond. This pattern only requires an output message. There is no input message.
- Solicit-response — the service initiates a message to the client and the client responds. To indicate this pattern, the output message from the service is written first in the WSDL document and the input message from the client is written second.

The patterns that do not have a response may still create request/acknowledge messages if required by the transport mechanism such as HTTP.

6.2.4 Service Registry

The service-oriented architecture, as illustrated in Figure 6.2, calls for a service registry for clients to be able to discover services. Of course in a simple situation, the client might already know about the location of the service, but in most situations where there are many services, a structured way of recognizing the location of services is needed.

Universal Description, Discovery and Integration (UDDI) is a discovery mechanism for Web services, which was introduced in 2001 and standardized through OASIS (Organization for the Advancement of Structured Information Standards). UDDI provides a specification for modeling information. It is targeted primarily towards business applications of Web services. UDDI registries are Web services that provide a location service in a service-oriented architecture. They can be globally accessible (public registries), or designed to be accessed by selected partners or just internally to an organization (private registries). Private registries may use the Internet but with limited access. There are publicly accessible UDDI registries provided by a few companies such as IBM and Microsoft. Such registries are targeted towards business-to-business interaction using Web services. For example, two businesses, one a manufacturer of products and one a purchaser of the products, might want to interact through a service-oriented architecture. The business wishing to purchase products might generate purchase orders sent to the manufacturer's Web services identified by a UDDI service registry.

The operation of a service-oriented architecture with a service registry is shown in Figure 6.11. After a registry is populated with Web service entries, the client can access the service registry to find out whether the desired Web service exists in the registry and if so where it is located. The service registry responds with the identification of the server and service capable of satisfying the needs of the client. Then, the client can access the server for the Web service interface. The server responds with a WSDL document describing the service and how to access it. The client can then send the Web service a request for an operation. The result of the operation is returned in a message from the Web service. All messages in this architecture are

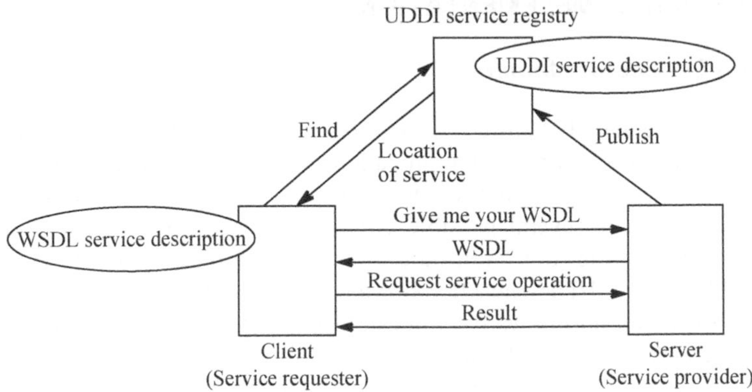

Figure 6.11 Steps to access a Web service in a service-oriented architecture.

SOAP messages. Note that the service registry itself has to be known both to the client and the service provider.

It would be feasible for the registry to return the WSDL interface document and then with this in hand, the client could make a request immediately to the Web service without asking for its WSDL interface document. That approach requires the service provider to send the WSDL document to the registry. It has the disadvantage that any changes made to the service that causes its WSDL document to be altered requires the WSDL document to be re-sent to the registry.

The final collection of software component in the Web services software stack consists of:

- HTTP transport
- SOAP message carrying XML documents
- WSDL
- UDDI used as Web service discovery mechanism

as illustrated with the associated protocols in Figure 6.12.

In a Globus Grid computing infrastructure, the function of a service registry is implemented in the information services group (Chapter 1, Section 1.6). A form of a service registry exists in Globus called an *index service*, which provides information on locally deployed services. We shall explore information services in a little more detail in Chapter 7.

Activity	Protocol/language
Service discovery	UDDI
Service description	WSDL
Service invocation	SOAP + XML
Service message transport	HTTP

Figure 6.12 Web services stack.

6.3 WEB SERVICE IMPLEMENTATION

6.3.1 Web Service Containers

Web services are generally "hosted" in a *Web service container*, that is, a software environment that provides the communication mechanisms to and from the Web services and their clients. A Web services container environment is illustrated in Figure 6.13. There are several possible environments that are designed for Web services, such as Apache Axis (Apache eXtensible Interaction System), IBM Websphere, and Microsoft .NET. The J2EE (Java 2 Enterprise Edition) server container is also a candidate for hosting Web services especially in business applications.

Apache Axis is available for free download (http://ws.apache.org/axis) and can be used for the Web service assignments described at the end of the chapter. Apache Axis requires an application server. Apache Axis can be installed on top of a servlet engine such as Apache Tomcat, which is suitable for the end-of-chapter assignments. However, Apache Axis could be installed on top of a fully fledged J2EE server.

Between the client and the service, it is convenient to have two additional components that act as intermediaries, one at the client end called *client stub* (also called a *client proxy*) and one at the service end called *server stub* (also called a *skeleton*) as illustrated in Figure 6.14. These stubs provide a structured way to handle the messaging and different client and server implementations. Interaction is between the client and its stub, between the client stub and server stub and between server and its server stub. The interaction between the stubs is across the network using SOAP messages. The client stub is responsible for taking a request from the client and converting the request into a SOAP message sent on the network. The client stub is also responsible for receiving responses on the network and converting to a suitable form for the client. The server stub is responsible for receiving a SOAP request from the client stub and converting it into a suitable form for the service. The server stub also converts the response from the service into a SOAP message for the client stub. The steps are:

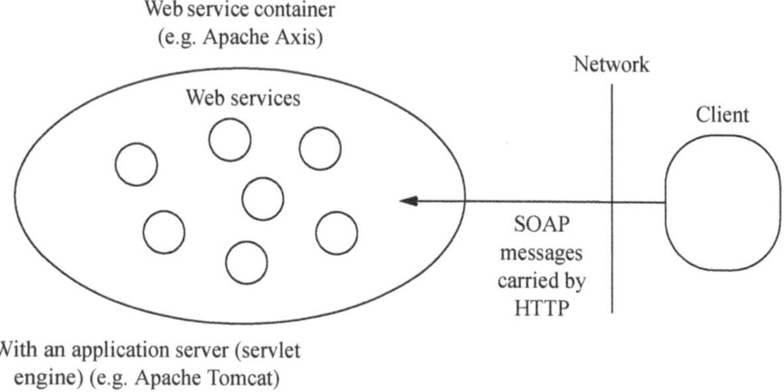

Figure 6.13 Web service environment.

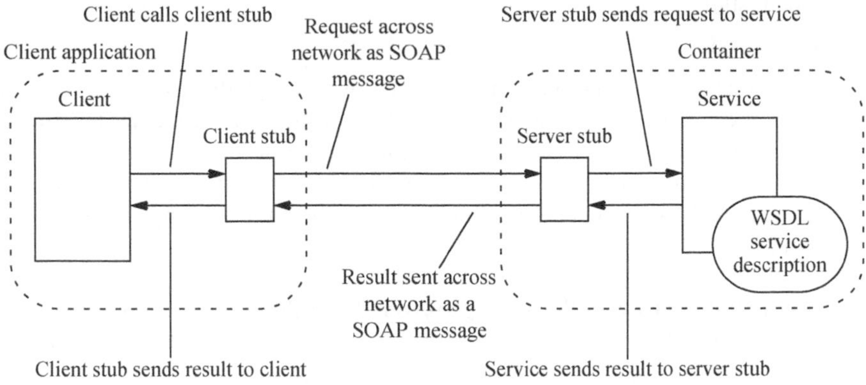

Figure 6.14 Client and Web service with stubs.

1. Client calls client stub
2. Client sends SOAP request across network
3. Server stub receives SOAP request
4. Server stub sends request to service
5. Service performs request
6. Service sends result to server stub
7. Server stub sends result across network as a SOAP message to client stub
8. Client stub receives SOAP request with result
9. Client stub sends result to client

The term *marshalling* describes the process of converting source data into a SOAP message, and the term *unmarshalling* describes the process of converting a SOAP message into data for the destination.

6.3.2 Building and Deploying a Service

One uses the traditional Computer Science term *building* a Web service to describe the process of compiling the Web service components ready for execution. The Web service then needs *deploying* in the container, that is, placing the files in the correct locations in the container, after which the service can be accessed by clients. There are several ways to create a Web service within a container and have it accessible by clients, that is, to build and deploy a Web service, depending upon where one starts in the building process. The fundamental components to create are:

- Web service code
- WSDL service description file
- Web service stub

and those to test or use the service are:

- Client stub
- Client code

One can start with writing the Web service, and move through the list of tasks to build and deploy the service, or start with a description of the service, the WSDL file, to create the stubs and a template (interface) for the Web service. Then implement in the interface. In the following, we will briefly outline the tools assuming that we are using Apache Axis environment, which has several tools for building and deploying a Web service.

Java Web Service (JWS) Deployment Facility. The simplest way in Axis to deplore a Web service is to use its *Java Web Service* (JWS) deployment facility. This facility could be used for a very simple Web service. To deploy in this facility, first, the Web service Java class file is created. All public methods in the service code will become available and accessible as service operations of the deployed Web service. Restrictions for JWS deployment include using data types known to Axis. The service code is stored in a file with a `.jws` extension rather than a `.java` extension and placed in a location known to Axis.

To take a very simple example, a Java class is shown below.

```
public class MyFunct {
  public int funct1(int x){
    int f = x + 2 * x;      // compute some function of x
    return f;
  }
}
```

This class has a single method called `funct1` that returns a function of the single input parameter, `x`. In our example, the file is called `MyFunct.jws`. In Axis, an environment variable is provided, called `CATALINA_HOME` that specifies the path to the Apache Tomcat servlet container home directory, for example `/usr/local/jakarta-tomcat-5.0.25`. The `.jws` file should be placed in `$CATALINA_HOME /webapps/axis/`.

Instant Deployment. In the Axis JWS facility, the `.jws` file is automatically compiled by the facility if necessary when the service is called by a client. Hence, one could actually reach it immediately as a Web service through the service URL, which would have a `.jws` extension. Axis converts the SOAP message for the methods of the service. However, instant deployment is only suitable for simple services and for testing. For more realistic situation, stubs are created. One would normally create a WSDL file. Also to assist in deployment, one might use a *Web service deployment descriptor* (WSDD) document. WSDD is an XML language for describing how to deploy a Web service. A WSDD document provides for greater flexibility than with the JWS "instant" deployment facility.

Creating Web Service Description Language (WSDL) document. There are a couple of tools provided by Apache to create the WSDL document for a service. The program `Java2WSDL` generates a WSDL document directly from a Java class or interface for the service. Axis also provides a way of obtaining a WSDL file from an already deployed service, by adding `?WSDL` to the URL of the service. For example, if the `MyFunct` Java class file is located at `http://localhost:8080/axis/MyFunct.jws` in an Apache Axis environment, setting your browser to point to

```
http://localhost:8080/axis/MyFunct.jws?wsdl
```

will display the WSDL file. The port number shown here (8080) may be different in an actual system. Port 8080 is commonly used as the alternative port to port 80 that is used for normal HTTP Web sites. Port 443 is the normal port for a secure HTTPS. Port 8443 is often used in this case.

Creating Stubs. Once the WSDL files has been created, the next step is to create the stubs. The Axis `WSDL2Java` tool can be used for both creating the server stubs (if not using instant deployment) and client stubs. We can couple the `?wsdl` facility with invoking `WSDL2Java` to create the required files for the client with the composite command such as

```
java   \
-classpath $AXISCLASSPATH org.apache.axis.wsdl.WSDL2Java   \
http://localhost:8080/axis/MyFunct.jws?wsdl
```

Files that are generated include the client stub, which is derived from the information in the binding element of the WSDL file. When `WSDL2Java` is used for the server side, a service stub and deploy/undeploy WSDD files are created. Complete information can be found in the Axis user's guide (Apache Software Foundation 2005).

Compiling Files. The files now need to be compiled with, for example, the command:

```
javac -classpath $AXISCLASSPATH    \
             localhost/axis/MyFunct_jws/*.java
```

Web Service Clients. Depending upon the actual details of the deployed service and Web service environment, a simple Java program can be used to access the service. An example for the `MyFunct` Web service is

```
import
localhost.axis.yourusername.MyFunct_jws.MyFunctServiceLocator;
import localhost.axis.MyFunct_jws.MyFunctService;
import localhost.axis.MyFunct_jws.MyFunct;
public class MyFunctClient {
public static void main(String args[]) throws
  Exception {
    MyFunctService service = new MyFunctServiceLocator();
    MyFunct myFunct = service.getMyFunct();
    int x = (new Integer(args[0])).intValue();
```

```
    System.out.println("The result of using" + args[0] + "
      in my function is " + myFunct.funct1(x));
  }
}
```

The client code is placed in a suitable directory, say called `.../Webservices` and compiled and executed as a normal Java program.

The previous description simply highlights the process. Much more detail can be found in the references quoted in Further Reading.

6.4 SUMMARY

This chapter introduced the following concepts:

- Service-oriented architecture
- Web services, including:
  - Registries
  - SOAP
  - WSDL
- Client-service implementations using Axis, including:
  - JWS facility
  - Creating WSDL and stubs

The topic is continued in Chapter 7, where stateful Web services are introduced and their application on a Grid computing environment.

FURTHER READING

A reading knowledge of XML is required to understand the deployment of Web services. There are several on-line XML tutorials including the W3Schools XML Tutorial at http://www.w3schools.com/xml/. The XML/SOAP standardization body is World Wide Web Consortium (W3C). Its home page is http://www.w3.org/XML/. Additional SOAP materials can be found at http://www.w3c.org/TR/soap. There are several books on XML and Web services, including by Graham et al. (2005), Benz and Durant (2003) and Erl (2004).

The associated course materials provided with this book at http://www.cs.uncc.edu/~abw/grid_computing uses Apache Axis tools. Details of this environment can be found at http://xml.apache.org/axis. Other hosting environments for Web services include Microsoft .NET and IBM Websphere.

BIBLIOGRAPHY

Apache Software Foundation 2005. Axis user's guide. http://ws.apache.org/axis/java/user-guide.html

Benz, B. with J. R. Durant. 2003. *XML programming bible*, Wiley: Indianapolis, Indiana.

Erl, T. 2004. *Service-oriented architecture: A field guide to integrating XML and Web services*, Upper Saddle River, NJ: Prentice Hall.

Graham, S, D. Davis, S. Simeonov, G. Daniels, P. Brittenham, Y. Nakamura, P. Fremantle, D. König, and C. Zentner. 2005. *Building Web services with Java: Making sense of XML, SOAP, WSDL, and UDDI*, 2nd ed., Indianapolis Indiana: SAMS Publishing.

House, J., M. Holliday, and R. Ruff. 2005. Assignment 1: A generic Web service compiling, deploying, and modifying a simple Web service. UNC–C/UNC–W Grid computing course, Fall 2005. http://www.cs.uncc.edu/~abw/ITCS4010F05/index.html

OASIS Advancing Open Standards for the Information Society. http://www.oasis-open.org/home/index.php

W3C SOAP Version 1.2 2007. Part 1: Messaging framework. W3C Recommendation. http://www.w3.org/TR/soap12-part1/

SELF-ASSESSMENT QUESTIONS

The following questions are multiple choice. Unless otherwise noted, there is only one correct answer for each question.

1. What is a remote procedure call?

(a) A procedure call that is not very friendly
(b) A procedure call on a local computer that is executed on a remote computer
(c) A procedure held in the disk memory of a computer
(d) A procedure call that is called and executed on a remote computer

2. How are Web services addressed?

(a) By URIs
(b) By programmer-defined Web service names
(c) By memory addresses
(d) By port numbers

3. What is meant by a service-oriented architecture?

(a) The concept of using services to get oriented to each other
(b) The way that services are organized within a container
(c) A client-server model in which servers publish their services in a registry and clients access the registry to find services
(d) A business model focused on providing service to customers

4. How does one determine what operations can be invoked on a service?

(a) Guess
(b) Each operation is given a predefined name agreed to by a standard of W3C

(c) Invoke the operation `tellMe()`
(d) Look at the WSDL interface document

5. What does WSDL stand for?

(a) Web Service Description Language
(b) Web Service Data Language
(c) Web Service Deployment Language
(d) Web Standard Description Language

6. What is a WSDL document?

(a) One that describes how to access a service and use it
(b) One that describes the data of a service
(c) One that describes how to use the client code
(d) None of the other answers

7. What can UDDI be used for?

(a) To transmit data between computers
(b) As a universal data description interface
(c) For modeling information in a Web services registry
(d) To encode the characters of the world

8. What is Apache Axis?

(a) A hosting environment for Web services
(b) A tool used by American Indians
(c) A compiler
(d) A type of make tool

9. What is SOAP?

(a) Something you use to clean yourself
(b) A protocol for passing XML documents
(c) A protocol for passing object-oriented programs
(d) Service-oriented architecture protocol
(e) None of the other answers

10. Which of the following contains all the services that have been deployed in a system?

(a) Class
(b) Shell
(c) Container
(d) Blob
(e) Bucket

11. What is a client stub?

(a) A way of offending a customer
(b) Code between the client code and the network
(c) A document that explains the client code
(d) None of the other answers

12. What is a .jws file?

(a) A Java Web Service file, a type of source file for a service that is automatically compiled if necessary when the service called
(b) A Java Web Service file, which is the final compiled service
(c) A Java Web Service file, which is the server stub
(d) A Job Worker Service file

13. What is meant by the term portType?

(a) The types of ports available to be used
(b) The type of data passed through a port
(c) A specific port chosen to be used by a container
(d) An abstract interface definition of a service

PROGRAMMING ASSIGNMENTS

These programming assignments require an environment for deploying Web services, such as the Apache Axis/Jakarta Tomcat servlet container. Details on how to establish this particular environment can be found at the Apache home page (http://ws.apache.org/axis/). The environment can be established on a personal Windows, Mac, or Linux system, or on an account on a centralized server.

6-1 Using Axis tools, create the WSDL file for the generic service given in Section 6.2.3 (Figure 6.10).

6-2 Follow the instructions given at `http://www.cs.uncc.edu/~abw/ITCS4010F05/assign1.doc` (steps 1 to 6) to deploy the prewritten Web service that performs the mathematical square operation. Compile the prewritten client to exercise the Web service.

6-3 Modify the Web service in Assignment 6-2 to add two additional methods, `isPrime` and `isEven`, as described at `http://www.cs.uncc.edu/~abw/ITCS4010F05/assign1.doc` (step 7). The `isPrime` method returns a Boolean indicating whether the number is prime (`TRUE` or `FALSE`) and the `isEven` method returns a Boolean indicating whether the number is even (`TRUE` or `FALSE`). Exercise the service with an appropriately written client.

6-4 Write and test a Web service that has two arguments, x and y, and returns the maximum of x and y.

6-5 Create a Web service such as in Assignment 6-2 on one networked computer and access it from another networked computer.

6-6 Create a Web page as a front-end to a client that can access any Web service. Provide text fields for the user to enter the service URL and arguments and to display the result returned from the Web service. Demonstrate your Web page with a Web service such as in previous assignments.

CHAPTER 7
System Infrastructure II: Grid Computing Services

In Chapter 6, Web service technology was outlined. Now we will move onto how that technology is applied to Grid computing, which includes the key question of how state should be implemented. It is very important that standardized approaches should be adopted, and we will describe effects towards standardization. This is an ongoing process with continual improvements and refinements. The chapter will continue a study of the Globus toolkit and in particular its use of Web service technology.

7.1 GRID COMPUTING AND STANDARDIZATION BODIES

Standardization is critical for the wide-scale adoption of Grid computing. The underlying network for Grid Computing has become the Internet rather than customized dedicated network for the most part. Even if a dedicated network is used, the network uses Internet Protocols. Prior to Grid computing, standardization bodies had come into existence for standardizing Internet and World-Wide Web protocols. IETF (Internet Engineering Task Force) was formed in 1985 for establishing Internet standards including the previously developed TCP/IP protocol. Other standardization bodies appeared with the ever-growing Internet. W3C (World Wide Web consortium) was founded by Tim Berners-Lee shortly after he conceived the World Wide Web in the early 1990s. W3C works on standardization of Web-related technologies including XML. A related Internet standardization body is OASIS (Organization for the Advancement of Structured Information Standards). OASIS began as a consortium called SGML Open in 1993 to promote the SGML markup language. It became OASIS in 1998 to focus on structured information standards including the XML markup language. Another body is DMTF (Distributed Management Task Force), which was created in 1992 for IT systems management infrastructure.

The Grid community started to form its own special interest bodies for adopting Grid computing standards in the late 1990s. In the US, the Grid Forum was formed in 1998. Later, Grid Forum was merged with the eGrid (European Grid) forum and the Asia-Pacific Grid community to form the international Global Grid Forum (GGF), which became the central forum for discussing and developing Grid computing standards. GGF had three meetings each year, numbering from the first meeting (GGF1 in Amsterdam in March 2001). GGF became the Open Grid Forum (OGF) in 2006, with a merger with an industry group called Enterprise Grid Alliance (EGA) and continued the tri-yearly meetings around the world, GGF 18 becoming OFG 18. These meetings grew to a very large endeavor, with participants across the world attending.

Once Web services were introduced, many standards began to be developed for use in Web service environments, most notably by the W3C organization. Web services exploit XML, which had been introduced in 1998, and SOAP, which had been ratified in 2000. Subsequent standards continued to be developed including WSDL described in Chapter 6, and large family of WS-* standards where * refers to the name of one of many standards. These standards were not developed specifically for Grid computing but it became natural to use them for a Grid computing environment with the adoption of Web services in Grid computing. Web services provide an easily identifiable interface through WSDL documents. and conveniently use Internet addressing (URLs). However, there are some issues with using Web services for Grid computing as we shall now address.

7.2 INTERACTING GRID COMPUTING COMPONENTS

7.2.1 Development of a Service-Oriented Approach

Grid computing infrastructure needs to address all aspects of operating a Grid computing environment including how security, job submission, file transfers, and information services are implemented. Individual components need to be provided for each area that interact with each other and with components at remote sites. How Grid computing infrastructure should be implemented has been evolving since its introduction in the mid-1990s. It has gone through several development cycles, which have still not finished and may never do so as dissatisfaction is found in each approach. Early Grid computing work focused on customized solutions, that is, projects developed in-house software to achieve Grid computing facilities. Some projects concentrated upon using a specific approach such as using an object-oriented approach. As described in Chapter 1, the Globus project started after the early I-Way experiment on Grid computing in 1995, and developed a software toolkit that has progressed through several versions. The Globus toolkit version 2 (GT 2) introduced in 2002 became widely adopted by the Grid community and provided a suite of software tools (APIs) for building a Grid environment. GT 2 became a de facto standard and was very highly regarded but it was not based upon any industry-wide standards. Many early Grid computing projects in the early 2000s used GT 2.x. For example, our undergraduate Grid computing course offered in 2004 and other Grid computing courses of that era used GT 2.x.

The early work of developing Grid computing infrastructure tools such as GT 2.x did not use Web services described in Chapter 6 as it predated Web services as we know it, which came into being later. The early experiences provided the impetus for later standardized work. The use of standardized technologies is critical to the wide adoption of Grid computing. With the advent of Web services, it was recognized that Web services provided the way forward for Grid computing to give a uniform interface for Grid computing components especially those that need access through the Internet. A key aspect of Grid services is XML, which provides a unified platform-independent way of describing information for Web services. As described in Chapter 2, XML has also been adopted in job description languages.

7.2.2 Stateful Web Services

Web services are usually *stateless*. They do not remember information from one invocation to the next. That works fine in many Web service applications. They do not need to know what happened with previous invocation by another client, nor should they. A simple example of a stateless Web service is the Web service introduced in Chapter 6 that performs a function using as input the arguments to the Web service and returning a value only based upon input arguments and the function. For example, the Web service `MathService` in Figure 7.1 has a method `inc` to add one to a supplied argument and return the result.

A *stateful* Web service has the ability to remember information from one invocation to the next and between invocations by different clients. This would be most useful in Grid computing applications to enable a sequence of actions to be done dependent upon past actions and results. A stateful Web service is illustrated in Figure 7.2. In this example, the Web service performs the `add` operation on a variable called `data` inside the Web service, and its value is retained between invocations by clients, the same client or different clients. Each client has access to the value of

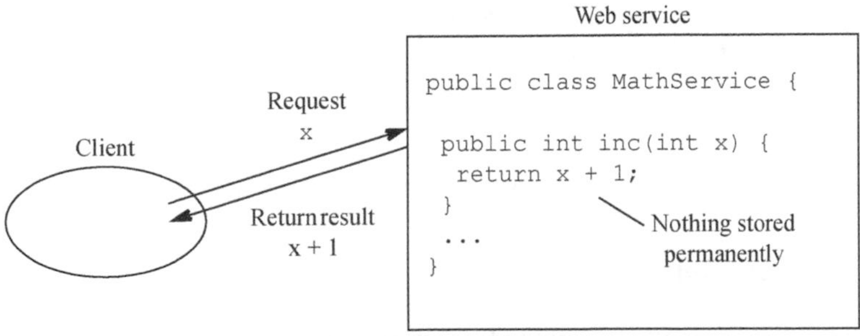

Figure 7.1 Stateless Web service.

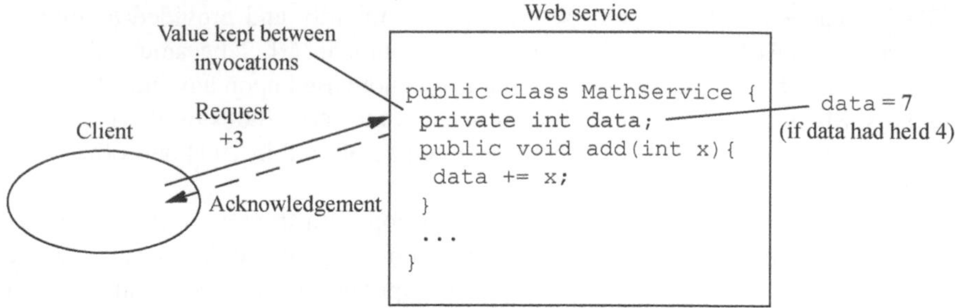

Figure 7.2 Stateful Web service.

`data`. The figure shows a client making a request to add 3 to `data`, which will become 7 if its original value was 4. Suppose there is another method called `getValue` that returns the current value of `data`. If the same or another client accesses the Web service using that method, it will get the value 7. In Figure 7.2, state is shown contained in a private variable within the Web service; however, this approach is not desirable and a better approach is to take the state out of the Web service to leave the Web service itself stateless, as we shall see.

Clearly the previous example is artificially simple although it is useful to explain the concept. A practical example of a stateful Web service would be where one accesses a large database as illustrated in Figure 7.3. Here, a client makes a request to access data from a large persistent database through a stateless Web service. The stateless Web service accesses the database. The result of the request, that is, the required stored data, is returned to the client through the Web service. The Web service could provide other methods such as transferring data from one database to another database. A key aspect here is that the state is contained in a *resource* that is separate from the Web service. The Web service becomes a front-end to the resource (or resources). We shall come back to this concept later.

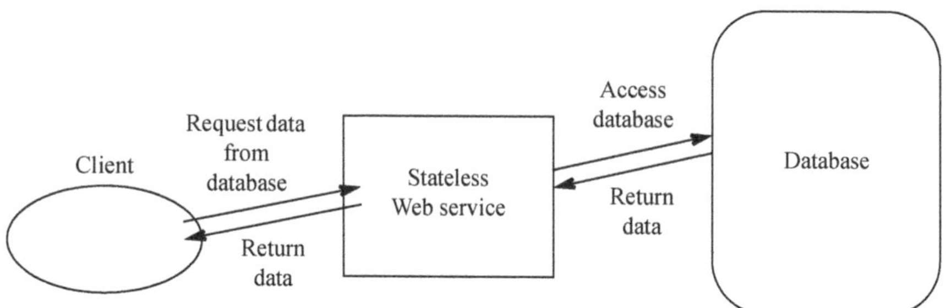

Figure 7.3 Stateful Web service as a front-end to a database.

7.2.3 Transient Services

Another feature originally thought to be required for Grid computing is the ability to make a service *transient*, that is, a service that can be created and destroyed. Web services are *non-transient* and generally always available for clients. They can be accessed by different clients on a need basis. They do not have the concept of dynamic service creation and destruction. Transient services, on the other hand, are created and potentially can be destroyed. Typically, although not necessarily, transient services would be created by specific clients and do not outlive their clients. Rather than make the service transient, one can make the stateful resource associated with the service transient or created dynamically. We will look at how transient services/resources might be created. In general, it has led to a somewhat object-oriented approach where instances are created using a factory service. Dynamic creation means that matters such as their lifetime also need to be addressed.

7.3 OPEN GRID SERVICES ARCHITECTURE (OGSA)

7.3.1 Purpose

Open Grid Services Architecture (OGSA) was originally proposed by Foster et al. in the seminal paper "The Physiology of the Grid" (2002) and was introduced at the Global Grid Forum GGF4 meeting in February 2002 as a Grid computing standard. OGSA defines standard mechanisms for creating, naming, and discovering service instances and addresses architectural issues relating to interoperable services for Grid computing. OGSA addresses all the fundamental services of Grid computing such as for job management, resource management, security services, and service discovery. It specifies standard interfaces for these services, that is, what each service should provide but it does not give implementation details. OGSA requires stateful services but does not say how that will be achieved. Following the traditional approach to system design, such implementation details are left for other standards.

7.3.2 Open Grid Services Infrastructure (OGSI)

The first attempt by the Grid computing community to specify how OGSA could be implemented was the *Open Grid Services Infrastructure* (OGSI) standard, which was introduced in 2002–2003 (final draft in 2003). OGSI specifies the way that clients interact with services (that is, service invocation, management of data, security mechanism, etc.). The approach taken in OGSI to implement a stateful Web service was to modify the Web Service Description Language, WSDL, to enable the state to be specified. The modified language was called *Grid Web Service Definition Language* (GWSDL).[1] OGSI also introduced and described the term *Grid service* as an extended Web service that conforms to its OGSI standard. GWSDL provided support for the extra features in Grid services that were not present in Web services. In addition to a means of representing state, OGSI included inheritance of portTypes (interfaces), a way of addressing services using so-called Grid Service References,

[1] The word "definition" comes from when Web Service Description Language was called Web Service Definition Language.

notifications, and a point-to-point message passing mechanism. More details of OGSI can be found at Tuecke et al. (2003).

With the appearance of OGSI, the Globus team moved onto implementing the OSGI standard, and released Globus version 3 in 2003. Version 3 was a complete implementation of OGSI and provided additional Web services, some built onto top of OGSI. Instances of services could be created dynamically. A few Grid computing projects migrated from Globus version 2 to Globus version 3, including our undergraduate Grid computing course in 2004–2005, but it soon became apparent that there were significant issues with OGSI. Within a year or so, the Grid community had found the approach not acceptable because:

- It significantly modified pure Web services.
- It required new tools. GWSDL required tools in addition to that used for Web services.
- It was too object oriented in approach.
- There was too much specified in one standard. It would be better broken down in a series of specifications.

A better way was needed.

7.3.3 WS-Resource Framework

While the Grid community had been exploring ways to embody state in Web services for Grid computing environments, the Web service community had also been working on approaches to stateful Web services. The two communities began to merge on an approach for stateful Web services. A group that included several industry partners (Computer Associates, Fujitsu, Hewlett-Packard, IBM) and the Globus Alliance (a group associated with Globus established in September 2003 for developing Grid technologies) proposed a specification called *WS-Resource Framework* (WSRF) in 2004. It was ratified by OASIS essentially to replace OGSI and make the implementation of a stateful Web service acceptable to both the Web services and Grid services communities. WSRF specifies how to make a Web service stateful and other features, without drifting too far from the original Web service concept.

The term Grid service, which originally was introduced with OGSI, continued for while. A broad meaning of Grid service is any service that conforms to interface conventions of a Grid computing infrastructure. A narrow WSRF meaning is a service that conforms to WSRF. The term seems to have lost favor, maybe because it is better to think of services in a Grid environment simply as regular Web services. OGSA also continues as an overall framework. OGSA requires a stateful Web service. WSRF specifies how that stateful Web service is implemented. Stateful Web services are extensions of the original stateless Web service and a WSRF Web service could be stateless by simply not using the state features.

A stateful Web service is obtained in WSRF by having a stateless service and stateful resources where the stateless Web service is a front-end to stateful *resources*; hence the name WS-Resource Framework.

Resource Properties. *Resource properties* is the name given to data items in the *resource*. They can consist of

- Data values about the current state of the service — results of calculations, etc.,
- Metadata — information about data
- Information about the whole resource — termination time, etc.

The service interface is described in WSDL. The WSDL file serves the same purpose as in the original stateless Web service. Using WSDL allows existing WSDL parsing tools to be used. A significant addition in the WSDL file is a specification of the resources. In WSRF, the Web service is described in a WSDL document and the resource is specified in a separate Resource Properties document (or merged into the WSDL document in the case of simple resources). A *WS-Resource* is a Web service and an associated resource.

Let us revisit the stateful Web service example shown previously in Figure 7.2. Figure 7.4 shows a WSRF version of this stateful Web service. In this example, there is a single resource property, `data`, acted upon by the `add` method. The `add` method adds an integer `x` given as an argument to `data`. This service implements an operation on a WSRF resource property and so WSDL will include definitions relating to the resource property.

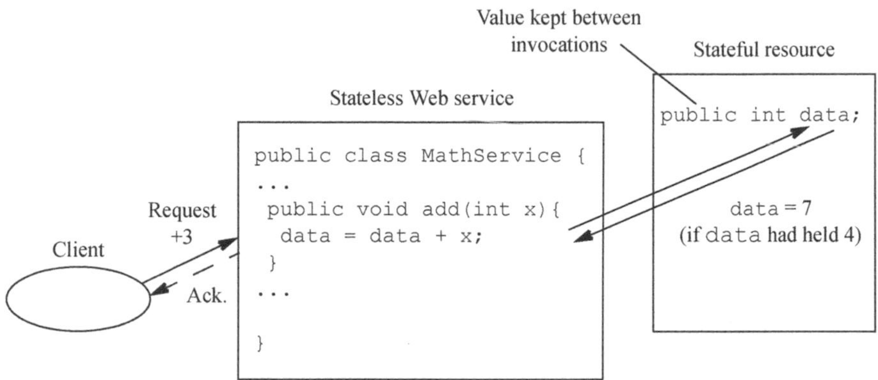

Figure 7.4 WSRF stateful Web service.

Endpoint References. Each WSRF service needs an addressing mechanism that includes addressing to the resources. Pure stateless Web services are addressed with URIs (Uniform Resource Identifiers). Typically, they are addressed by URLs (Uniform Resource Locators) as used for Web sites. URLs are a subset of URIs. The WSRF service addressing mechanism is defined in the *WS-addressing* standard and uses a term called an *endpoint reference* (EPR), which is an XML document that contains various information about the service and resource. Specifically, the endpoint reference includes both the service address (URI) and a resource identification called a *key*. The key is a number. The service that accesses the resources and the resources are paired together. The client makes a request to the (stateless) Web service, addressing the service using the EPR. The URI part of the

EPR identifies the service. The Web service selects the resource using the resource key inside the EPR, as illustrated in Figure 7.5.

An endpoint reference has required and optional entries. An endpoint reference could be used simply to address a Web service without an associated resource, i.e., a stateless Web service. In that case, the EPR would have the form

```
<wsa:EndpointReference>
     <wsa:Address>
          http://www.cs.uncc.edu/~abw/MyWebService
     </wsa:Address>
</wsa:EndpointReference>
```

where the `Address` element is required, the prefix `wsa` is the WS-Addressing namepace (`http://www.w3.org/2005/08/addressing/wsdl`), and `http://www.cs.uncc.edu/~abw/MyWebService` is the URL of the Web service.

With the inclusion of a Web service resource (i.e., *WS-Resource*), an EPR is used to address the Web service and an associated resource. The following is an example.

```
<wsa:EndpointReference>
     <wsa:Address>
          http://www.cs.uncc.edu/~abw/MyWebService
     </wsa:Address>
     <wsa:ReferenceParameters>
          <myrp:ID> 234 </myrp:ID>
     </wsa:ReferenceParameters>
</wsa:EndpointReference>
```

The resource identifier (`234` in the example above) is specified within the `ReferenceParameters` element of the EPR. Note that it is not specified within the WSDL document.

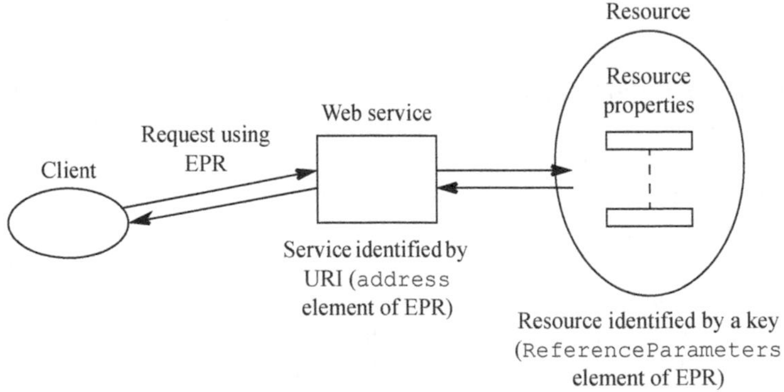

Figure 7.5 Endpoint reference (EPR).

In the WS-Addressing standard, the contents of `wsa:Address` and `wsa:ReferenceProperties` elements must appear in the SOAP header:

```
<soap:Envelop>
<soap:Header>
    ⋮
<wsa:To>http://www.cs.uncc.edu/~abw/MyWebService</wsa:To>
<resourceID>234</resourceID>
    ⋮
</soap:Header>
<soap:Body>
    ⋮
</soap:Body>
</soap:Envelop>
```

WSRF Specifications. WSRF is actually a collection of four specifications (standards):

- *WS-ResourceProperties* — specifies how resource properties are defined and accessed
- *WS-ResourceLifetime* — specifies mechanisms to manage resource lifetimes
- *WS-ServiceGroup* — specifies how to group services or WS-Resources together
- *WS-BaseFaults* — specifies how to report faults

Additional related WS-* standards include:

- *WS-Notification* — collection of specifications that specify how to configure services as notification producers or consumers.
- *WS-Addressing* — specifies how to address Web services. Provides a way to address a Web service/resource pair. Defines the endpoint reference.

7.3.4 Generic Stateful WSRF Service

We will briefly describe how to program a simple WSRF service. The following section is based upon detailed code provided by Sotomayor and Childers (2006). Here, the code has been simplified, including not having specific operations or a `types` element for defining the data format of the messages. However, as we shall see, a `types` element will be still needed to define the resource properties. The reader is referred to Sotomayor and Childers (2006) for a more complete discussion of stateful Globus 4 services.

WSDL Code. In Chapter 6, a generic Web service was outlined that performed a function `funct1` that has one integer argument `arg1` and returns an integer `result` based only upon the supplied argument. Now, that simple Web service will be extended so that the function acts upon an integer, `data`, as illustrated in Figure 7.6. The actual function is still undefined at the WSDL level. The integer `data` is separated out as a resource property.

The format of the stateless service WSDL 1.1 document was given in Chapter 6, Section 6.2.3, and this carries over for a WSRF WSDL document. But in addition, the properties of the resources need to be specified.

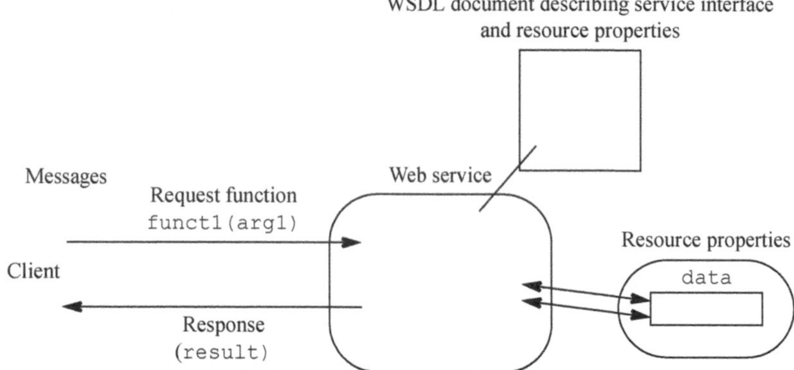

Figure 7.6 Generic stateful Web service with a WSDL document and a stored value.

Namespaces. A number of extra namespaces are introduced into the WSDL document to cater for the extra WSRF elements.

Messages. There are no changes to the message element from that used in Chapter 6, i.e.,

```
<!-- M E S S A G E S -->
<message name="Funct1In">
    <part name="in0" type="xsd:int"/>
</message>
<message name="Funct1Out">
    <part name="out0" type="xsd:int"/>
</message>
```

Resource Properties. Resource properties are specified in the `types` elements.

```
<types>
...
<!-- RESOURCE PROPERTIES -->
<xsd:element name="data" type="xsd:int"/>
<xsd:element name="FunctionResourceProperties">
<xsd:complexType>
    <xsd:sequence>
        <xsd:element ref="tns:data"
            minOccurs="1" maxOccurs="1"/>
```

```
      </xsd:sequence>
   </xsd:complexType>
   </xsd:element>
   </xsd:schema>
   ...
   </types>
```

`data` is declared as occurring just once.

PortType. The portType element is the same as the stateless version except that `ResourceProperties`. is included as an attribute:

```
<!-- P O R T T Y P E -->
<portType name="GenericService"
wsdlpp:extends="wsrpw:GetResourceProperty"
wsrp:ResourceProperties="tns:FunctionResourceProperties">
<operation name="funct1">
  <input message="tns:Funct1In"/>
  <output message="tns:Funct1Out"/>
</operation>
</portType>
```

Bindings. In a traditional WSDL document (Chapter 6), bindings provide how the abstract interface maps to concrete message protocols. This is not needed in Globus 4.x systems because bindings are automatically inserted by Globus when the service is built using Globus tools. Hence, there is no `binding` element in the code. Also there is no `service` element providing a network address in the presented code.

Service Code. The WSRF service has two major parts:

- Resource properties
- Service code

For simple service and resources such as the first WSRF service described in Sotomayor and Childers (2006, Chapter 6), the service methods and resource properties can be combined into one file, in essence service code that consists of

```
public class MyFunct {
private int data          // resource property
  ...
  public int funct1(int x){
    int f = ... ;          // compute some function of x
    return f;
  }
  ...
}
```

The actual code is somewhat more complex as it has to incorporate a GT 4 class for the way that the resource property is to be handled, which is outside the scope of this book. Sotomayor and Childers (2006) provide a full treatment of the different ways resource properties can be handled.

Deploying GT 4 Services. Globus version 4 uses the Apache Axis Web service container internally and the basic steps for deploying a GT4 service are similar in concept to that described for Apache Axis in Chapter 6. The package of files needed for deployment are the WSDL service interface file, the service code, WSDD (Web Service Deployment Descriptor) deployment file, the JNDI (Java Naming Directory Interface) configuration file, and GT4 build files. The package can be built using the Ant (Another Neat Tool) build tool, creating a so-called `gar` file (Grid archive). Globus provides a build script `globus-build-service.sh` that contains the `ant` and `bash` commands to build the service. A Windows version in Python is provided.

Once the `gar` file has been created, the Globus tool `globus-deploy-gar` is used to deploy the service. Then, the GT4 container has to be started/re-started to complete the process and make the service accessible. The GT4 container is started with the command `globus-start-container`. The option `-nosec` will start the container with no security, that is, without needing certificates using the `http` protocol rather than the `https` protocol. Figure 7.7 shows a sample output from the command `globus-start-container -nosec`. In this example, the host computer has the Internet address 166.82.130.77 with the alternative HTTP port number 8080 but this can be changed using the `-p` option in the `globus-start-container` command.

Running the GT4 container only requires the Globus core common runtime component and not the full Globus toolkit. Versions of the Globus core are provided in Java, C, and Python. The Java GT4 core can be installed and services deployed used on any machine with appropriate versions of JDK (Java Development Kit) and

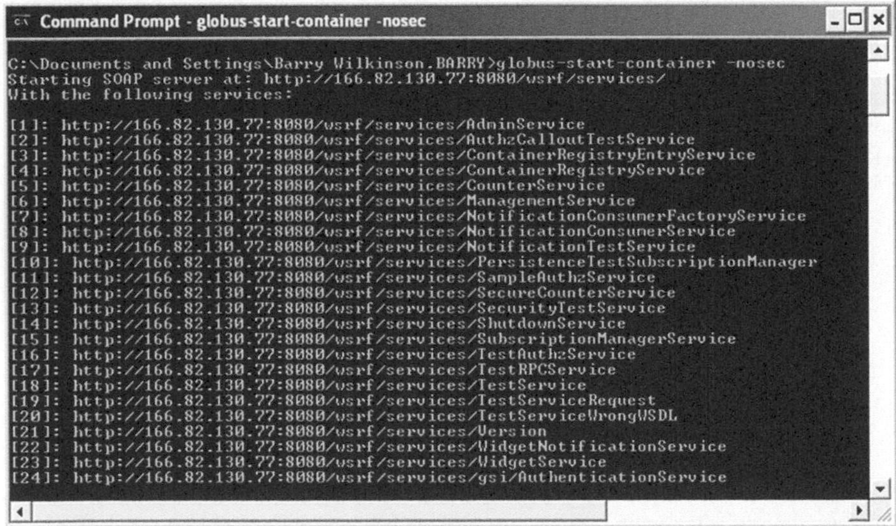

Figure 7.7 Globus 4.0 container with services.

Ant (for Windows, Python). Links to installation details are included in assignments at the end of the chapter that call for deploying GT4 services. Deploying your own Globus services is useful to understand the inner workings of Globus components and for creating your own Globus service based application. When a container is started, it will list all services deployed in the container. Hence, if a shared container is running a centralized server, one would see everyone's personally deployed services. Each service must have a unique name. It is possible to write a script to automatically add usernames to service names to make them unique but all work is held up every time one restarts the container to redeploy services. Using multiple containers is also problematic. The number of containers maintained on a single server is limited to available memory, generally to just a few containers. Hence, being able to install the GT4 core on a personal computer is attractive and enables individual experimentation. An Internet connection is needed to download the software but is not needed to install and test services locally. Without an external Internet connection, the local host address (e.g., 192.168.1.100) or a local network address (e.g., 127.0.0.1) will be displayed in Figure 7.7 instead of an external network address (166.82.130.77 in Figure 7.7).

Once a service is deployed in a GT4 container and the container is running, the service can be used by a client in much the same way as with stateless Web service in Chapter 6. A URI will be used to locate the services, for example `http://166.82.130.77:8080/wsrf/services/WidgetService` in Figure 7.7. However, an endpoint reference has to be obtained. A snippet of code for that is

```
import classes ...
public class Client {
public static void main(String[] args) {
ServiceAddressingLocator locator =
                         new ServiceAddressingLocator();
...
String myserviceURI = args[0];
EndpointReferenceType endpoint =
                         new EndpointReferenceType();
endpoint.setAddress(new Address(myserviceURI));
FunctPortType myservice
               = locator.getMathPortTypePort(endpoint);

// use service function
myservice.funct1(...);
...
```

The full code and its explanation can be found at Sotomayor and Childers (2006).

7.3.5 Additional Features of WSRF/GT4 Services

Sotomayor and Childers (2006) discuss many advanced features of GT4 services in detail which are beyond the scope of this book. Here, we will briefly outline some concepts.

Configuring and Managing Resources. Previously, there was a single resource although it could have multiple resource properties inside the resource. Both the service code and the resource code were held in a single file. Having one file is not the preferred way except for a simple service. Ideally, there should be separate classes, and classes are provided for different arrangements, including having one class for the arrangement combining the service and resource, which is called `ReflectionResourceProperty` class.

Resources are managed by a *resource home*, which provides resource management functions including locating the resources. The resource home class for a combined service-resource pair is called `ServiceResourceHome` class. The resource home class for a single resource that is separate from the service is called `SingletonResourceHome` class.

Ideally, we would like to have multiple resources. In WSRF, resources can be created (but *not* the services themselves) using the WS-Resource factory pattern—a traditional object-oriented approach to creating resources using a factory service. The factory service is responsible for creating WS-resources and interacts with the resource home that subsequently will manage the resources. As mentioned earlier, each resource assigned a unique key, which together with service URI identifies WS-resource pair (endpoint reference). The factory service is requested to create a resource using the `createResource` method, and uses the resource home to create resource. The `createResource` method will return an EPR of the newly created WS-Resource. Notice that the client making the request to the resource factory needs to know the location of the resource factory. When a service needs access to the resource it will contact the resource home to find the resource.

Lifecycle Mechanisms. Mechanisms are available in WSRF to specify how long a resource exists. GT4 provides mechanisms to specify when a resource is automatically destroyed. It can be destroyed immediately by a client invoking `destroy` operation of a WSRF service. Notice that the factory is responsible for creating a resource, but a service destroys it. A resource can also be destroyed by the GT4 command `globus-wsrf-destroy` on the command line, e.g.,

```
globus-wsrf-destroy -e EPRfile.epr
```

where `EPRfile.epr` is a file that contains the ERP of the resource. Alternatively, destruction can be scheduled sometime in the future. The termination time is exposed as a resource property and can be set with GT4 command `wsrf-set-termination-time`, e.g.,

```
wsrf-set-termination-time -e EPRfile.epr 100
```

would set the termination time as 100 seconds.

Lease-Based Lifecycle. In lease-based management, resources must be kept alive otherwise the resource dies after the set lifetime. Interested parties (clients) must renew the lease by updating the termination time or the resource will be destroyed. This approach guarantees cleanup without having to use a destroy operation explicitly.

Notifications. *WS-notification* defines mechanisms for notifying clients when something interesting happens, for example when a resource property reaches a certain value. One could use polling but this would be very inefficient. Examples of interesting changes include:

- Changes to resource property values
- Methods added
- Methods removed
- Resources destroyed

Clients need to subscribe to receive notifications and then will be informed of any changes as illustrated in Figure 7.8. Subscriptions are for a particular topic. Resources can be implemented with GT 4 classes that provide for automatic notifications whenever property changes.

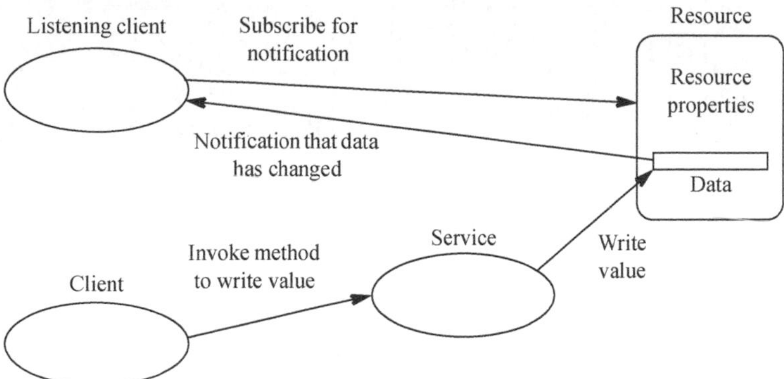

Figure 7.8 Automatic notifications.

7.3.6 Information Services

Information services collect information from various sources in a Grid environment. They can provide information on running jobs and on the availability of Grid resources (computers, etc.). One of the special features of Grid computing is the possibility of discovering remote resources in the Grid. Globus 4 information services collectively is called the *Monitoring and Discovering System* (MDS4 in GT 4) and consists of a set of three WSRF information components:

- Index service
- Trigger service
- WebMDS

from which a framework can be constructed for collecting and using information. The three components are part of the full GT4 package. Strangely, they are not provided in the GT4 core although index and trigger services are WSRF services hosted in the GT 4 container. These services will display if installed as illustrated in Figure 7.9, shown here with a local network IP address, 192.168.1.100. This particular customized Globus core installation comes from following the Globus "sticky-note" tutorial for illustrating features of Globus services (Ananthakrishnan, et al. 2005).

```
C:\WINDOWS\system32\cmd.exe - bin\globus-start-container -nosec
C:\tutorial\gt-install>bin\globus-start-container -nosec
Starting SOAP server at: http://192.168.1.100:8080/wsrf/services/
With the following services:

[1]: http://192.168.1.100:8080/wsrf/services/InMemoryServiceGroupEntry
[2]: http://192.168.1.100:8080/wsrf/services/TriggerFactoryService
[3]: http://192.168.1.100:8080/wsrf/services/IndexFactoryService
[4]: http://192.168.1.100:8080/wsrf/services/Version
[5]: http://192.168.1.100:8080/wsrf/services/IndexService
[6]: http://192.168.1.100:8080/wsrf/services/NotificationConsumerService
[7]: http://192.168.1.100:8080/wsrf/services/DefaultTriggerServiceEntry
[8]: http://192.168.1.100:8080/wsrf/services/TriggerServiceEntry
[9]: http://192.168.1.100:8080/wsrf/services/IndexServiceEntry
[10]: http://192.168.1.100:8080/wsrf/services/StickyNoteService
[11]: http://192.168.1.100:8080/wsrf/services/AdminService
[12]: http://192.168.1.100:8080/wsrf/services/DefaultIndexService
[13]: http://192.168.1.100:8080/wsrf/services/DefaultIndexServiceEntry
[14]: http://192.168.1.100:8080/wsrf/services/DefaultTriggerService
[15]: http://192.168.1.100:8080/wsrf/services/ShutdownService
[16]: http://192.168.1.100:8080/wsrf/services/ContainerRegistryService
[17]: http://192.168.1.100:8080/wsrf/services/TriggerService
[18]: http://192.168.1.100:8080/wsrf/services/gsi/AuthenticationService
[19]: http://192.168.1.100:8080/wsrf/services/InMemoryServiceGroupFactory
[20]: http://192.168.1.100:8080/wsrf/services/InMemoryServiceGroup
[21]: http://192.168.1.100:8080/wsrf/services/ContainerRegistryEntryService
[22]: http://192.168.1.100:8080/wsrf/services/SubscriptionManagerService
```

Default index service

Default trigger service

Figure 7.9 Index and trigger services in a GT4 container.

Index Service. The GT4 index service uses the query and notification features described in Section 7.3.5 to obtain resource property information from WS-resources, as illustrated in Figure 7.10. Note that the index service does *not* act as a local service registry listing the services—it act as registry for the resources. As mentioned, the resource home is involved in creating resources. The `add` method of the appropriate resource home class is used to create resources. This method can be overridden to include registering resource with the index service, as illustrated in Figure 7.11.

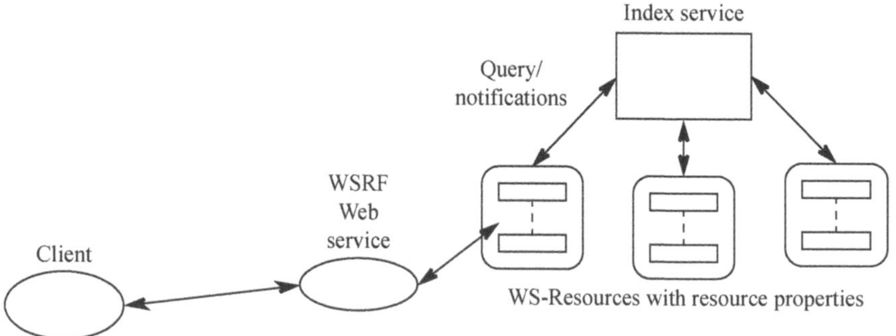

Figure 7.10 Index service.

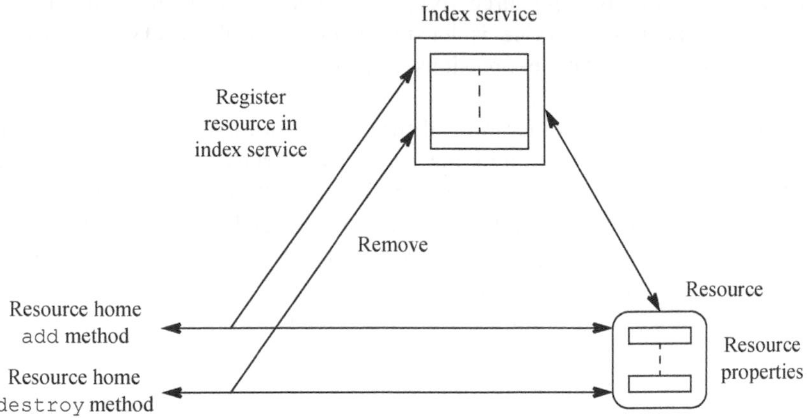

Figure 7.11 Registering resource in an index service.

Resource properties can be implemented in different ways but the information they contain is converted to an XML document for exchange with other components. This document is called a *resource properties document.* The index service also stores the resource property information in an XML format. A resource properties document can be queried on the command line with the GT4 command `wsrf-query`. This command can be used to query the index service if the service exposes the `QueryResourceProperties` portType (Sotomayor and Childers 2006). `wsrf-query` uses an XPath expression[2] as final argument to select specific entries in the resource properties document. For example:

```
wsrf-query -s http//localhost:8080/MyService '/*'
```

will select all elements. (The XPath expression `'/*'` is actually optional and the default.) The result might be

```
<ns1:MyServiceResourceProperties xmlns:ns1= " ... ">
     <ns1:data>234</ns1:data>
</ns1:MyServiceResourceProperties>
```

The `-s` option in the `wsrf-query` command indicates the service URL. One can use the `-e` option with a supplied file containing the endpoint reference instead. A single resource property can be returned using the GT4 command `wsrf-get-property`, or specific multiple properties using the GT4 command `wsrf-get-properties`, both by specifying the properties by their fully qualified names.

An index service can be browsed with the GT4 command such as:

```
wsrf-query -s   \
http://localhost:8080/wsrf/services/DefaultIndexService '/*'
```

[2] See Chapter 9, Section 9.2.2, for more Xpath expressions for selecting XML elements.

which will list all the resource properties and their values in an XML format, where the index service is located at http://localhost:8080/wsrf/services/DefaultIndexService. XPath expressions can be used to select specific resource properties.

Index services support hierarchical structures. Information from various index services can be aggregated into a higher-level index service, as illustrated in Figure 7.12. A *community index service* might contain information propagated from all the index services of the virtual organization and provide a global view of the state of the Grid. Since the resource properties include those associated with regular Globus services such as GRAM services, the community index service can be searched for monitoring and discovery within the Grid environment.

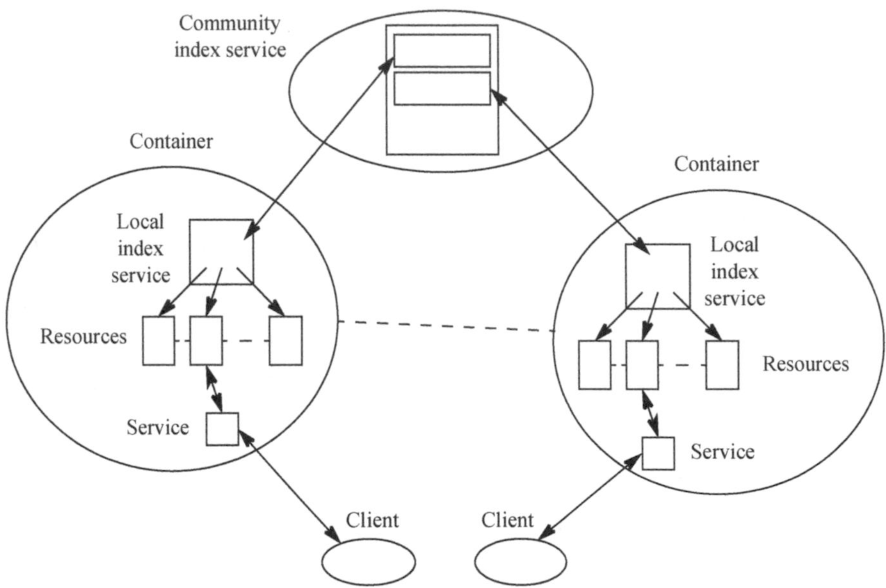

Figure 7.12 Hierarchical index services.

Trigger Service. The *trigger service* responds to specific conditions occurring within the Grid environment and first appeared in GT4. It subscribes to an index service (or another source of WSRF information) to be notified of changes, which are compared to the provided XPath expression. When a match occurs, prescribed actions take place such as sending an email or updating a Web page.

WebMDS. WebMDS (Web Monitoring and Discovering System) is a servlet that provides a Web-based interface to display XML-based information such as resource property information, and as such can be a front-end to index services.

7.4 SUMMARY

This chapter introduced the following:

- Grid computing standardization
- Stateful Web services
- Open Grid Service Architecture (OGSA)
- WS-Resource Framework (WSRF)
- Globus components
- WSRF service implementation
- Information services
- Globus index service

FURTHER READING

The principal source for coding Globus toolkit WSRF services is Sotomayor and Childers (2006). This book provides a very detailed account on how to code such services with many complete code examples. Details of coding WSRF services not in the context of Globus can be found in Graham et al. (2005).

Grid computing software development is a moving target. WSRF was developed to overcome the previous OGSI/GWSDL attempt at defining a stateful Web service. But even WSRF has its critics and we can expect more developments. A topic not covered here is the *Representational State Transfer* (RESTful) Web Service approach.

BIBLIOGRAPHY

Ananthakrishnan, R., C. Bacon, L. Childers, J. Gawor, J. Insley. 2005. How to build a service using GT4. Globus Alliance. http://www.globus.org/toolkit/tutorials/BAS/GT4BuildAServiceV19.pdf

Banks, T. 2005. OASIS Web Service Resource Framework (WSRF) – Primer, Committee draft Dec. 1. http://docs.oasis-open.org/wsrf/wsrf-primer-1.2-primer-cd-01.pdf

Foster, I, C. Hesselman, and S. Tuecke. 2001. The anatomy of the Grid enabling scalable virtual organizations. *Int. J. High Performance Computing Applications*. Also appeared as Chapter 6 in Berman, F., G. Fox, and T. Hey, eds. 2003. *Grid computing: Making the global infrastructure a reality.* Chichester England: John Wiley & Sons.

Foster, I., C. Kesselman, J. M. Nick, and S. Tuecke. 2002. The physiology of the Grid. *Global Grid Forum*. Also appeared as Chapter 8 in Berman, F., G. Fox, and T. Hey, eds. 2003. *Grid computing: Making the global infrastructure a reality.* Chichester England: John Wiley & Sons.

Foster, I., H. Kishimoto, A. Savva, eds., D. Berry, A. Djaoui, A. Grimshaw, B. Horn, F. Maciel, F. Siebenlist, R. Subramaniam, J. Treadwell, and J. Von Reich. 2005. The open Grid services architecture, version 1.0, Jan. 29. http://www.gridforum.org/documents/GWD-I-E/GFD-I.030.pdf

Graham, S, D. Davis, S. Simeonov, T. Boubez, G. Daniels, P. Brittenham, Y. Nakamura, P. Fremantle, D. König, and C. Zentner. 2005. *Building Web services with Java: Making sense of XML, SOAP, WSDL, and UDDI.* 2nd ed. Indianapolis Indiana: SAMS Publishing.

Joshy, J., and C. Fellenstein. 2004. *Grid computing.* Upper Saddle River, NJ: Prentice Hall.

Sotomayor, B., and L. Childers. 2006. *Globus toolkit 4: Programming Java services.* San Francisco: Morgan Kaufmann.

Tuecke, S., I. Foster, J. Frey, S. Graham, C. Kesselman, T. Maquire, T. Sandholm, D. Snelling, and P. Vanderbil. 2003. Open Grid Services Infrastructure version 1.0 (OGSI). http://www-unix.globus.org/toolkit/draft-ggf-ogsi-gridservice-33_2003-06-27.pdf

Wikipedia entry: Distributed Management Task Force. http://en.wikipedia.org/wiki/DMTF. http://en.wikipedia.org/wiki/DMTF

Wikipedia entry: OASIS (organization) http://en.wikipedia.org/wiki/OASIS_(organization)

Wikipedia entry: World Wide Web Consortium. http://en.wikipedia.org/wiki/World_Wide_Web_Consortium

SELF-ASSESSMENT QUESTIONS

The following questions are multiple choice. Unless otherwise noted, there is only one correct answer for each question.

1. What is a stateless Web service?

(a) A Web service that cannot remember prior events
(b) A Web service without local variables
(c) A Web service that is not associated with a particular state in the country
(d) A Web service that can remember prior events

2. What is a non-transient service?

(a) An instance of a service that does not receive data
(b) An instance of a service that outlives its client
(c) An instance of a service that generates stateful data
(d) An instance of a service that generates stable data

3. What is the Open Grid Services Architecture (OGSA)?

(a) A standard for defining a structure for interoperable Grid computing services but not its implementation
(b) A standard for defining a structure for interoperable Grid computing services that includes its implementation
(c) An open source implementation of Grid computing middleware
(d) A Grid computing environment that is open to all

4. What is a (WS) notification in Grid services?

(a) A mechanism of the GGF committee to inform the community of changes to the (WS) standard
(b) A mechanism for client to inform a Grid service of its existence
(c) A mechanism to inform a client of changes in the service or resource
(d) A final demand to pay your taxes

5. What is the WS-Resource Framework (WSRF)?

(a) A way of connecting a Web service to a resource without specifying its implementation
(b) A framework for Web service reliability
(c) A standard that provides for Web services embedded within a computing resource
(d) A standard that provides for stateful Web services

6. When one issues the command:

```
globus-start-container -p 8081 -nosec
```

what does one see?

(a) 10 pages of errors messages
(b) Nothing because the `-p` option with port `8081` cannot be used with `-nosec` (no security)
(c) A list of services deployed by all users in the container
(d) A message that states whether port 8081 can be used without Globus security
(e) A list of the services that only you have deployed in the container

7. In WSRF, what is a factory service?

(a) A service designed to build products
(b) A service designed to create services
(c) A service designed to create instances of services
(d) A service designed to create WS-resources
(e) None of the other answers

8. Name one basic difference between a Web service and a WSRF Grid service

(a) None
(b) A Grid service is accessed from a Grid
(c) A Grid service can be stateful
(d) The connections between Web services are in the form of a Web and the connections between Grid services are in the form of a Grid

9. Name one program (tool) that is used to deploy a GT 4 service.

(a) ant
(b) vi
(c) cc
(d) Java

10. How is a stateful WSRF Web service located?

(a) By a programmer-defined service name
(b) By a URI or URL

(c) By an endpoint reference
(d) By a port number

11. What is the purpose of the Globus index service?

(a) To hold index variables for arrays
(b) To operate on an array of functions
(c) To maintain a list of available services
(d) To maintain a list of available resources

PROGRAMMING ASSIGNMENTS

These programming assignments require a Globus 4.x core environment for deploying services. It is not necessary to have the full Globus installation, only the core. Details on how to establish the core can be found at (Sotomayor and Childers 2006) or the Globus home page (http://www.globus.org/). Also refer to assignments at the book home page (http://cs.uncc.edu/~abw/GridComputingBook). The software packages needed are appropriate versions of JDK, ant, Python, and the Globus core. The environment can be established on a personal Windows, Mac, or Linux system, or on an account on a centralized server.

7-1 Download the files for the Math service provided at the on-line tutorial "The Globus Toolkit 4 Programmers Tutorial" (http://gdp.globus.org/gt4-tutorial/) or at (Sotomayor and Childers 2006) for the first math service with the Java interface:

```
public interface Math {
  public void add(int a);
  public void subtract(int a);
  public int getValueRP();
}
```

Follow the instructions provided at http://gdp.globus.org/gt4-tutorial/ or at (Sotomayor and Childers 2006) to build and deploy the service and test it with the provided client.

7-2 Modify the Math service and associated files given in Assignment 7-1 so the service supports the multiplication operation. To do this task, you will need to modify Service code (`MathService.java`) and WSDL file (`Math.wsdl`).

CHAPTER 8
User-Friendly Interfaces

The focus of this chapter is graphical user interfaces (GUIs), which are preferable to command-line interfaces for many non-Computer Science users. Graphical user interfaces are important if Grid computing platforms are to become accepted to a broader community of users. The chapter concentrates upon two areas:

- Graphical tools for generating workflows, and
- Web-based portal design for access Grid resources.

Workflows can be interconnected job submission components and in that context follow on directly from Chapter 3 on job schedulers. Indeed, simple workflows were constructed in Condor in Chapter 3. The workflow components in a Grid workflow can also be clients to Web services (Chapters 6 and 7).

Web-based portals were purposely introduced early in the book (Chapter 1) to give the "big picture" from which lower level details followed in subsequent chapters. Now, we will study the internal design of portals. It is relatively straightforward to create customized portal using a portal toolkit or framework, and we will focus on one such framework called GridSphere.

8.1 INTRODUCTION

Clearly, users would like to see a simple convenient but flexible interface to a Grid platform. We are all familiar with Web interfaces to access the Internet for Internet-based activities. The GUI Grid computing Web-based portal user interface introduced in Chapter 1 provided a simple interface for the central activities of a Grid user.

Many production Grids use such Grid portals, but also many Grid activities fall back upon the command-line interfaces that were described subsequently in the book. When one is using a traditional Linux command-line interface, one might wonder what is so different from just making an `ssh` connection to remote computers. This comparison has been made before. In Chapter 2, Section 2.1, user actions to run jobs were described and how a Grid environment should provide more features than is easily done with `ssh` remote connections, including automatic discovery of resources. In Chapter 5, Section 5.1.2, a case was made for a unified security Grid environment in the context of single sign-on and the security environment.

Although issuing a `Globusrun-ws` command on the command line to submit a job and `ssh`-ing into a remote computer and issuing commands to run a job are very similar on the surface. It is actually significantly different underneath as the user does not personally log on a remote machine on the command line to run a Globus job. The remote connection happens behind the scenes using proxies. We can point to a number of improvements in implementation in the Grid version including single sign-on, which is significant if one wants to submit jobs to different remote computers and to do third-party transfers, that is, to transfer a file from one remote computer to another from a third. Doing all this on the command line requires users to open up a console window and know the commands in detail. Windows users especially are not used to even opening a console window to run application, let alone issuing complex commands. So it is very desirable to offer users a better user interface, the type of interface they have come to expect on personal computers with windows, pull-down menus, drag-and-drop, and double clicking on icons to launch applications. In this chapter, we will describe such graphical user interfaces (GUIs). We will begin with graphical interfaces for workflows and then move on to Web-based interfaces for general Grid actions such as submitting and monitoring jobs.

8.2 GRID COMPUTING WORKFLOW EDITORS

8.2.1 Workflows

A computational workflow describes a sequence of tasks that need to be done to perform an overall computation. Each task has inputs and outputs (results). The results of one task are used by other tasks as inputs, creating dependencies between the tasks. A simple workflow might be a series of sequential tasks as one might find sequential procedural programming. It may be necessary to allow tasks to complete before performing the subsequent tasks in the sequence, depending upon the dependencies between the tasks. In Condor (Chapter 3, Section 3.2.2), such workflows were created in Condor using a directed acyclic graph that was encoded into a DAG file. However, this a very primitive approach. Dependencies may be very complex. The results of multiple jobs need to be staged as input for multiple subsequent jobs and the inputs and outputs may not be immediately compatible and may require transformations to be acceptable. The tasks themselves may be significant applications. This is especially true for scientific applications. For example, bioinformatic applications might need to use multiple specialized programs to process data. Sometimes, they access very large, possibly distributed, databases. Sometimes, they use visualization tools at particular sites to view the results and assist in their analysis.

As we saw in Chapter 7, Grid computing has moved to using a Web service approach for much of its middleware, so that invoking a command such as a GRAM job submission command is in fact invoking a Web service. Distributed applications themselves could be exposed as Web services. Hence, a Grid workflow typically becomes a form workflow of interconnected Web services. There are languages for constructing Web service workflows. IBM proposed the XML *Web Services Flow Language* (WSFL) targeted towards constructing Web service workflows for business applications in the early 2000s. Microsoft also proposed an XML business Web service orchestration language called XLANG., which is an extension of WSDL. Both were superseded by the *Web Services Business Process Execution Language* (WS-BPEL or BPEL), an OASIS standard introduced in 2004. WS-BPEL is targeted towards business applications. Whatever the underlying representation of the workflow, we are interested in raising the level for the users so that they do not need to write workflow descriptions in a language. This leads to a workflow editor.

8.2.2 Workflow Editor Features

There have been a very large number of projects for creating workflow tools for Grid computing. A number of XML languages have also appeared for Grid computing workflows. Twenty-one Grid workflow projects are listed by von Laszewski (2006). A very detailed taxonomy of workflow editors can be found in (Yu and Buyya 2005). The following are some desirable features for the workflow editor:

- A workflow editor should provide the ability for the user to construct a workflow of interconnected local and remote components easily.
- Workflows lend themselves to pictorial representation and ideally, workflow editors should have a graphical drag-and-drop interface.
- The pictorial representation should be separated from the concrete implementation.
- An XML workflow description is desirable.
- The components of the workflow should be able to be grouped together into reusable components for other workflows.

Workflow editors with graphical interfaces include Taverna (Taverna project) and Kepler (Kepler project). The Taverna workflow editor is focused on distributed components (including services) running on local and remote machines. It has its own graphical interface. Kepler has a broader focus of scientific workflows but can now handle distributed computing, including Grid computing. It leverages a powerful open-source scientific workflow editor called Ptolemy II developed at the University of California–Berkeley, and uses the Ptolemy II graphical interface (called Vergil) directly. Ptolemy is concerned with modeling concurrent processes in real-time embedded systems. Its development can be traced back to the mid-1980s. Work on Ptolemy II began in 1996. Version 7.0 was introduced in 2008. In Ptolemy, the individual components of the workflow are called *actors* (from early work elsewhere). Actors communicate between themselves through messages. More details on Ptolemy can be found at (Brooks et al. 2008).

In the next section, we shall focus on a Grid computing workflow editor called GridNexus, which has all the features listed above and is very easy to use. We use GridNexus in our Grid computing class.

8.2.3 GridNexus

GridNexus was developed at the University of North Carolina–Wilmington in early-mid-2000s specially for constructing Grid computing workflows (Brown et al. 2005). As with Kepler, GridNexus is based upon Ptolemy II and uses the Ptolemy II graphical interface. Many of the graphical user interface features of GridNexus are inherited from Ptolemy II, including some terminology, onto which some new features are introduced for a Grid computing environment, such as Web service clients, etc. Also new in GridNexus, the workflow is separated from its execution with the use of an XML scripting language called JXPL[1] that was designed as part of the GridNexus project. The GridNexus graphical interface is derived from Ptolemy II and has a top command ribbon and three frames, two on the left and a main drawing frame on the right (a *palette* or *canvas*) as illustrated in Figure 8.1. The upper left frame displays a library of modules that can be used in the workflow. The user constructs the workflow by drag-and-drop and drawing actions on the canvas, pulling modules from the upper left frame module library. Once modules are placed onto the canvas, module outputs can be dragged to module inputs. The inputs and outputs available will depend upon the modules selected. The lower left frame displays a thumbnail of the actual workflow being constructed on the canvas.

A workflow is executed by selecting the run icon on the top ribbon. This will first create a description of the workflow in the JXPL language. The JXPL file so produced is then read and executed by a JXPL processor. When the workflow has Web services, these components will be accessed from wherever they reside, as illustrated in Figure 8.2. Note that if a Web service is part of the workflow, as illustrated, a GridNexus Web service client will be invoked to access the Web service.

Modules in the workflow are categorized as *sources* that have no inputs but produce output data, *transformers* that convert input value(s) into an output, *sinks* that take input value(s) to perform some operation but do not produce output on the workflow. There are various library modules that are sources, sinks, or transformers. Many of the basic components are derived directly for Ptolmey II. The constant source module provides a constant value, which can be set by the user (by double clicking the constant component on the workflow). A large number of arithmetic transformers exist, including addition, subtraction, multiplication, division, and modulo, and trigonometric and mathematical functions.

JxplDisplay is a GridNexus-specific sink module for creating the JXPL script for the workflow and causing it to be executed. The JXPL script without it being

---

[1] JXPL is not an abbreviation for anything now although it may have been originally.

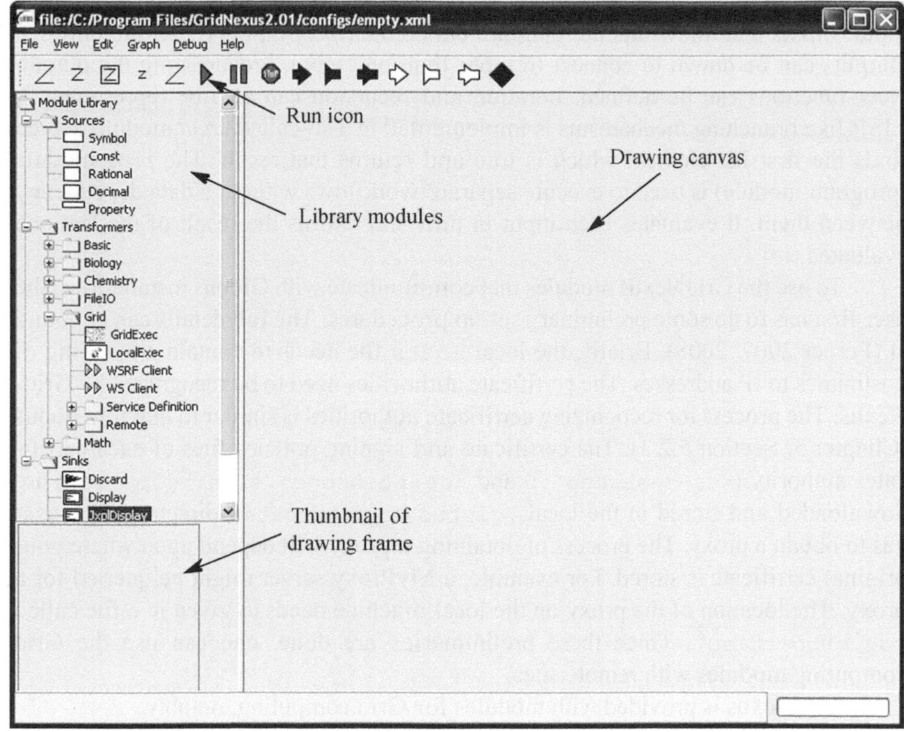

Figure 8.1 GridNexus user interface.

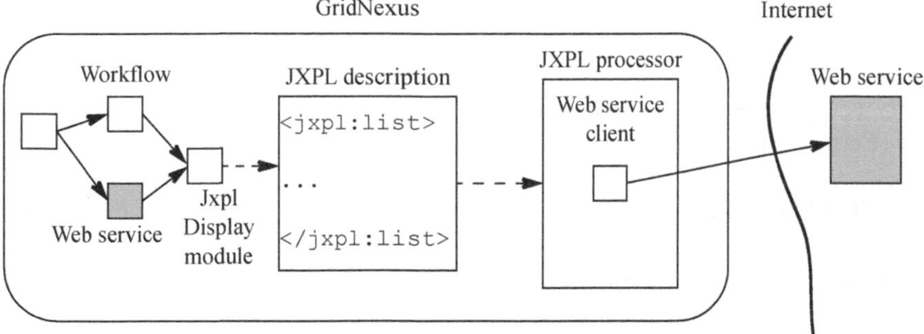

Figure 8.2 GridNexus accessing remote service.

executed can be obtained by double-clicking the JxplDisplay module and deselecting "Evaluate Jxpl." This is useful for debugging a workflow.

The modules of a workflow can be grouped together into a so-called *composite* and considered as a single entity and used in other workflows. A composite will have named inputs and outputs. A user can save composites in a user library that appears on the upper-left frame (see Figure 8.1). Composites come directly from Ptolemy II.

Other features include allowing more than one source to be merged into one input. Inputs that allow this merging are called *multiport* inputs (from Ptolemy II). Outputs can be drawn to connect to more than one input, broadcasting the output. New functions can be defined. Iteration and recursion can also be represented A LISP-like branching mechanisms is implemented in a so-called *cond* module, which finds the first list element which is true and returns that result. The *prog* module (program module) is used to execute separate workflows without a data dependency between them. It evaluates each input in turn and returns the result of the last one evaluated.

To use the GridNexus modules that communicate with Globus installations, the user first has to do some preliminary set-up procedures. The full details can be found at (Ferner 2007, 2008). Briefly, the local `hosts` file needs to contain a mapping of hostnames to IP addresses. The certificate authorities need to be recognized by GridNexus. The process for recognizing certificate authorities is similar to that for Globus (Chapter 5, Section 5.2.3). The certificate and signing policies files of each certificates authority (`<ca-hashcode>.0` and `<ca-hashcode>.signing_policy`) are downloaded and stored in the local `.globus/certificates` directory. The user has to obtain a proxy. The process of obtaining a proxy will depend upon where your original certificate is stored. For example, a MyProxy server might be queried for a proxy. The location of the proxy on the local machine needs to given in a file called `cog.properties`[2]. Once these preliminaries are done, one can use the Grid computing modules with remote sites.

GridNexus is provided with modules for Grid computing, notably:

- GridExec for submitting GRAM jobs (GRAM client)
- GridFTP and SFTP modules for file transfers
- WS client module
- WSRF (grid service) client module

which can be found under `Module library -> Transformer -> Grid`. Let us briefly review various modules and their use. The following materials are derived from (Ferner 2007, 2008).

GRAM Client. GridNexus has a module called `GridExec` for submitting Globus GRAM jobs. Figure 8.3 shows a simple workflow using `GridExec`. This particular workflow causes the program `/bin/echo` to be executed with the arguments `hello world`. This corresponds to:

```
globusrun-ws -submit -F coit-grid01.uncc.edu:8440  \
                   -c /bin/echo hello world
```

A JDD job description file is constructed with default settings for output, i.e.,

[2] CoG (Commodity Grid) Kit is a set of libraries to simplify the development of applications that use Grid computing components. CoG Kit is actually used for the development of some Globus components. See Chapter 9 Section 9.4.1 for more details on CoG Kit.

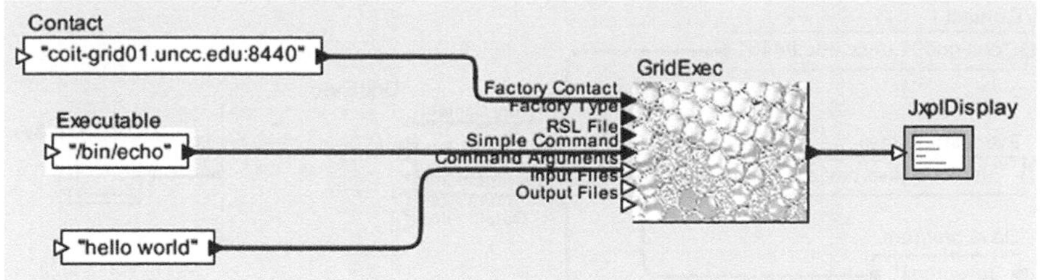

Figure 8.3 Invoking the GridExec module.

```
<job>
    <executable>/bin/echo</executable>
    <argument>hello</argument>
    <argument>world</argument>
    <stdout>${GLOBUS_USER_HOME}/stdout</stdout>
    <stderr>${GLOBUS_USER_HOME}/stderr</stderr>
</job>
```

for the Globus `globusrun-ws` command and essentially causes the command

```
globusrun-ws -submit -F coit-grid01.uncc.edu:8440  \
                 -f echo.xml
```

to be executed where `echo.xml` is the job description file for the job. GridExec will also accept the job description file directly (RSL File input in Figure 8.3). A specific job scheduler can be specified with the `Factory Type` input, for example SGE, Condor, LSF, PBS, etc., specified as a string otherwise the default is the Fork job manager.

The job could be a Java program in which case the executable specified is the Java interpreter and the first argument is the Java class file, as illustrated in Figure 8.4. In this figure, `/usr/local/java/bin/java` is the Java interpreter and `javaprog1` is the Java program class file. The program argument is `89`. Notice that the two arguments for the Java interpreter, `javaprog1` and `89`, are input using separate constant boxes that are merged into the `Command Argument` multiport input. Multiport inputs are identified by white triangles whereas those inputs that do not accept more than one input are identified by black triangles, see Figure 8.4. A characteristic of the GridNexus multiport is that the order that the inputs are drawn on the workflow is very important as this determines the order that the inputs enter the multiport. So for Figure 8.4, the output of `Cont3` (the Java class file) must be drawn to the `Command argument` input first, then the other `Const` boxes which hold the arguments for the Java program.

File Transfers. In the workflows of Figures 8.3 and 8.4, specific input and output files can also be specified using the `Input Files` and `Output Files` inputs. For example, suppose we want to transfer result of the program in Figure 8.4 to a remote location, that location could be given on the `Output Files` input. This

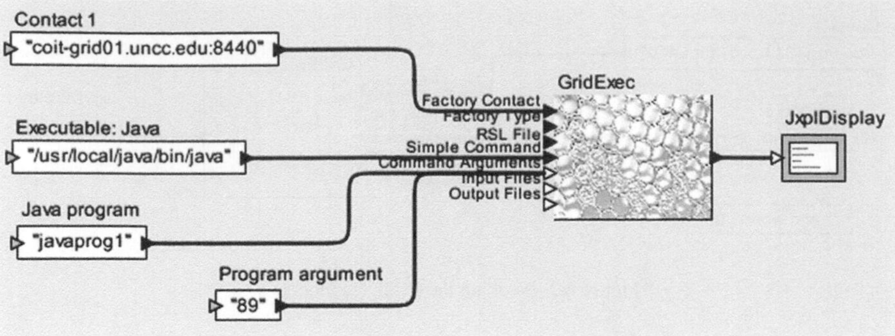

Figure 8.4 Invoking the GridExec module for a Java program.

will have the effect of submitting a job description file with an output file transfer. An example is

```
<job>
  <executable>/usr/local/java/bin/java</executable>
  <argument>javaprog1</argument>
  <argument>89</argument>
  <stderr>${GLOBUS_USER_HOME}/stderr</stderr>
  <fileStageOut>
   <transfer>
    <sourceUrl>file:///nsf-home/abw/stdout</sourceUrl>
    <destinationUrl>
      gsiftp://torvalds.cis.uncw.edu/home/grid/abw/stdout
   </destinationUrl>
   </transfer>
  </fileStageOut>
</job>
```

Such file staging requires GridFTP, which is not available in Globus 4.0 for a Windows platform. GridNexus has a module for file transfers using SFTP, which is available in Windows but requires public key authentication.

Ferner (2008) introduced a way to file transfers in GridNexus using a separate GridNexus module for the file transfer. Figure 8.5 shows an example derived from (Ferner 2008). In this example, first the Java program is executed using the upper GridExec module (same construction as in Figure 8.4). Next, in the lower GridExec module, a Globus `globusrun-ws` command is issued to execute the `globusrun-url-copy` command to transfer the output file to another site, in this example torvald.cis.uncw.edu. This is equivalent to:

```
globusrun-ws -submit -F coit-grid01.uncc.edu:8440 -c  \
globusrun-url-copy file:///nsf-home/abw/stdout        \
gsiftp://torvalds.cis.uncw.edu/home/grid/abw/stdout
```

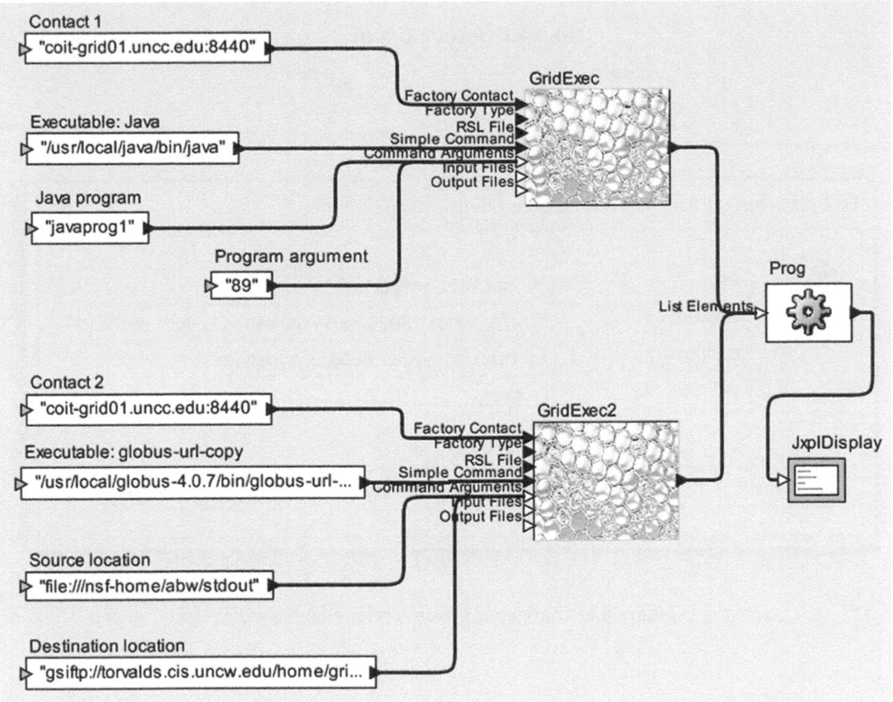

Figure 8.5 Output staging. (Derived from (Ferner 2008).)

WS Client Module. The WS client module acts a client to a Web service with a given WSDL interface document. Once dragged onto the workflow, it is configured from a provided WSDL document. An example is shown in Figure 8.6. The URL of the Web service WSDL in this case is:

```
http://coit-grid02.uncc.edu:8080/ ... /MyMath.jws?wsdl
```

In this particular case, the Web service here is deployed with the `.jws` instant deployment facility and the Apache Axis `?WSDL` query facility is used to automatically obtain the WSDL file from an already deployed Web service (see Chapter 6, Section 6.3.2).

The `MyMath` service has one method, `squared`, which will return the square of the input. This service is described by House, Holliday, and Ruff (2005). Once the "Commit" button is clicked, the WSDL document should be read and the service name should be displayed with its named input as illustrated in Figure 8.6. Prior to that, no name will be displayed. To test the service, one might configure the workflow as shown in Figure 8.7.

WSRF Client. The GridNexus WSRF module acts as a client for a WSRF-compliant (Grid) service. With WSRF Grid services, unfortunately it is not possible

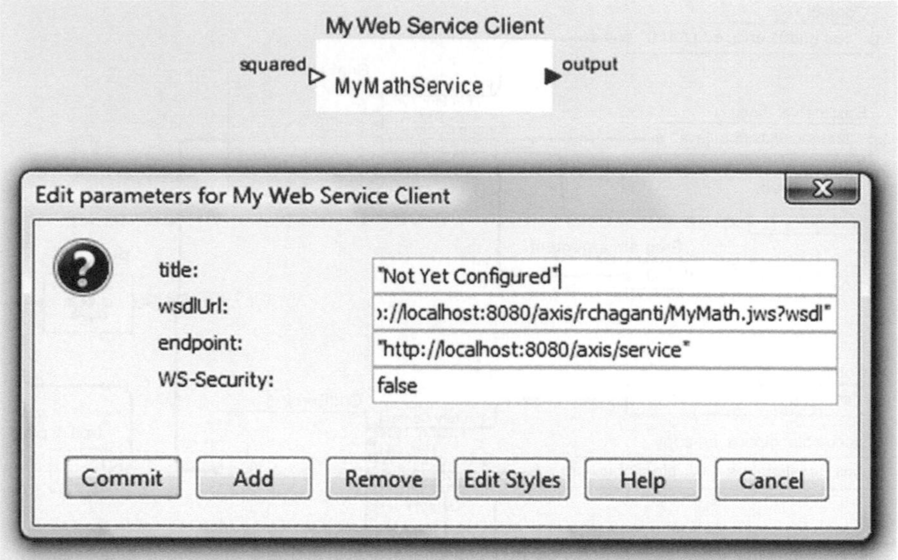

Figure 8.6 Configuring a Web service client module.

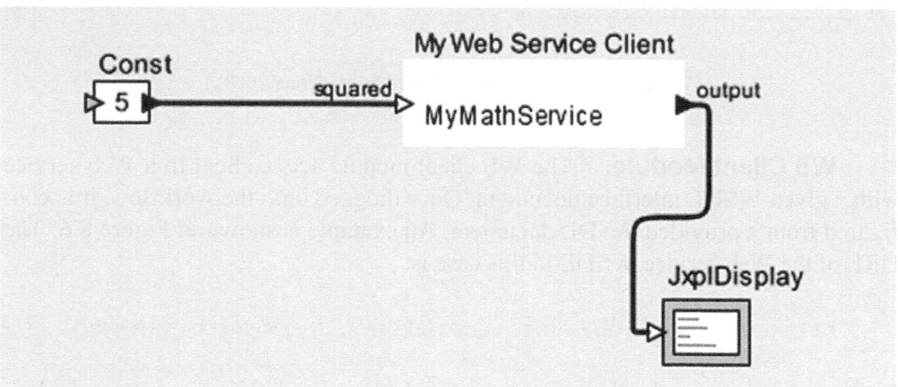

Figure 8.7 Testing a Web service client module.

to obtain all the necessary interface information just from the WSDL as in the case of a regular stateless Web service. The WSRF module requires the following information to be entered:

- The URL of the service, or factory URL
- The location of the `AddressingLocator` class of the service

to be able to configure itself to work with the service JAR package. With these two items entered, the input name(s) should be displayed. An example of a configured WSRF client is shown in Figure 8.8. Then the service can be used in a workflow such as shown in Figure 8.9, as done by Ferner (2007). The `const` module is used to

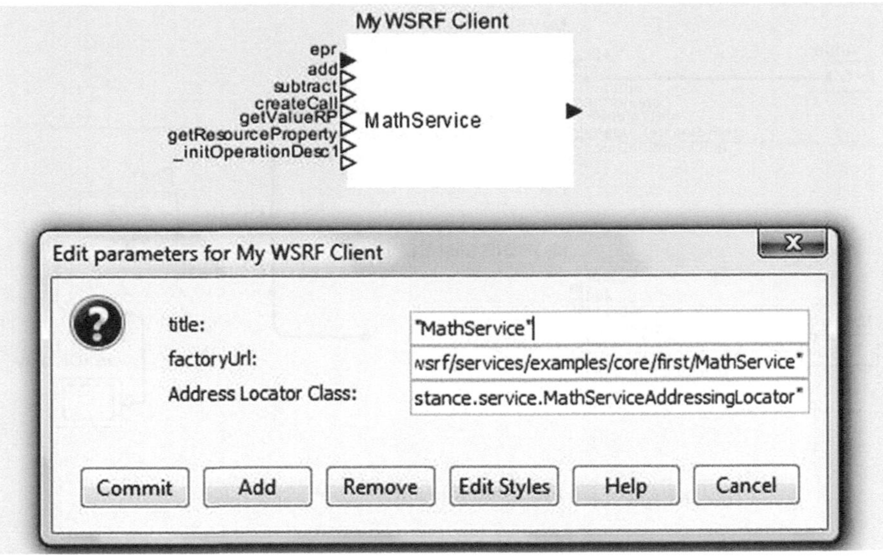

Figure 8.8 GridNexus WSRF client set-up.

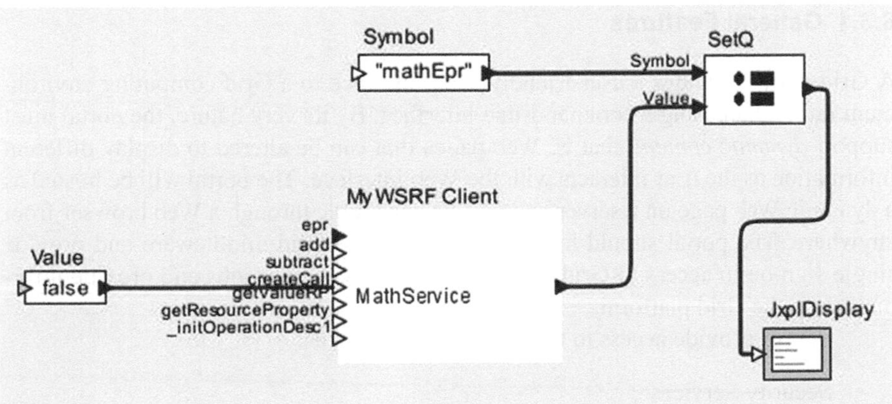

Figure 8.9 Invoking WSRF client.

invoke the `create` method, which has no arguments. Something has to be provided on an input to invoke the associated method. In GridNexus, `false` is used. This approach has a problem for a method whose argument might be `false`. The `SetQ` module assigns a value to a symbol and is used here to set a value to a symbol called `MathEpr`, which then will be displayed as such from the `JxplDisplay` module. The EPR symbol is then available subsequently to invoke the `destroy` method.

Figure 8.10 shows a workflow consisting of two WSRF Web services. This workflow uses a `Prog` module to execute both service clients, an approach used in the GridNexus assignments written by Ferner (2007, 2008).

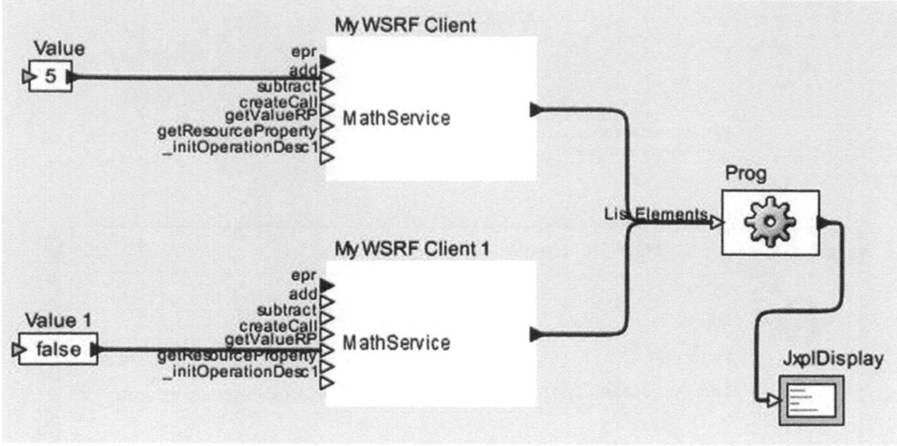

Figure 8.10 GridNexus workflow with two WSRF services.

8.3 GRID PORTALS

8.3.1 General Features

A Grid portal provides a user-friendly Web interface to a Grid computing environment rather than using a command-line interface. By its very nature, the portal must support *dynamic content*, that is, Web pages that can be altered to display different information as the user interacts with the Web interface. The portal will be hosted as a dynamic Web page on a server and will be accessible through a Web browser from anywhere. The portal should hide the details of the Grid middleware and provide single sign-on to access to Grid computing services, submit jobs and provide information on the Grid platform.

Portals provide access to familiar services such as:

- Security Services

  Management of certificates

- Remote File Management

  Access to files and directories
  Moving files

- Remote job management

  Job submission
  Workflow management

- Grid information services

  Static information (machine type, etc.)
  Dynamic information (machine load, etc.)

A common generic layout of a portal is shown in Figure 8.11. Users can select tabs and menu items in much the same way as one would navigate a commercial Web site. Typically, an active window is in the center. Some early portals would display floating windows after users made their selections. Some portals would have facilities to display graphical output. Feeds from Internet sources are also a possibility. Mostly, users will fill in HTML text fields and click on buttons. To run a job, for example, information is entered such as the name of the executable, the arguments, names of input and output files, etc., as was introduced in Chapter 1.

The portal usually incorporates user registration and proxy management components. The most notable proxy management component is the MyProxy server. Portals also provide access to information—anything related to tasks at hand, including communication with people in the virtual organization. In fact, some portals started simply as informational portals.

A Grid portal comes in one of two forms:

- A general-purpose portal as a front-end to access a Grid computing platform in non-specific application domains.
- A portal tailored to a particular application domain.

A portal may have a broad scientific domain. Sometimes, the term *science portal* is used to emphasize its use in a science domain. The term *gateway* is also used to describe a portal, for example a *science gateway* for scientists. Scientists in this context could mean many scientific disciplines such as physics, chemistry, biology, etc. A portal may be tuned to a specific area, for example bioinformatics. An application-specific portal should provide access to specific software packages and databases of the application domain. For example a bioinformatics portal might provide access to BLAST software and protein databases.

Any self-respecting Grid computing project has a portal. Table 8.1 lists a few of the many scientific portals. There are in fact literally hundreds of portals around the world. TeraGrid lists 32 separate portals just associated with TeraGrid (TeraGrid).

Most Grid portals are created with a *portal toolkit.* Portal toolkits provide a software framework and components to put together a portal easily. The software components can be reused, and potentially components developed by others can be

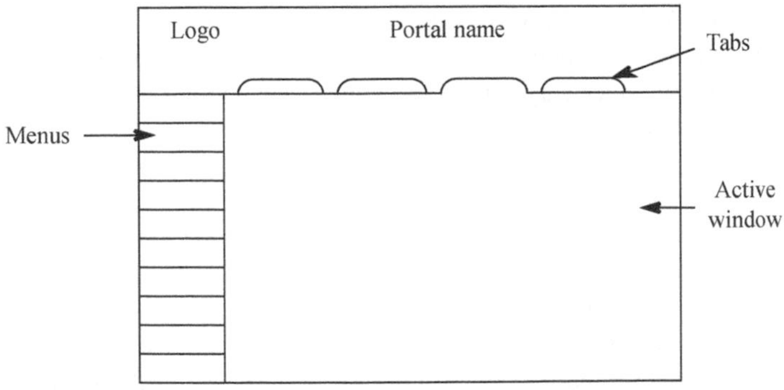

Figure 8.11 Generic layout of a Grid portal.

TABLE 8.1 EXTREMELY SMALL SAMPLE OF SCIENCE PORTALS IN 2000–2008

Name	URL
Biomedical Information Research Network (BIRN)	https://portal.nbirn.net/gridsphere/gridsphere
Chronos system (Geoscience)	http://portal.chronos.org/gridsphere/gridsphere
GEONgrid (Geoscience)	https://portal.geongrid.org/gridsphere/gridsphere
CIMA X-Ray Crystallography Common Instrument Middleware Architect for Grid enabling instruments	http://cimaportal.indiana.edu:8080/gridsphere/gridsphere
LEAD (Linked Environments for Atmospheric Discovery)	https://portal.leadproject.org/gridsphere/gridsphere
NEESGrid (Network for Earthquake Engineering Simulation)	https://central.nees.org/login.php
RENCI science portal (Bioinformatics)	https://portal.renci.org/portal/
TACC User portal (Texas Advanced Computing Center)	https://portal.tacc.utexas.edu/gridsphere/gridsphere

incorporated. Ideally, standard interfaces should exist. The *presentation layer* that the user sees should be separated in the software construction from the back-end. Before looking at specific portal toolkits, let us first look at some available technologies for constructing a portal toolkits.

8.3.2 Available Technologies

Dynamic Content. Dynamic content refers to a Web page display that can be altered while it is being viewed usually due to user requests, as opposed to static content in which the Web page simply displays fixed information without the possibility of it changing under varying circumstances. Dynamic content can be done at the client side, that is, after the page is downloaded, by having code embedded in the HTML page that is executed by the browser to process user input and alter the Web page. The JavaScript language was designed specifically for such embedded code and all browsers support JavaScript. An excellent introduction to JavaScript is by Knuckles (2001). An alternate way of providing dynamic content is at the server side, that is, the processing to create new information on a Web page is done on the server prior to it being sent to the client. When a user makes a request within a Web page, the server receives this request and sends the appropriately altered HTML page, which is then displayed. Such server-side dynamic content is commonly used when the request requires access to a remote database or other resource only available at the server side. Client-side and server-side methods are often used together.

Grid portals can use client-side dynamic content but generally require server-side dynamic content. Server-side dynamic content can use regular programming languages such as C and Java to create the HTML pages that are sent to the client, but there are also technologies specifically for generating server-side dynamic content including CGI (Common Gateway Interface), PHP (originally Personal Home Page, now Hypertext Preprocessor), ASP.NET (Active Server Pages .NET framework), Java Servlets, and JSP (Java Server Pages). CGI is the oldest and provides a standard protocol between Web servers and client applications. PHP and ASP.NET are non-Java technologies. PHP is a scripting language specifically designed for providing server-side dynamic content. ASP.NET is a Web application framework developed by Microsoft that can provide dynamic content and a successor to ASP (Active Server Pages). However, many Grid portals focus on Java implementations using Java Servlets and JSP. GridSphere, which we mentioned in Chapter 1 as our course portal, uses Java servlets and JSP. Hence, let us look at these technologies in more detail. Being Java based, it is also platform independent and not tied to one operating system or manufacturer.

Java Servlets. Java servlets are Java objects that receive requests from Web clients and generate responses. The Servlet interface in `javax.servlet` package defines required methods that must be implemented for client-servlet interaction. A *servlet container* is a form of Web server that provides an environment for servlets and map URLs to specific servlets. Another term used is *servlet engine,* which supports servlets. Apache Tomcat is a servlet engine. Servlets can maintain state across server transactions by various means. More information about servlets can be found at (Wikipedia Java Servlets).

Java Server Pages (JSP). Using Java servlets alone would typically require the invoked Java programs to create the HTML using `println` statements. Java Server Pages (JSP) technology is a complementary SUN technology to Java servlets that is used to create Java servlet code from static content. A JSP file is an HTML page with embedded JSP tags. The JSP tags provide for creating the Java servlet code, which is created automatically from the JSP file by a JSP compiler. The final servlet code itself might be fully compiled machine-executable code or Java byte code executed by a Java virtual machine (JVM).

Figure 8.12 shows the interaction of a client with a servlet engine, JSP files, and servlets. The client makes a request for a Web page. Pages with the extension `.jsp` indicate a JSP page. Using a just-in-time approach, the first time the `.jsp` page is accessed, the servlet engine will take the `.jsp` page and create the servlet code. The servlet methods will be called and create the HTML page, which is directed to the Web client for display. Should the page be requested again, the servlet code is already created and can be executed immediately.

JSP additions to the HTML page are in the form of JSP tags provided for inserting code and related actions. Three tags are available for inserting code. The *declaration* tag (`<%! ... %>`) is used to declare variables and methods, i.e.:

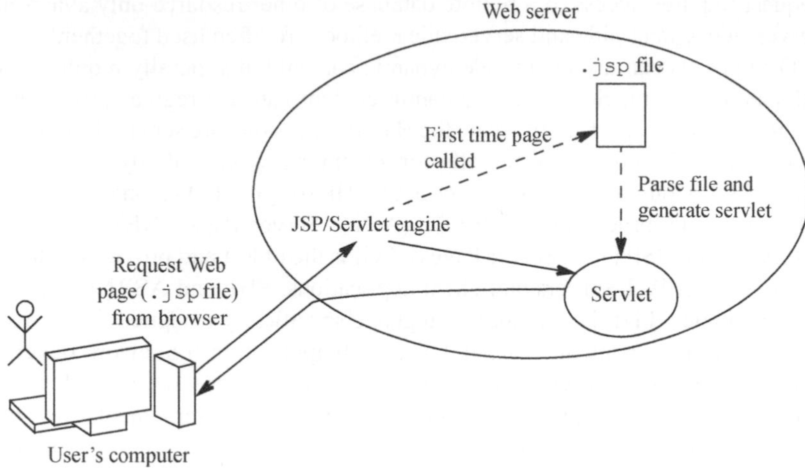

Figure 8.12 JSP/Java servlet environment.

```
<%!
     Java variable declarations or/and
     Java methods
%>
```

The *scriptlet* tag (`<% ... %>`) can also be used to include Java code including variable declarations and methods, but also is broader to include any Java code fragment, i.e.:

```
<%!
     Any Java code fragment
%>
```

The *expression* tag (`<%= ... %>`) will compute a Java expression and convert the result into a string that is inserted in-line into the HTML code, i.e.,

```
HTML code <%= Java expression %> HTML code
```

There are two other tags in JSP, the *directive* tag and the *action* tag. The *directive* tag (`<%@ directive ... %>`) enables additional information to be provided, where `directive` describes the type of directive tag, notably `page`, `include`, or `taglib`. These allow, among other things, one to include additional HTML files, JSP files, and Java packages, and a tag library to extend the functionality of tags. These additional files enable reuse and separation of code functions. Examples are:

```
<%@ page import = "java.util.*" %>
<%@ include file = "shared/template.html" %>
<%@ taglib uri = "htp://java.sun.com/portlet"
                  prefix = "portlet"%>
```

The *action* tag `<jsp: ... >` is used to invoke server-side JavaBeans, transfer control to another page, and support for applets.

Our description of JSP is just a brief overview. The full details of JSP is outside the scope of this textbook and the reader is referred to the SUN home page for JSP documentation for more information.

Commodity Grid (CoG) Kits. Commodity Grid (CoG) kits were conceived during the early development of Globus from 1996 onwards (von Laszewski et al. 2001). They had the objective of combining commodity software technologies with Grid components (hence the name "commodity Grid kits") and providing a higher-level interface to Grid components. The so-called commodity technologies referred to here are accepted software components and standard protocols and should not be confused with hardware commodity components. Software components include common libraries, programming languages such as Java, C, and Python, and standard distributed computing frameworks and Internet protocols. The first and most prevalent CoG kit is the Java CoG kit, which provides Java APIs that enable, for example, one to submit and monitor Globus jobs and transfer files by CoG kit calls within a Java application program. This avoids the use of lower-level Globus APIs, which are complex and change from one version of Globus to another. CoG kit development continued during the same period as Globus. In fact, Java CoG kit is used in GT 3.2 and GT 4 for the Java-based GSI, GridFTP, MyProxy, and GRAM. A part of the CoG kit, known as *JGlobus* is now included in the Globus 4 distribution. CoG kits encourage reuse of code avoiding duplication of effort, and hopefully isolate the application programming from changes in the underlying Grid middleware, which changed rapidly during the early 2000s and continues to date.

Figure 8.13 shows a sample CoG kit program to transfer files (Villalobos 2007). Java CoG kit also provides facilities for graphical interfaces targeted at portal design, and became an important component in early portals. It also developed workflow facilities (von Laszewski and Hategan 2005), including an XML-based workfow specification language and execution engine called Karajan.

8.3.3 Early Grid Portals

Several portals and portal toolkits have been developed since the mid-1990s. The early ones began before standards were developed for portal design and used existing technologies such as Java servlets, JSP, and CoG kit. These portals are called first generation Grid portals by Li and Baker (2005), who describe first generation portals as mainly using a three-tier architecture. The client browser is in tier one, which communicates with a Web server and proxy credential server in tier two, which communicates with an application manager in tier three. The application manager manages Grid resources, which are also in tier three. Usually, there needs to be a persistent storage or a database to hold static and dynamic information about the Grid environment. In Li and Baker's description, the application manager may use an events archive to store the state of the application, which can be monitored by the Web browser. Some examples of general-purpose portal toolkits in this period include:

```
filetransfer.java
import org.globus.cog.abstraction.impl.common.AbstractionFactory;
import org.globus.cog.abstraction.impl.common.task.ServiceContactImpl;
import org.globus.cog.abstraction.interfaces.FileResource;
import org.globus.cog.abstraction.interfaces.SecurityContext;
import org.globus.cog.abstraction.interfaces.ServiceContact;
import org.ietf.jgss.GSSCredential;

public class filetransfer {
  public static void main(String args[]){
     filetransfer trans = new filetransfer();
     try {
        trans.transferFile();
     } catch(Exception e){

     }
  }
  public void transferFile() throws Exception{
     FileResource client = null;
     client = AbstractionFactory.newFileResource("gridftp");
     SecurityContext securityContext=
     AbstractionFactory.newSecurityContext("gridftp");
     securityContext.setCredentials(null);
     client.setSecurityContext(securityContext);

     ServiceContact serviceContact=
     new ServiceContactImpl("coit-grid03.uncc.edu:2811");

     client.setServiceContact(serviceContact);

     client.start();      //Start the client
     client.getFile
  ("/home/username/stdout.txt","home/username/stdout.local.txt");
     client.stop();       //Stop the client
}
```

Figure 8.13 CoG kit program to transfer files.

- Grid Portal Development Kit (GPDK) — used JSP for the presentation layer and JavaBeans and Java CoG back-end. GPDK is now not supported.
- Grid Portal Toolkit (Gridport) — used HTML for the presentation layer and Perl/CGI.
- Ninf/Gridspeed portals — used JSP/Java servlets for the presentation layer and Java CoG kit back-end.

All of the above interfaced to a MyProxy server and the Globus toolkit. Such general purpose portal toolkits can be used to create a portal, for either a broad non-specific application Grid computing platform or for narrower application domains. Examples using early Grid portal toolkits include the NPACI Hotpage Grid portal based upon GridPort to provide access to NPACI resources (Thomas and Boisseau 2003).

8.3.4 Development of Grid Portals with Portlets

Although called toolkits, most early portal toolkits had little ability, if any, to be truly customized and were not very flexible. They were often tied to specific programming tools and Grid software, such as Globus 2.4. The specific programming structure was not suitable for users to develop new portals themselves. The APIs were not standardized. In contrast, the portal implementation should be flexible, meet Grid industry standards, and be able to be extended using parts developed by others. After Grid computing embraced Web services as a core technology, the portal framework needed to interact with these services. Clearly, a separation of functionality of the presentation layer from the back-end logic is desirable. A software engineering approach called the Model-View-Controller (MVC) originally developed in the late 1970s is attractive. In the MVC model, there is a model component, view component, and a controller component. The view component is responsible for the user's visual interface. The view component has to interface with the controller component and is typically implemented with JSP. The controller component is responsible for handling the input/output and responding to requests and events and typically implemented using servlets. The model component is responsible for manipulating the data and domain-specific information at the back-end and also interfaces the view and controller components.

Portlets. A general approach for portal design developed in the early-mid 2000s is to use software components called *portlets* for the presentation layer of a portal (the visual interface with the user). Each portlet provides a specific functionality and an area within the portal as illustrated in Figure 8.14. Each portlet might be associated with a particular Grid service depending upon the functionality. The portal can use as many portlets as is needed to create the required functionality. There can be multiple portlets within a single tabbed window. Portlets provide for all the func-

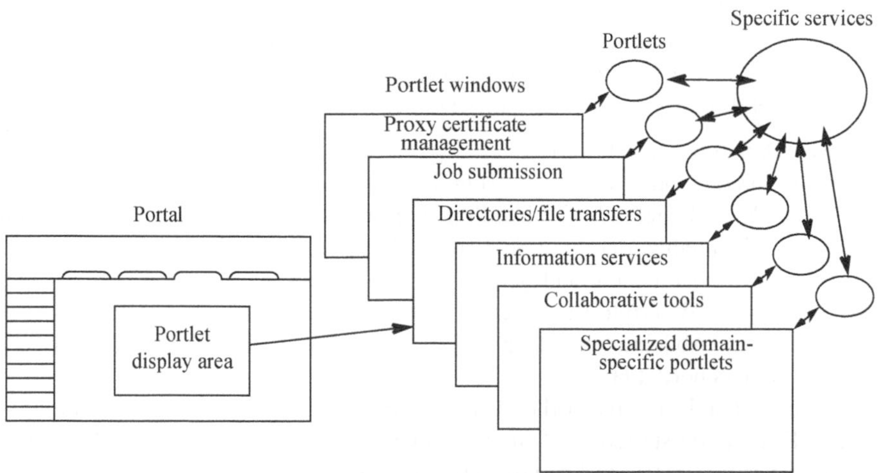

Figure 8.14 Portlets.

tionality that would be expected to access Grid resources, including:

- Proxy certificate management
- Job submission and run-time management
- Remote file transfers
- Access to information services (resource status, etc.)
- Collaborative tools (email, chat, discussion boards, etc.)

and also depending upon the application, specialized portlets for interfacing to domain specific applications.

The portlets themselves can be compared to servlets and require a similar environment called a *portlet container* managed by a *portlet server*. In general, portlets do not communicate with each other, only with the user and the services they front-end. They only provide for the presentation-level. With the portlet approach, it should be easy to reconfigure user's view. Before universally agreed standards were adopted, several venders developed portlet APIs including IBM's Websphere portlet APIs. The open-source Apache Jetspeed project embodied portlets. However, different portlets from different sources should be able to be plugged into portal. Central to its broad adoption and to enable third-party plug-in's is the development of an accepted portlet standard.

JSR 168. After the early experiences of portal designs in the late 1990s, an effort was made in the 2000–2002 period to develop a Java portlet specification, leading to the Java Specification Request JSR 168 released in October 2003. JSR 168 is also called Java Portlet Specification version 1.0 and is based upon the Apache Jetspeed portlets.

Portlet code generally has the following structure:

1. Initialize
2. Render portlet
3. For a request received:
    - Accept request and perform required back-end actions
    - Render display according to result
4. Finalize and destroy portlet

The JSP 168 method for Step 1 is `init()`. If the portlet is to be rendered, the appropriate JSR 168 method is `doView()`. Handling a request is illustrated in Figure 8.15. The two JSR 168 methods for handling a request are `processAction()` and `render()`. The method `processAction()` takes two objects `ActionRequest` and `ActionResponse` for input and output, respectively. Apart from the portlet code, XML deployment descriptors are needed. More information about the JRS 168 APIs can be found at (Abdelnur and Hepper 2003) and (SUN 2003). We shall describe a specific JSR 168 portlet later.

Other related work on Grid tools was going on during the period. The National Science Foundation started the NSF Middleware Initiative (NMI) in 2001 initially over three years to create and deploy network services focusing on Grid and related software. It provided a centralized location for important Grid software including

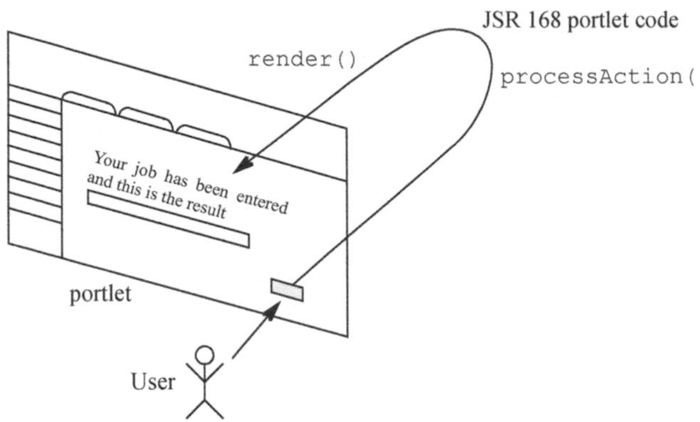

Figure 8.15 JSR 168 portlet activities.

Globus, Condor, MPI-G2, and portal software. In 2003, a new collaborative project called the Open Grid Computing Environment Consortium (OGCE) was established to create collaborations and shareable components for portals. The OGCE portal release 2 consisted of a core set of JSR 168 compatible Grid portlets. This portal is independent of the specific container. Two portal containers were supported, uPortal and GridSphere. Originally, GridSphere and OGCE2 together would be described as a OGCE2/GridSphere portal, but subsequently it was simply referred to as GridSphere.

GridSphere. GridSphere is a portal framework that became extremely widely adopted. Many subsequent production Grid portals are based upon GridSphere as evidenced from the URLs of the portals listed in Table 8.1. Our Grid course portal is based upon GridSphere. The GridSphere portal can be identified for sure by the portal URL when the default directory is used (... `/gridsphere/gridsphere`). The portal may use the default HTTP port 80 or an alternative non-privileged HTTP port 8080 so as not to conflict with a regular Web server. Production Grid portals often use the secure HTTPS, which would use the default port 443, or an alternative port such as 8440. The port chosen must not conflict with the Globus container if running on the same server, which typically runs on port 8443 by default.

The GridSphere core portlets include:

- Login
- Locale, profile, and layout personalization
- Administration portlets for creation of users, groups, portlet management
- Localization support (French, English, Spanish, German, Dutch, Czech, Polish, Hungarian, Italian, Arabic, Japanese, and Chinese)

onto which many other portlets can be installed from various sources, for example the MyProxy server portlet, Globus job submission and control portlets, information services portlets, collaborative tools such as Sakai, etc.

GridSphere can be installed easily on a personal computer (Windows, Mac, or Linux) and it is convenient to do so to develop one's own portlets, see for example, an assignment in our course (Villalobos, Land, and Wilkinson 2007). Portlets in GridSphere are deployed into a servlet engine, for example Tomcat as shown in Figure 8.16. First, one would install the Tomcat servlet engine. Then one would install GridSphere. The default location for the installed GridSphere is `http://localhost:8080/gridsphere/gridsphere` or simply `http://localhost:8080/gridsphere`. At this location is a set-up screen to create the portal administrator account. Once configured, one gets the login screen as shown in Figure 8.17. Many production portals use this login page. Note the neat language selection at the top right side of the login page, which will change the login language and alter the layout to match language customs if necessary. Once logged in, a row of tabs will display, one for each portlet or portlet group installed.

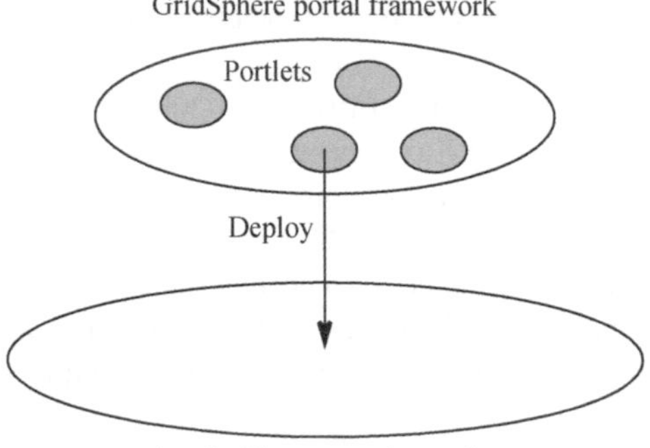

Figure 8.16 GridSphere portlets deployed into a servlet engine (Tomcat).

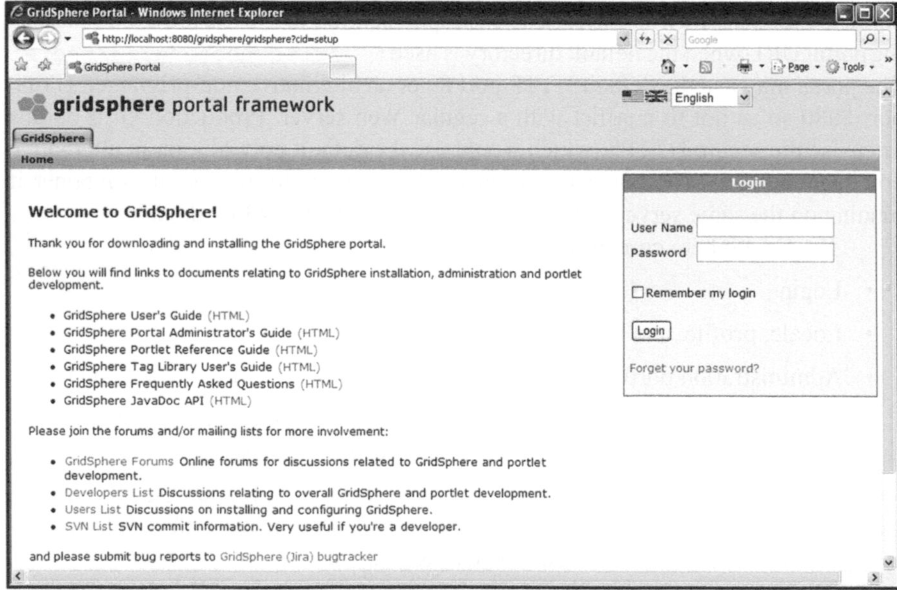

Figure 8.17 Default GridSphere login page.

Creating and Deploying a JSR 168 portlet. It is extremely easy to create one's own portlet in GridSphere using sample code as a template, as described in (Villalobos, Land, and Wilkinson 2007). The following is directly from this source. We will briefly go over the steps. Suppose one wanted to create a simple portlet as a front-end to a Java program. To make it concrete, suppose the Java program would just tell whether a number was even or odd. Although trivial, this will illustrate the steps required, which would be applicable to larger realistic Java programs. Suppose the portlet should look as shown in Figure 8.18. This layout can be described by an HTML table consisting of a single column and four rows as shown in Figure 8.19. The first column holds text (which may wrap around depending upon the desired width of the table). The second row holds where a user will enter a number. The bottom row holds an HTML button named "Get Answer" for the user to indicate that the result should be computed and displayed. The third row holds the result, text indicating whether the number is even or odd. This would only display after the result is requested.

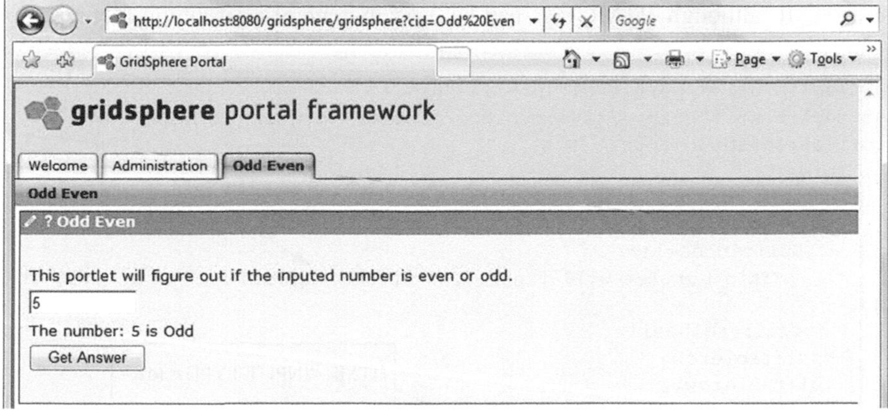

Figure 8.18 Final portlet displayed in GridSphere.

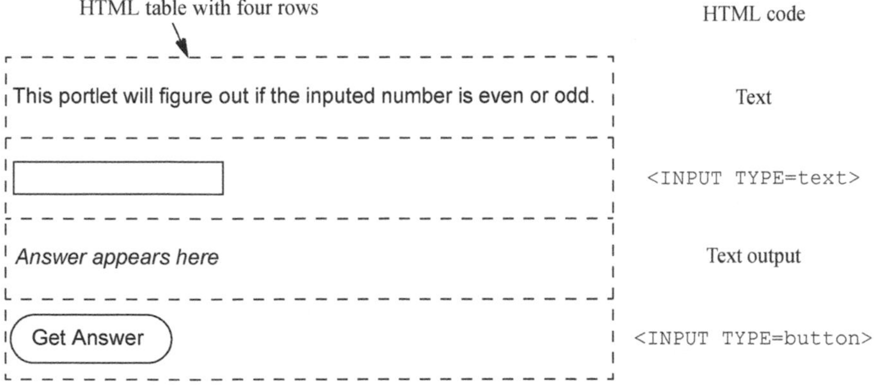

Figure 8.19 Portlet layout.

GridSphere provides an Ant[3] script to create the directory structure and sample files needed for a portlet. The portlet designer then simply provides the Java source file that does the required evaluation and modifies the supplied portlet deployment descriptor files accordingly. Suppose the project directory is called `oddevenportlet`. All files relating to the portlet are placed in specific folders within this directory. The `src` directory holds the source file. The `build` directory holds the corresponding class file. The `webapp` directory holds the deployment descriptor files. In each case, there is a specific sub-directory structure. More details can be found at (Villalobos, Land, and Wilkinson 2007).

The files needed to deploy a particular portlet are:

- Java source file for the portlet actions
- HTML/JSP file describing the portlet layout
- Three portlet deployment descriptor files

HTML/JSP File. The layout is defined by a JSP file called `MainPage.jsp` as shown in Figure 8.20. This file uses tags from the `PortletUI` library instead of pure HTML although HTML tags can be used in general as it simply defines the

```
<%@ taglib uri="/portletUI" prefix="ui" %>
<%@ taglib uri="http://java.sun.com/portlet" prefix="portlet" %>
<portlet:defineObjects/>
<ui:form>
  <ui:table width="500">
    <ui:tablerow>
      <ui:tablecell>
        This portlet will figure out if the inputed number is even or
odd.
      </ui:tablecell>
    </ui:tablerow>
    <ui:tablerow>
      <ui:tablecell>
        <ui:textfield size="10" beanId="valueTF1"/>
      </ui:tablecell>
    </ui:tablerow>
    <ui:tablerow>
      <ui:tablecell>
        <ui:text beanId="answer"/>
      </ui:tablecell>
    </ui:tablerow>
    <ui:tablerow>
      <ui:tablecell>
        <ui:actionsubmit action="action" value="Get Answer"/>
      </ui:tablecell>
    </ui:tablerow>
  </ui:table>
</ui:form>
```

Text

HTML <INPUT TYPE = text>

Text output

HTML <INPUT TYPE = button>

Figure 8.20 JSP code Java source code, `MainPage.jsp`.

[3] A Java-based build tool from the Apache project similar to make but using XML scripts.

HTML layout for the portlet. The JSP tags `<%@ taglib uri="/portletUI" prefix="ui" %>` and `<%@ taglib uri="http://java.sun.com/portlet" prefix="portlet" %>` declare that custom tags are used as defined in a tab library given by the URI, and provides prefixes names (`ui` and `portlet`). The portlet tag library `<portlet:defineObjects/>` establishes `renderRequest`, `renderResponse` and `portletConfig` objects. The table width is defined as 500 pixels.

Source File. The Java source file causes the actions required when the user interacts with the portlet. The Java source code is shown in Figure 8.21. This program

```
package edu.uncc.abw.portlets;
import org.gridlab.gridsphere.provider.*;
import javax.servlet.UnavailableException;
import javax.portlet.*;
import java.io.IOException;
import java.io.*;

public class OddEven extends ActionPortlet {
   private static final String DISPLAY_PAGE = "MainPage.jsp";
   public void init(PortletConfig config) throws PortletException {
      super.init(config);
      DEFAULT_VIEW_PAGE = "prepare";
   }

   public void action(ActionFormEvent event) throws PortletException {
      TextFieldBean value1 = event.getTextFieldBean("valueTF1");
      TextBean answer = event.getTextBean("answer");
      int val = Integer.parseInt( value1.getValue() );

      if(value1.getValue() == null ){
         answer.setValue("");
      }
      else {
         if( isEven(val) ){
            answer.setValue("The number: "+value1.getValue()+" is Even");
         }else{
            answer.setValue("The number: "+value1.getValue()+" is Odd");
         }
      }
      setNextState(event.getActionRequest(), DISPLAY_PAGE);
   }

   public void prepare(RenderFormEvent event) throws PortletException {
      setNextState(event.getRenderRequest(), DISPLAY_PAGE);
   }

   public boolean isEven(int val){
      return val % 2 == 0;
   }
}
```

Figure 8.21 Java source code for portlet, `OddEven.java`.

calls the method `init()` to initialize the JSR 168 portlet, and provides two methods for handling user requests, `action()` and `prepare()`. The method `action()` performs the required computational actions upon a request and creates data for the HTML text box. The method `prepare()` renders the portal. The method `isEven()` is a helper method for `action()`. The HTML/JSP file is identified with `DISPLAY_PAGE` and used in the rendering routine.

Deployment Descriptor Files. There are three deployment descriptor files:

- `portlet.xml` — JSR 168 standard, describing the portlet
- `layout.xml` — GridSphere file describing layout of portal window
- `group.xml` — GridSphere file describing a collection of portlets

These files are very simple XML files that do not require namespaces and can be easily deciphered, see next. There are other deployment files that are generated automatically during deployment but the three listed are the only ones that the user would alter.

The `portlet.xml` file is shown in Figure 8.22 and identifies the Java program class file, in this case `edu.uncc.abw.portlet.Oddeven.java.class`. This class file will be found in the build directory once the Java source program is compiled and the portlet is deployed.

```
<?xml version="1.0" encoding="UTF-8"?>
<portlet-app
  xmlns="http://java.sun.com/xml/ns/portlet/portlet-app_1_0.xsd"
  version="1.0"
  xmlns:xsi="http://www.w3.org/2001/XMLSchema-instance"
  xsi:schemaLocation=
    "http://java.sun.com/xml/ns/portlet/portlet-app_1_0.xsd">
  <portlet>
    <description xml:lang="en">Odd Even Portlet</description>
    <portlet-name>OddEven</portlet-name>
    <display-name xml:lang="en">Odd Even Portlet</display-name>
    <portlet-class>edu.uncc.abw.portlets.OddEven</portlet-class>
    <expiration-cache>60</expiration-cache>
    <supports>
      <mime-type>text/html</mime-type>
      <portlet-mode>edit</portlet-mode>
      <portlet-mode>help</portlet-mode>
    </supports>
    <supported-locale>en</supported-locale>
    <portlet-info>
      <title>Odd Even</title>
      <short-title>Odd Even</short-title>
      <keywords>odd even</keywords>
    </portlet-info>
  </portlet>
</portlet-app>
```


Figure 8.22 `portlet.xml`.

The layout file, `layout.xml` is shown in Figure 8.23 and specifies the portlet tab layout and also needs the path to the Java program class file. The layout is described in terms of a table with a specified number of columns and rows in the same manner as HTML pages. This particular layout has one row and one column, that is, there is a single portlet within the portal window. Multiple portlets could be constructed within a portal window, each in one cell of a specified layout table and each with its own associated portlet files. Note that the HTML layout required for each portlet is defined elsewhere in the JSP page associated with the portlet.

The group file, `group.xml`, is shown in Figure 8.24 and specifies the group membership name and role for the portlet. Portlets are assigned a particular group for managing portlets.

```
<portlet-tabbed-pane>
  <portlet-tab label="Odd Even">
    <title lang="en">Odd Even</title>
    <portlet-tabbed-pane style="sub-menu">
      <portlet-tab label="oddeventab">
      <title lang="en">Odd Even</title>
        <table-layout>
          <row-layout>
            <column-layout>
              <portlet-frame label="Odd Even">
                <portlet-class>
                  edu.uncc.abw.portlets.OddEven
                </portlet-class>
              </portlet-frame>
            </column-layout>
          </row-layout>
        </table-layout>
      </portlet-tab>
    </portlet-tabbed-pane>
  </portlet-tab>
</portlet-tabbed-pane>
```


Figure 8.23 `layout.xml`.

```
<?xml version="1.0" encoding="UTF-8"?>
<portlet-group>
  <group-name>userPortlets</group-name>
  <group-description>The demo group</group-description>
  <group-visibility>PUBLIC</group-visibility>
  <portlet-role-info>
    <portlet-class>edu.uncc.abw.portlets.OddEven</portlet-class>
    <required-role>USER</required-role>
  </portlet-role-info>
</portlet-group>
```


Figure 8.24 `group.xml`.

Building and Configuring. Once all files are in place, the portlet is compiled and deployed with the provided Ant script. Then, Tomcat must be restarted and the portlet group membership configured using the GridSphere portlet group manager portlet. The new Odd Even portlet tab will appear and the portlet will be available for use (Figure 8.18) .

JSR 286 and WSRP. JSR 286 Portlets Specification v2.0 is an updated version of JSR 168 released in June 2008 after about five years of review and development. JSR 286 is (thankfully) backward compatible with JSR 168 portlets (SUN 2008). JSR 168 portlets can be deployed in JSR 286 portlet containers. JSR 286 incorporates inter-portlet communication, which was absent in JSR 168. During the same period of the JSR 286 development, OASIS was working on a standard for defining a Web service interface called *Web Services for Remote Portlets* (WSRP) for interacting with "presentation-oriented Web services." Version 1 was introduced in 2003 with WSDL for its interface description. WSRP Version 2 was introduced in 2008. JSR 286 includes an alignment with WSRP version 2.

8.4 SUMMARY

Within the focus of providing user-friendly interfaces, this chapter has two major topics—workflow editors with a graphical user interface and Web-based Grid portals. The GridNexus workflow was described in the section on workflow editors. With the Grid portal section, several relevant technologies were introduced for creating server-side dynamic content in Web pages, notably

- Java servlets
- Java server pages (JSP)
- Commodity Grid Kits
- JSR 168 Portlets

Then the GridSphere portlet toolkit was described in some detail including sample files that need to be created for a portlet, to lay the groundwork for possible work creating your own portlets within the Gridsphere/JSR 168/JSR 286 framework.

FURTHER READING

Much of the further reading material for this chapter can be found online through the home pages of the various technologies that are cited as references.

Published work on GridNexus include (Brown et al. 2005). Further information on using GridNexus can also be found from installing GridNexus and selecting help, which provides a number of tutorial examples.

An early description of CoG kit can be found in (von Laszewski et al. 2001). The CoG kit is also described in a chapter of the book *Making the Global Infrastructure a Reality* (von Laszewski 2003). Historical information on first generation portal toolkits can be found in chapters of the same book (Novotny 2003) and (Thomas and Boisseau 2003). Gridsphere, when installed, provides extensive materials in HTML format including user's and administrator's guides, portlet reference guide, and a tag library user's guide. Creating your own portlets within the Gridsphere framework is also explained in detail in (Villalobos, Land, and Wilkinson 2007).

BIBLIOGRAPHY

Abdelnur, A., and S. Hepper. 2003 Java™ Portlet specification version 1.0. Sun Microsystems, Inc. Oct. 7.

Alameda, J., M. Christie, G. Fox, J. Futrelle, D. Gannon, M. Hategan, G. von Laszewski, M. A. Nacar, M. Pierce, E. Roberts, C. Severance, and M. Thomas. 2007. The open Grid computing environments collaboration: Portlets and services for science gateways *Concurrency and Computation Practice and Experience* 19 (6):921–942.

Brooks, C., E. A. Lee, X. Liu, S. Neuendorffer, Y. Zhao, and H. Zheng. 2008. Heterogeneous concurrent modeling and design in Java (Volume 1: Introduction to Ptolemy II). EECS Department, University of California, Berkeley, Technical Report No. UCB/EECS-2008-28, April 1. http://www.eecs.berkeley.edu/Pubs/TechRpts/2008/EECS-2008-28.pdf

Brown, J. L., C. S. Ferner, T. C. Hudson, A. E. Stapleton, R. J. Vetter., T. Carland, A. Martin, J. Martin, A. Rawls, W. J. Shipman, and M. Wood. 2005. GridNexus: A Grid services scientific workflow system. *Int. Journal of Computer Information Science (IJCIS)* 6 (2): 72–82.

Ferner, C. 2007. Assignment 5 Using GridNexus to create workflows that use Web and Grid services. ITCS 4146/5146: Grid computing course (Spring 2007). http://www.cs.uncc.edu/~abw/ITCS4146S07/assign5S07.pdf.

Ferner, C. 2008. Assignment 3 GridNexus job submission. ITC 4146/5146: Grid Computing course (Fall 2008). http://www.cs.uncc.edu/~abw/ITCS4146F08/assign3F08.pdf.

Gridsphere portal framework. http://www.gridsphere.org/gridsphere/gridsphere

GridNexus UNC–Wilmington Grid computing project. http://www.gridnexus.org/

Grid Workflow Forum. http://www.gridworkflow.org/snips/gridworkflow/space/start

House, J., M. Holliday, and R. Ruff. 2005. Assignment 1: A generic Web service compiling, deploying, and modifying a simple Web service. UNC–C/UNC–W Grid computing course, Fall 2005. http://www.cs.uncc.edu/~abw/ITCS4010F05/index.html

Kepler Project. http://kepler-project.org/

Knuckles, C. D. 2001. *Introduction to interactive programming on the Internet using HTML and JavaScript.* New York: John Wiley & Sons.

Li, M. and M. Baker. 2005. *The Grid core technologies*. Chichester England: Wiley.

Novotny, J. 2003. The Grid protal development kit. In *Making the global infrastructure a reality*, ed. F. Berman, G. C. Fox, and A. J. G. Hey, Chap. 27, 657–673. Chichester England: Wiley.

Open Grid computing environments portal and gateway toolkit. http://www.collab-ogce.org/ogce/index.php/Main_Page

Sotomayor, B., and L. Childers. 2006. *Globus toolkit 4: Programming Java services*. San Francisco: Morgan Kaufmann.

SUN. JavaServer Pages Technology - Documentation. http://java.sun.com/products/jsp/docs.html

SUN. 2003. Introduction to JSR 168 – The Java portlet specification. http://developer.sun.com

SUN. 2008. Introducing Java portlet specifications: JSR 168 and JSR 286. Sun Developer Network. http://developers.sun.com/portalserver/reference/techart/jsr168/

Taverna project website http://taverna.sourceforge.net

TeraGrid. http://www.teragrid.org

Thomas, M. P., and J. R. Boisseau. 2003. Building Grid computing portals: The NPACI Grid portal toolkit. In *Making the global infrastructure a reality*, ed. F. Berman, G. C. Fox, and A. J. G. Hey, Chap. 28, 657–673. Chichester England: Wiley.

uPortal: Evolving portal implementation from participating universities and partners. http://uportal.org/

Villalobos, J. 2007. Creating Java CoG programs that connect Grid services. http://www.cs.uncc.edu/~abw/ITCS4146S07/Cog_assign.doc.

Villalobos, J., Land, J., and B. Wilkinson. 2007. Portlet development: creating a simple portlet. ITCS 4146 Grid Computing course (Spring 2007). http://www.cs.uncc.edu/~abw/ITCS4146S07/assign6aS07.doc.

von Laszewski, G., I. Foster, J. Gawor, and P. Lane. 2001. A Java commodity Grid kit. *Concurrency and Computation: Practice and Experience* 13 (8–9):643–662. http://www.mcs.anl.gov/~gregor/papers/vonLaszewski--cog-cpe-final.pdf.

von Laszewski, G., and M. Hategan. 2005. Java CoG kit workflow concepts. *Journal of Grid Computing* 3 (3–4):239–258. http://www.mcs.anl.gov/~gregor/papers/vonLaszewski-workflow-jgc.pdf.

von Laszewski, G., J. Gawor, S. Krishnan, and K. Jackson. 2003. Commodity Grid kits – middleware for building Grid computing environments in Grid computing. In *Making the global infrastructure a reality*, ed. F. Berman, G. C. Fox, and A. J. G. Hey, Chap. 26, 639–656. Chichester England: Wiley. http://www.mcs.anl.gov/~gregor/papers/vonLaszewski--grid2002book.pdf

Web Services for remote portlets specification v2.0. 2008. OASIS Standard. http://docs.oasis-open.org/wsrp/v2/wsrp-2.0-spec.html.

Wikipedia. ASP.NET. http://en.wikipedia.org/wiki/ASP.NET.

Wikipedia. Common gateway interface. http://en.wikipedia.org/wiki/Common_Gateway_Interface.

Wikipedia. Java servlets. http://en.wikipedia.org/wiki/Java_Servlet.

Wikipedia. PHP. http://en.wikipedia.org/wiki/PHP.

Yu, J., and R. Buyya. 2005. A taxonomy of workflow management systems for Grid computing. *Journal of Grid Computing* 3 (3–4):171–200.

SELF-ASSESSMENT QUESTIONS

The following questions are multiple choice. Unless otherwise noted, there is only one correct answer for each question.

1. What is GridNexus?

(a) A Grid portal
(b) A software package to create workflows

(c) A term used to describe an amorphous Grid
(d) A software package to run Grid jobs

2. What is JXPL?

(a) A job description language
(b) An XML language for JSR 168 portlets
(c) The XML language used in GridNexus to describe workflow
(d) A Java XML programming language

3. What is the purpose of the GridNexus *JxplDisplay* module?

(a) To display the workflow
(b) To compute the JXPL document for the workflow and optionally evaluate it
(c) To display the JXPL manual
(d) To display all the input for the workflow

4. What is the purpose of the GridNexus *WS Client* module?

(a) Provides a module for a local or remote Web Service
(b) Provides a module for a client that accesses a local or remote Web Service
(c) Provides a module for a Web service that accesses a client
(d) None of the other answers

5. How is the GridNexus *WS Client* configured?

(a) It self-configures
(b) The user provides a title or name
(c) The user provides the URL of the Web service
(d) The user provides the URL of the Web service WSDL document
(e) None of the other answers

6. What is the difference between the GridNexus *WS Client* and the GridNexus *WSRF Client*?

(a) Nothing significant
(b) WS client interfaces to a Web Service and WSRF client interfaces to a Grid service
(c) WSRF client is a backward-compatible improvement to the WS client
(d) WS client is a backward-compatible improvement to the WSRF client

7. What is the purpose of the GridNexus *GridExec* module?

(a) Submit a job to GRAM
(b) Create a Grid execution service
(c) Perform an `ssh` connection to a remote computer
(d) Execute a Grid service

8. What is the GridNexus *composite* module?

(a) A workflow with every module connecting to every other module
(b) A module composed of every type of module that exists in GridNexus
(c) A workflow considered as one unit with a set of inputs and a set of outputs
(d) A module that two modules with the output of the first module connecting to the input of the second module

9. What is a Grid portal?

(a) Software for moving information between Grids
(b) A database of Grid resources
(c) The interface between Grid services
(d) A Web-based application for a user to access Grid services and resources

10. What is meant by *dynamic content* in the context of the implementation of Web pages?

(a) A Web page that keeps changing at regular intervals
(b) A Web page that changes in response to changing information and client requests
(c) A Web page using dynamic memory
(d) A Web page that has to be refreshed at regular intervals

11. What is a Java servlet?

(a) A Java program (object) that handles requests from Web clients
(b) A Web service written in Java
(c) A small Java-based server
(d) A Java program that hands out small objects

12. What is a Java Server Page (JSP)?

(a) Java code that implements a Web service
(b) A SUN technology used in conjunction with servlets to create Web pages with dynamic content
(c) Java code that can implement a server
(d) An XML language used to describe Java server programs

13. What is Commodity Grid (CoG) kit?

(a) A toolkit to create a Grid platform for Grids on commodity cluster hardware
(b) A toolkit specifically for trading commodities on the Grid
(c) The lowest level software beneath Globus to access the Grid resources
(d) A toolkit providing higher-level interfaces to Grid components than basic Globus APIs

14. What is a portlet?

(a) A pull-down menu in a Grid portal
(b) A group of portals
(c) A small Grid portal
(d) A tabbed window within a portal
(e) A back-end component in a portal such as a Web service
(f) Software component used for an area within a portal providing a presentation-level interface associated with some functionality

15. In a JSR 168 portlet environment, suppose the file `layout.xml` describes a table with two rows and two columns. Describe its effect?

(a) It creates a layout for a portal window that has four cells arranged as two rows and two columns, one cell for each portlet.

(b) It creates a layout for a portlet that has four cells arranged as two rows and two columns.
(c) It creates a layout for a complete portal that has four cells arranged as two rows and two columns. The lower left cell can hold selectable menu items. The upper left cell can hold a logo.
(d) It defines a user input layout.

16. What is one difference between JSR 168 and JSR 286?

(a) Nothing. They are the same.
(b) 120,000 new APIs.
(c) JSR 286 includes facilities for inter-portlet communication.
(d) They are completely different standards with nothing in common.

PROGRAMMING ASSIGNMENTS

GridNexus — Some of these assignments are closely based upon (Ferner 2008). More details can be found at this reference. The first step is to install GridNexus. GridNexus can be installed on a personal computer (Windows, Mac, or Linux). Instructions on installing the software is provided at the GridNexus site (http://www.gridnexus.org). GridNexus is Java-based and requires Java 1.5 or later.

8-1 Create an arithmetic workflow in GridNexus to compute the following:

$$f(A, B, C, D) = A*B + C/A$$

where $A = 2$, $B = 4$, and $C = -9$. Examine the JXPL output and confirm the output is correct.

Preliminary Setup To Access Remote Resources. Assignments that access remote Grid resources or remote services require the appropriate accounts on the remote resources and credentials set up on the local GridNexus installation. The specifics will depend upon the installation. Briefly, for GridNexus installed on a local computer running Windows to access a remote Linux server:

1. ***Certificate Authority files.*** Locate the directory `.globus\certificates` in `C:\Documents and Settings\<USERNAME>` on the local computer. If it does not exist, create it. Obtain the certificate authority certificate file and its signing policy file. They are usually located on the server at `/etc/gridsecurity/certificates`. The two files will have names `<cert_hash>.0` and `<cert_hash>.signing_policy`, where `<cert_hash>` is the hash code for the certificate authority (an 8-digit hexadecimal number). Store the files in `.globus\certificates`. Files from multiple certificate authorities may be stored locally if you want GridNexus to handle certificates signed by multiple certificate authorities.

2. **`hosts`** ***file.*** Locate the `hosts` file on the computer you have installed GridNexus. It will be found at `C:\WINDOWS\system32\drivers\etc\hosts` on a Windows

system. Add IP address - server name mappings for the remote computers you will be using. For example:

152.15.98.24 coit-grid01.uncc.edu
152.15.98.25 coit-grid02.uncc.edu
152.15.98.26 coit-grid03.uncc.edu
152.15.98.27 coit-grid04.uncc.edu

3. ***Obtain Proxy.*** This step will depend upon where your certificate is being held, from which the proxy is created. If your certificate is on a remote computer with Globus installed, log onto the remote computer and obtain a new proxy by executing the command `grid-proxy-init`. The proxy will be stored on the remote computer at `/tmp/x509up_u<user_ID>`. The full path and name of the file can be found from the command `grid-proxy-info`. (If the MyProxy server is being used, one would use a MyProxy command to obtain the proxy, see Chapter 5, Section 5.3). Transfer the file from the remote server to the local computer at `C:\Documents and Settings\<USERNAME>\.globus\x509up_<USERNAME>`. Confirm that the file `cog.properties` exists in the `.globus` directory and it contains the full path of your proxy, i.e.: `proxy=/Documents and Settings/<USERNAME>/.globus/x509up_<USERNAME>`. Correct as necessary.

Assignments Executing Remote Jobs:

8-2 Create the workflow illustrated in Figure 8.3 to execute the program `/bin/echo` on the remote computer with the arguments `hello world`. Confirm JXPL output is correct, which will indicate where the output has been sent. The actual echo output will be at `$GLOBUS_USER_HOME/stdout` on the remote computer. Log onto the remote computer and verify the output is correct.

8-3 Write a Java program to implement the functionality of `/bin/echo` and store the class file on the remote computer. Establish the program works remotely. Create the workflow to execute this Java program through GridNexus (see Figure 8.4)

8-4 Continuing from Assignment 8-3, add to the workflow so that the output will be re-directed back to the local computer (see Figure 8.5).

8-5 Repeat Assignments 8-3 and 8-4 but by writing an RSL file to effect the execution and file staging and using this RSL file in the workflow as input to GridExec.

GridSphere — The following assignments require GridSphere to be installed. This can be done on a personal computer (Windows, Mac, or Linux). Instructions on installing the software are provided in the referenced links.

8-6 Follow the instructions given for the assignment at http://www.cs.uncc.edu/~abw/ITCS4146S07/assign6aS07.doc to install GridSphere and deploy the simple odd-even portlet.

8-7 Deploy a portlet that allows the user to perform add, subtract, multiply, and divide operations on two numbers entered by the user. The portlet may have any reasonable layout. Figure 8.25 shows one sample layout created by Ramya Chaganti, a Computer Science MS student at UNC–Charlotte in 2008.

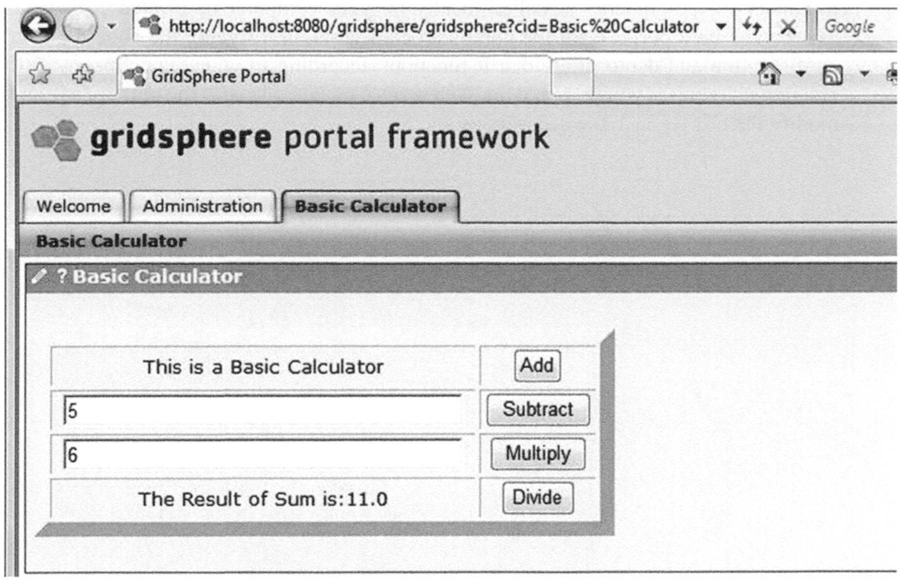

Figure 8.25 One layout for a calculator portlet.

8-8 Follow the instructions given for the assignment at http://www.cs.uncc.edu/~abw/ITCS4146S07/Cog_assign.doc to install the CoG kit libraries, test the installation and compile and run the sample code to submit jobs and transfer files.

8-9 This assignment brings together knowledge of Web service deployment and portlet design. Create a portlet interface for a Web service. The specific design is not specified but it must have the following components:

- A portlet that is hosted in Gridsphere and provides the user interface
- A local Java client program
- A Web service

as illustrated in Figure 8.26. The Web service may be a WSRF service such as provided

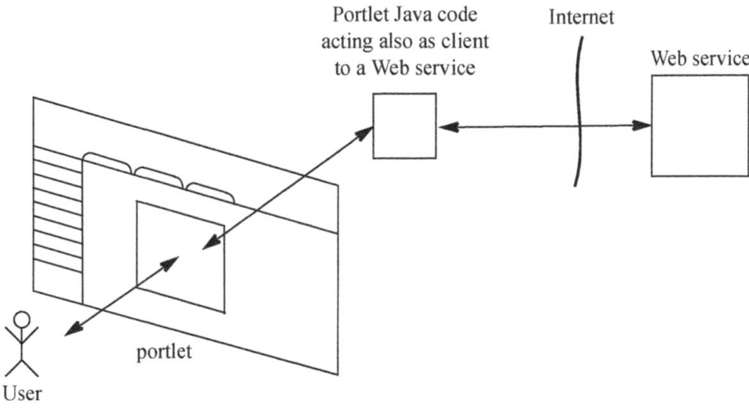

Figure 8.26 Portlet acting as a front-end to a Web service (Assignment 8-9).

in Chapter 7 assignments, installed locally on your computer or located elsewhere. There must be a purpose for the combined configuration. Develop the code for your configuration and demonstrate that it functions according to your requirements. This assignment requires you to merge the Web service client code with the portlet code and modify the `build.xml` file accordingly.

CHAPTER 9
Grid-Enabling Applications

In previous chapters, we described how to execute jobs through a command-line interface and graphical user interfaces (GUIs), but for the most part, the actual structure of the jobs submitted have been assumed to be ready for the Grid platform. In this chapter, we explore which jobs and applications are suitable for a Grid platform, how to modify them if they are not, and how to execute them on a Grid platform.

9.1 INTRODUCTION

9.1.1 Definition of Grid Enabling

The term *Grid enabling* an application is a poorly defined and understood term. It does *not* mean simply executing a job on a Grid platform. Many computer programs designed for a single computer can be shipped to a remote Grid site and executed in little more than was possible after making an `ssh` connection to a remote computer. Grid enabling implies that something special had to be done to make the program run on the Grid platform, i.e., it was enabled in some way to utilize distributed nature of the Grid platform. With that in mind, a simple definition of Grid enabling an application is:

> *Being able to create an application that uses the distributed resources available on a Grid platform more effectively than simply executing the application on a single, local or remote resource.*

A more detailed definition that matches our view of Grid-enabling applications is (Nolan 2008):

> *Grid Enabling refers to the adaptation or development of a program to provide the capability of interfacing with a grid middleware in order to schedule and utilize resources from a dynamic and distributed pool of "grid resources" in a manner that effectively meets the program's needs.*

How one Grid enables an application is still in the research domain without a standard approach. Here, various potential approaches will be described. A particular type of application might require a different approach so let us first briefly outline the applications that might suit a Grid platform.

9.1.2 Types of Jobs to Grid Enable

Clearly, one can run simple batch jobs on a Grid resource using commands such as Globus command `globusrun-ws` as described in Chapters 2 and 3. But all this would do is to cause a program executable to be executed on a remote or local Grid machine. In the presence of a job scheduler, the scheduler would schedule the job to be executed. Although this way is perhaps the most common way of using a Grid platform, it does not make most use of the possibilities that exist in a distributed computing platform nor the potential collaborative nature of computing that Grid computing can offer. A distributed computing platform such as a Grid platform offers the possibility of executing multiple programs. The programs can be executed on separate computers in the Grid with the potential of reduced time of overall execution time.

Multiple computers can be employed in two ways:

- Multiple computers can be used separately to solve individual problems
- Multiple computers can be used collectively to solve a single problem

We shall look at each of the above areas in more detail in subsequent sections. In Section 9.2, we will consider *parameter sweep*, which comes under using multiple computers separately to solve individual problems, but as a special case where the individual problems are instances of the same problem with different inputs. In Section 9.3, we will look at various ways of porting an existing program on multiple Grid computers without re-writing the application to any great extent. This section includes exposing an application as a service, which also can be a framework for creating a new distributed application. Hence, it can be seen as both a technique for making a single application accessible of a Grid platform and also as a technique for using multiple Grid computers to solve a single problem. In Section 9.4, we look at writing an application specifically for a Grid using Grid middleware and higher-level middleware independent APIs. In Section 9.5, we introduce applying multiple Grid computers to solve a single problem using parallel programming techniques, specifically using message-passing libraries.

9.2 PARAMETER SWEEP

9.2.1 Parameter Sweep Applications

It turns out that in some domain areas, scientists need to run the same program many times but with different input data. Such problems are known as the *parameter sweep applications (PSAs)*. The scientist attempts to "sweep" across a parameter space with different values of input parameter values in search of a solution. There are several reasons for doing a parameter sweep. There may be no closed mathematical solution and a trial-and-error approach is performed. Human intervention may be involved in a search or design space. Examples appear in many areas of science and engineering. For example, a scientist might wish to search for a new drug and needs to try different formulations that might best fit with a particular protein. An aircraft design engineer might be studying the effects of different aerodynamic designs on the performance of an aircraft. Sometimes, it is aesthetic design process and there are many possible alternative designs and a human has to choose. Sometimes, it is a learning process in which the design engineer wishes to understand the effects of changing various parameters. Typically, there will be many parameters that can be altered and there might be a vast combination of parameter values. Ideally, some automated way of doing parameter sweep is needed one that includes both specifying the parameter sweep and a way of scheduling the individual sweeps across the Grid platform.

9.2.2 Implementing Parameter Sweep

Parameter sweep can be simply achieved by submitting multiple job description files, one for each set of parameters, but that is not very efficient. Parameter sweep applications are so important that research projects on Grid tools have been devoted to making them efficient on a Grid and providing for them in job description languages. It appears explicitly in the RSL-1/RSL-2/JDD job description languages described in Chapter 2, Section 2.2.2, with the count element. For example, inserting

```
<count> 5 </count>
```

in the RSL-2/JDD job description file would cause five instances of the job to be submitted. This in itself would simply cause five identical executables to be submitted as jobs. Four would be pointless unless either the code itself selected different actions for each instance, or different input and output files were selected for each instance in the job description file. Macro-substitution can be used with the job description language to effect changes to the job description file.

JSDL version 1 does not have a specific provision for parameter sweep, but it has been extended unofficially to incorporate features for parameter sweep. The JSDL 1.0 specification allows for extensions although such extensions limit the interoperability. The extensions to JSDL 1.0 were discussed at OGF 19 in 2007 and received further consideration at several subsequent OGF meetings. Two forms of parameter sweep creation have been identified—enumeration in a list and numerically related arguments.

Arguments Enumerated in a List. The approach here is to have additional JSDL elements to select from existing job description elements and replace the body of the selected elements with values from a list. This requires two additional elements:

- `Parameter` — to select the element
- `Value` — to list the values

These new elements are contained within an `Assignment` element for each assignment. Multiple and nested assignments can be made to satisfy various scenarios, that is making multiple simultaneous substitutions in different combinations. The approach is shown in Figure 9.1. Each instance of the parameters sweep would select the next value from a list. Although shown modifying the `Argument` element, it could modify any element including that specifying the executable (which would not be true parameter sweep). The final result would in essence be a JSDL job description file for each instance of the job with the substitution as illustrated in Figure 9.2.

Selection Expression. Figure 9.1 shows an expression that selects an XML element. The selection can be done using the XML Path language (XPath). The XPath language provides a way to select an XML element in an XML document. XPath itself is *not* an XML language. It was mentioned in Chapter 7, Section 7.3.6, for selecting an XML element from an XML document associated with an index service but not fully explained. Briefly, suppose the XML document has the form

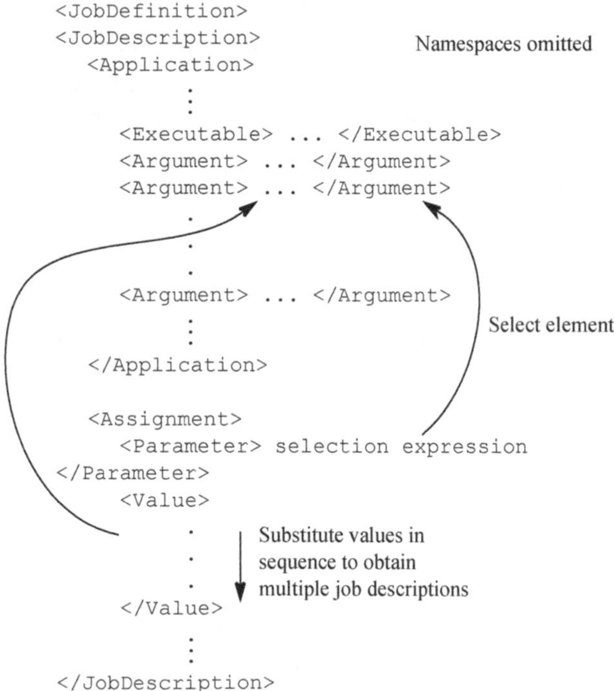

Figure 9.1 Parameter sweep element selection and substitution.

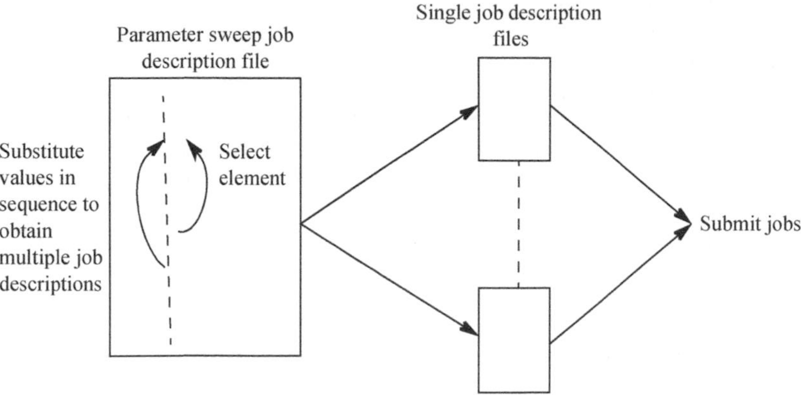

Figure 9.2 Parameter sweep element selection and substitution.

```
<a>
      <b>
            <c>
            </c>
      </b>
</a>
```

the XPath expression to identify the c element (i.e., `<c> ... </c>`) would be /a/b/c. This is the path to the element (hence the name XPath). The construction here is the same as a path to a file held in a subdirectory. XPath allows for much more expressive forms to select an element. For example suppose there are multiple tags called `<c>` as in

```
<a>
      <b>
            <c> ... </c>
            <c> ... </c>
            <c> ... </c>
                 ⋮
            <c> ... </c>
      </b>
</a>
```

The XPath expression to select the third c element would be /a/b/c[3]. Leading double forward slashes (//) can be used to indicate an element anywhere in the document. Therefore /a/b/c[3] could be reduced to //c[3] if it is known that the only c elements are within the b element, but this form of XPath expression has to be used with care as it actually means all third c elements of their parents anywhere in the document.

To take an example for a parameter sweep, consider the JSDL job

```
<jsdl:JobDefinition>
<jsdl:JobDescription>
<jsdl:Application>
<jsdl-posix:POSIXApplication>
  <jsdl-posix:Executable>/bin/echo</jsdl-posix:Executable>
  <jsdl-posix:Argument>Hello</jsdl-posix:Argument>
  <jsdl-posix:Argument>Fred</jsdl-posix:Argument>
</jsdl-posix:POSIXApplication>
</jsdl:Application>
</jsdl:JobDescription>
</jsdl:JobDefinition>
```

To alter the second argument to be Bob, Alice, and Tom (three sweeps), the job description might be

```
<jsdl:JobDefinition>
<jsdl:JobDescription>
<jsdl:Application>
<jsdl-posix:POSIXApplication>
  <jsdl-posix:Executable>/bin/echo</jsdl-posix:Executable>
    <jsdl-posix:Argument>Hello</jsdl-posix:Argument>
    <jsdl-posix:Argument>Fred</jsdl-posix:Argument>
</jsdl-posix:POSIXApplication>
</jsdl:Application>
</jsdl:JobDescription>
<sweep:Sweep>
  <sweep:Assignment>
    <sweep:Parameter>
      //jsdl-posix:Argument[2]
    </sweep:Parameter>
    <sweepfunc:Values>
      <sweepfunc:Value>Bob</sweepfunc:Value>
      <sweepfunc:Value>Alice</sweepfunc:Value>
      <sweepfunc:Value>Tom</sweepfunc:Value>
    </sweepfunc:Values>
  </sweep:Assignment>
</sweep:Sweep>
</jsdl:JobDefinition>
```

The XPath expression `//jsdl-posix:Argument[2]` selects the second `jsdl-posix:Argument` element. The output from the `echo` program during the sweep would be:

```
Hello Bob
Hello Alice
Hello Tom
```

(or possibly in different orders depending upon how the jobs are scheduled on the machines that run the jobs).

Multiple assignment elements can be used to alter more than one argument in each sweep. The `Assignment` elements can also be nested to create combinations of arguments.

Numerically Related Arguments. Job description languages such as JSDL can be extended to increment an integer argument automatically with a `for`-like construct. The `for` construct would specify the values of an argument, which would substitute in a similar fashion to the previous substitutions—essentially a macro-substitution. This approach is used in XPML (eXtensible Parametric Modeling Language) for the Gridbus resource broker (The GridBus Project). First, a `parameter` element is used in the XPML job description document to specify the argument values, for example

```
<parameter name="arg1" type="integer" domain="range">
    <range from="1" to="99" type="step" interval="2"/>
</parameter>
```

where the argument has the name `arg1`. The values for `arg1` in this particular case are 1, 3, 5, ... 9. The argument `arg1` would occur later within the `execute` element, for example

```
<execute>
    <command value=" ... "/>
    <arg value="$arg1"/>
    ...
</execute>
```

One value of `arg1` substitutes for each sweep. DRMAA also has facilities to do bulk job submission.

Coupled with actually specifying sweep of parameter in a job description document, the jobs need to be efficiently scheduled across Grid resources. Ideally, they are run on different compute resources at the same time, otherwise we would not get the speed-up advantage. Parameter sweep might be regarded as the simplest way of using multiple Grid computers. But it may not be applicable. Now let us move on to increasingly more complex ways that multiple computers can be used in different situations.

9.3 USING AN EXISTING PROGRAM ON MULTIPLE GRID COMPUTERS

9.3.1 Data Partitioning

In some problems, the data needed for solving a problem can be divided into parts and processed separately to solve the problem. This is known as *data partitioning*. Once the data is divided, individual computers can be used with each part. There are problems that particularly lend themselves for this approach, especially those problems that search large databases. For example, BLAST (Basic Local Alignment

Search Tool) is applied to very large databases in bioinformatics to find statistical matches between gene sequences. A BLAST user might submit a sequence query that is then compared to a very large database of known sequences in order to discover relationships or to match the sequence to a gene family.

If there is just one sequence from the user, the database might be partitioned into parts and different computers work on different parts of the database with the same sequence as illustrated in Figure 9.3 (Bangalore 2007). Of course, this assumes that the database can be easily split into parts. There will be a significant cost in sending large database partitions to different sites but once done, they can be reused. There will be communications at the edges of two databases to match the sequence across the partitions if the database partitions do not overlap.

An alternative approach if the user or users are submitting many queries is to submit each query to a different computer holding or having access to the whole database as illustrated in Figure 9.4 (Bangalore 2007). This approach is in essence a form of parameter sweep that we described earlier, where each input sequence is one

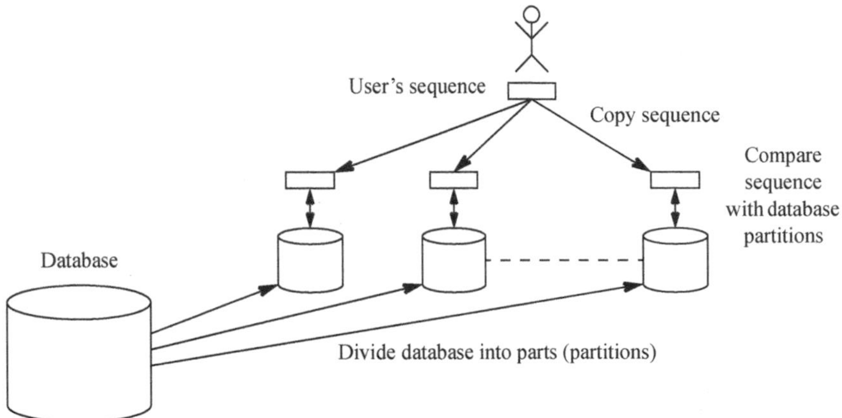

Figure 9.3 Partitioning BLAST database.

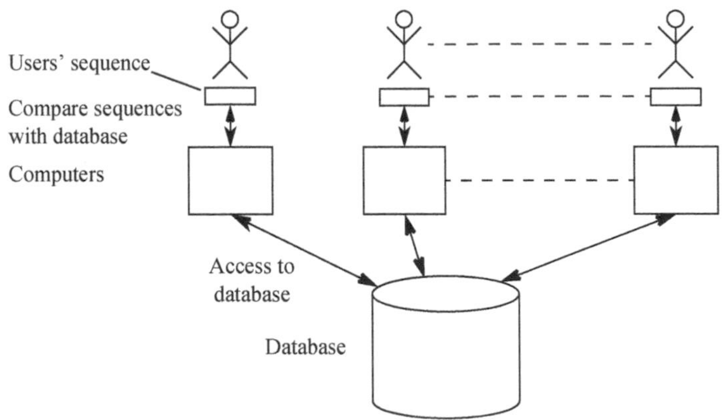

Figure 9.4 Multiple queries against a single BLAST database.

sweep. In a distributed system, significant network communications to access the database would result unless the whole database is already at each site.

The two simple approaches described for speeding up a BLAST search on a multiprocessor or distributed system do not require the BLAST application to be re-written or altered.

9.3.2 Deploying Legacy Code

A dictionary definition of legacy is "anything handed down from, or as from, ancestor".[1] In that vein, *legacy programs* refer to existing programs that are written and designed for previous computer systems but are still being used. Often they are written in traditional established programming languages such as C, C++, or Fortran, for a single computer and have dependencies with the target computer platform and operating system, i.e., they are not immediately portable to newer computer systems. Legacy code may be very old and use old programming practices. The source code, if available, may not be supported. However using legacy code may be attractive from a user's perspective, to continue with proven programs that they are already familiar with in their work. In fact, it may actually be necessary when newer programs do not exist.

Running legacy code on any platform that it was not designed for is a challenge; running legacy code of a Grid is even more so. First one has to establish that it can run on the particular Grid resource or the code can be modified so that it will. Grid resources are often heterogeneous. The best one could hope for such programs is to have binaries that match the remote compute resource and its operating system. With documented source code, re-writing is a possibility. Some applications may not be available in documented source code and may be pre-packaged by the manufacturer so rewriting may not be an option.

One project that addresses porting legacy code onto a Grid is the *Grid Enabling Legacy Software* (GriddLeS) project. That project focuses on the file handling and overloads existing file handling routines to redirect requests to remote locations if necessary.

9.3.3 Exposing an Application as a Service

Grid computing has embraced Web service technology for its implementation and it is natural to consider Web service technology for accessing applications on a Grid platform. An approach is to wrap the application code to produce a Web service, as illustrated in Figure 9.5. Wrapping means that application is not accessed directly but through the service interface. The Web service can be accessed through its URL from anywhere. An example of providing a service interface for the BLAST application is described by Bangalore and Afgan (2008). Another example is the toolkit for deploying applications are Grid services described by Guan, Velusamy, and Bangalore (2005).

---

[1] *Webster's New World College Dictionary*, 3rd ed., s.v. "Legacy, 2."

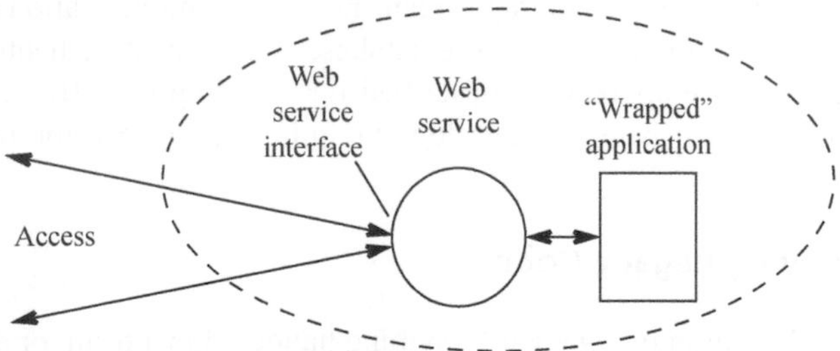

Figure 9.5 Web service wrapper approach.

The Web service could invoke the program, provide it with inputs and receive the results. If the Web service is written in Java, the service could issue a command in a separate process using the `exec` method of the current `Runtime` object with a construction

```
Runtime runtime = Runtime.getRuntime();
Process process = runtime.exec("<command>" )
```

where `<command>` is the command to issue, capturing output with

```
OutputStream stdout = process.getOutputStream();
    ...
```

This approach is explored in assignments at the end of the chapter.

The Web service approach described so far only provides an interface for an application. The application still has to be capable of executing fully within the particular environment. All input files must to be available. Input and output staging may be needed. The wrapped application might be accessed through a customized portlet in a portal as illustrated in Figure 9.6, which extends the concepts introduced in Chapter 8.

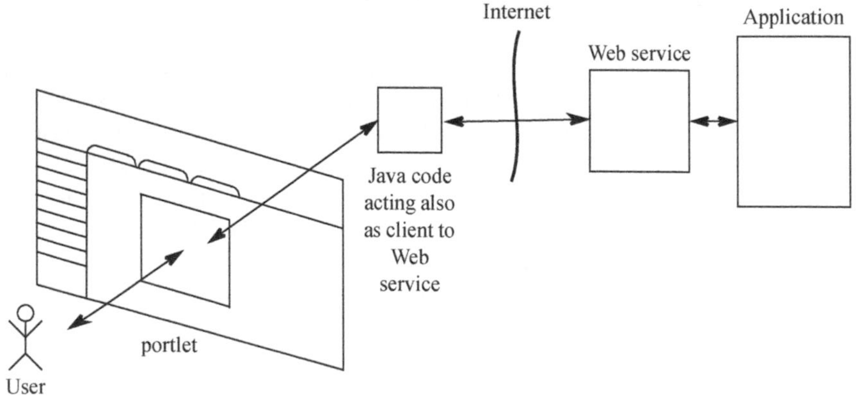

Figure 9.6 Portlet acting as a front-end to a wrapped application.

The approach could be extended to have multiple interacting services that make up a single application or multiple cooperating applications in a workflow. In keeping with the primary goal of Grid computing of providing a geographically distributed collaborative computing environment, some Grid applications are also physically distributed and fit well with the use of distributed services—for example, an application that uses remote sensors or equipment, or one that uses distributed databases. An application might use a specialized shared resource such as a shared hardware accelerator that is at one location. A hardware accelerator typically improves the speed to perform particular algorithms by orders of magnitude. They appear in specific domain area, such as BLAST accelerators in bioinformatics. One approach for distributed components is to use Web services as front-ends for both remote resources such as hardware accelerators and software components of remote applications as illustrated in Figure 9.7.

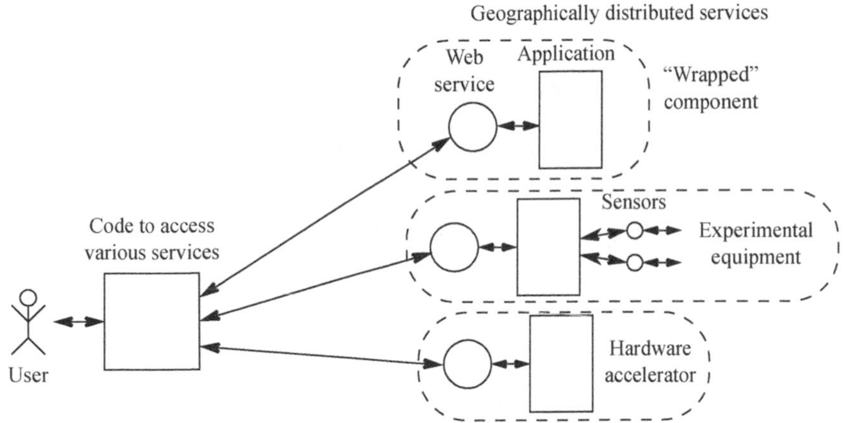

Figure 9.7 Application with physically distributed components.

9.4 WRITING AN APPLICATION SPECIFICALLY FOR A GRID

An application can be written specially to make use of the facilities that a Grid platform has to offer completely from the beginning. This implies initiating actions such as transferring files between locations as needed, locating suitable resources for the job, and starting new jobs from within an application. This approach is the most difficult of all and especially so if one uses Grid middleware APIs, which change from one version of the software to another without any universally agreed standards. Some progress has been made in raising the level and attempting to develop standards and higher-level APIs that everyone might use, but the state of this work has a lot to be desired.

9.4.1 Using Grid Middleware APIs

A Grid-enabling approach is to incorporate Grid middleware APIs in the application code for external operations such as file input/output. More advanced activities might take advantage of Grid middleware operations such as starting and monitoring jobs, and monitoring and discovery of Grid resources.

Using Globus APIs. Globus does provide APIs for invoking Globus services and routines, but there is an extremely steep learning curve for the user. There are literally hundreds, if not thousands, of C and Java routines listed at the Globus site. Kaiser (2004, 2005) and van Nieuwpoort (van Nieuwpoort) provide code to C++ copy a file using Globus APIs—*over 50 lines of code* and this is needed just to copy a file from one location to another.

Using CoG kit APIs. CoG kit was introduced in Chapter 8, Section 8.3.2, in the context of implementing portals. As mentioned, CoG kit is designed to raise the level of APIs to higher than Globus APIs. It is used in Globus. Figure 8.13 (page 240) shows the code to transfer files using CoG kit routines, which is much less complex than using Globus APIs but still requires setting up the Globus context.

9.4.2 Higher-Level Middleware-Independent APIs

Clearly it is desirable to have a higher level of abstraction than using middleware APIs not only because of the complexity of these routines, but also because Grid middleware changes very often. Globus typically is revised every couple of months and has major changes every year or two. Sometimes, very significant changes occur. Globus version 3 was abandoned after only about a year. There are significant compatibility issues between Globus version 4.0 and Globus version 4.2. It is extremely difficult to keep up with the changes. In addition, Globus is not the only Grid middleware. We have focussed on Globus but there are efforts around the world to provide Grid middleware, including UNICORE (Uniform Interface to Computing Resources), which started in Germany about the same time as Globus and continues to be developed, and gLite (Lightweight Middleware for Grid computing), which is part of the EGEE (Enabling Grids for E-sciencE) collaborative. To give an indication of the rapid changes that occur, gLite 3.0.2 Update 43 was released May 22, 2008. gLite 3.1 Update 27 was released July 3, 2008, six weeks later.

The concept of higher-level APIs above the Grid middleware is illustrated in Figure 9.8. These higher-level APIs should expose a simple interface that is not tied to specific version of Grid middleware or even Grid middleware family at all. The *Grid Application Toolkit* (GAT) developed in the 2003–2005 period (Kaiser 2004, 2005) followed this approach. An example of GAT code to copy a file can be found at (Kaiser 2005). Subsequently, an effort was made by the Grid community to standardize these APIs leading to the *Simple API for Grid Applications* (SAGA) project. An example of SAGA code to copy a file can be found at (Kielmann 2006), requiring just a few lines of regular C++ code. In this code, there is no mention of the underlying Grid middleware and the code is applicable to any such middleware.

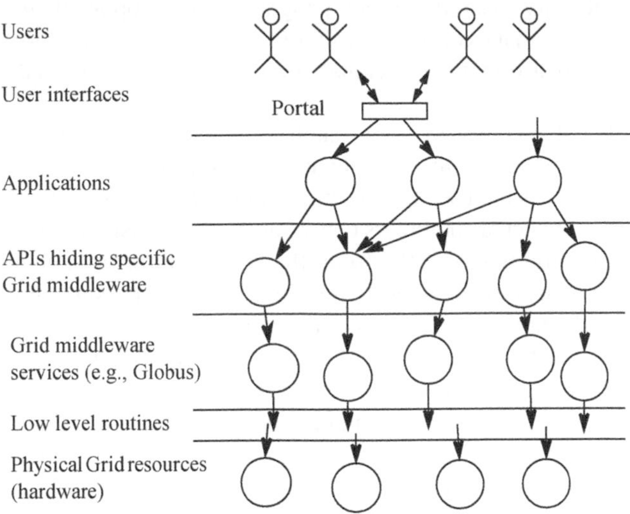

Figure 9.8 Layered approach to Grid middleware.

9.5 Using Multiple Grid Computers to Solve a Single Problem

9.5.1 Parallel Programming

Grid computing offers the possibility of using multiple computers on a single problem to decrease the time it takes to solve the problem. The potential of using multiple computers collectively has long been known (since the 1950s) and it is an obvious way of getting increased speed of execution. Suppose a problem can be divided into p equal parts, with each part executing on a separate computer and at the same time. Then, the overall time to execute the problem using p computers simultaneously would be $1/p$th of the time on a single computer. Of course, this is the ideal situation. Problems cannot usually be divided into equal parts that can be executed simultaneously and independently, and there are interprocess communication costs. But there is potential of significant increase of speed, and it is a basis of high performance computing. The term *parallel programming* is used to describe constructing programs or program sequences that operate together and collectively on separate computing resources to solve a problem and is also a very old term (1950s).

There are two general architectural approaches for constructing a system with multiple computers or processors:

- Shared memory system
- Distributed memory system

Shared memory systems, as the name suggests, has a common memory accessible by all processors. The processors would use the shared memory for data that each can access and for passing information between them. The most common shared memory programming technique is using threads either directly or through higher-level languages and libraries. Shared memory programming is applicable to multiprocessor computer systems that have physical shared memory such as multi-core computer systems. In a Grid platform, the individual computers might use shared memory programming.

In distributed memory systems, each computer has its own memory. The computer systems are physically separate and interconnected through a communication medium. Messages are used to exchange information between the computers. A cluster of locally connected computers comes under this classification. The programming model used follows directly from the architectural arrangement of using messages, i.e., message-passing programming. Message-passing programming is applicable to Grid computing platform, although there are significant technical and performance issues when ported onto a Grid, which we shall discuss later. The full treatment of message-passing programming is outside the scope of this book, but we will introduce the essential concepts and then describe how they might be applied to a Grid computing platform. A Grid computing platform has some superficial similarities to a local cluster but has some significant differences.

9.5.2 Message-Passing Approach

Message passing programming is generally done using library routines. The programmer explicitly inserts these routines into their code as illustrated in Figure 9.9. The separate programs on each processor operate as communicating processes. First, it is necessary to have a suitable message-passing software environment and libraries.

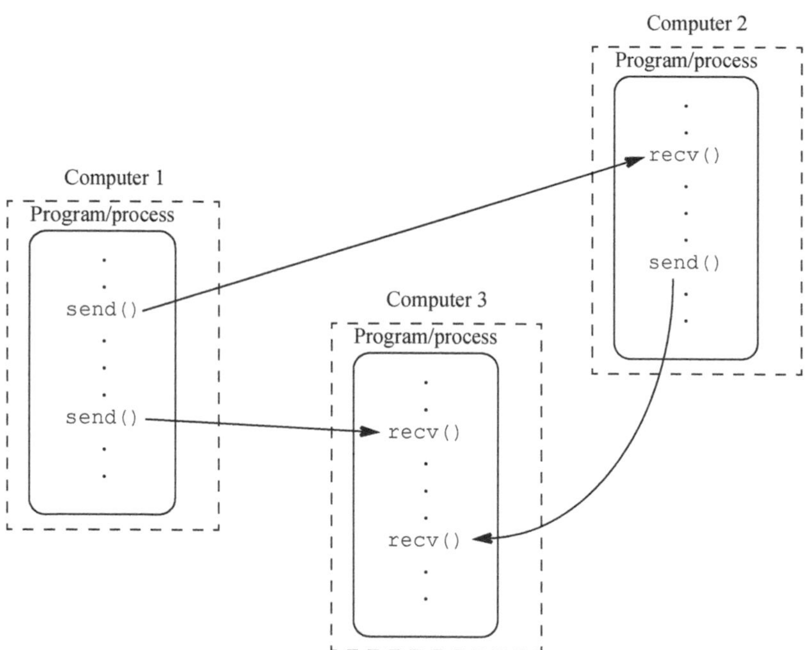

Figure 9.9 Message passing concept using library routines.

Perhaps the first highly successful message passing suite of libraries is PVM (Parallel Virtual Machine) developed at Oak Ridge National Laboratories in the late 1980s by Sunderam (Sunderam 1990). PVM became widely used in the early 1990s. It provided a fully implemented set of library routines for message passing.

Subsequently, a standard specification for message-passing APIs was developed called MPI (Message Passing Interface), which supplanted PVM in the mid-1990s. MPI was developed by the MPI Forum, a collective with more than 40 industrial and research organizations. It has been very widely accepted. Whereas PVM was both a specification of APIs and its implementation, MPI only specified the APIs (routine names, arguments, required actions, and return value) in C, C++, and Fortran. As with all standards, the actual implementation was left to others. The first version of the MPI standard was finalized in 1994. Version 2 of MPI was introduced in 1997 with a greatly expanded scope, most notably to offer new operations that would be efficient, such as one-sided message passing. MPI version 1 has about 126 routines and MPI version 2 increased the number of routines to about 156 routines. MPI-2 did depreciate some MPI-1 routines but even so, the standard is daunting. However, in most applications only a few MPI routines are actually needed. It has been suggested that many MPI programs can be written using only about six different routines.

There are several widely available implementations of MPI. MPICH, which started a collaborative project between Argonne National Laboratory and Mississippi State University, is one of the most widely used implementation of MPI. Another implementation of MPI called LAM-MPI has also been used quite widely although it is not being developed further. Not all early implementations of MPI version 2 supported every feature in the MPI version 2 standard. MPICH-2, a complete redesign of the original MPICH-1 has full support for the MPI version 2 standard. The open MPI project (OpenMPI), which started in 2004, has about 26 participating companies and research institutions and provides an open source implementation of MPI version 2.

9.5.3 Message-Passing Interface (MPI)

In the following, we will describe MPI but only in sufficient detail to explain important characteristics and factors for Grid enabling. More details can be found at (Wilkinson and Allen 2005).

Programming Model. In MPI version 1, the basic programming model is that each process executes the same code. The user writes a single program that is compiled to suit the processors and all processes are started together. This does not mean that all processors necessarily do exactly the same actions. Control statements can make individual processes do different things, as we shall see shortly.

MPI introduces the concept of a communication domain or context it calls a *communicator* (a non-obvious name). A communicator defines the scope of a communication and initially all processes are enrolled in a communicator called `MPI_COMM_WORLD`. The name of the communicator appears in all MPI message passing routines. The communicator concept enables one to constrain message-passing to be between agreed processes. It also enables library routines written to be others to be incorporated into a program without conflicts in the message passing, by assigning different communicators in the library routines. Many MPI programs, however, do not need additional communicators[2].

The processes within a defined communicator are given a number called a *rank* starting from zero onwards. The programmer will use control constructs, typically `IF` statements, to direct processes to perform specific actions. For example, one could have

```
if (rank == 0) ...      /* do this */;
if (rank == 1) ...      /* do this */;
   ⋮
```

Usually the computation is constructed as a master-slave model in which one process (the master) performs one set of actions and all the other processes (the slaves) perform identical actions although on different data, i.e.,

```
if (rank == 0) ...      /* master do this */;
else ...                /* all slaves do this */;
```

A sample "Hello World" MPI program is shown in Figure 9.10. This program will send the message "Hello World" from the master process (rank = 0) to each of the other processes (rank ≠ 0). Then, all processes execute a `println` statement. In

```
#include <stdio.h>
#include <string.h>
#include "mpi.h"
main(int argc, char **argv ) {
   char message[20];
   int i,rank, size, type=99;
   MPI_Status status;
   MPI_Init(&argc, &argv);
   MPI_Comm_size(MPI_COMM_WORLD,&size);
   MPI_Comm_rank(MPI_COMM_WORLD,&rank);
   if(rank == 0) {
      strcpy(message, "Hello, world");
      for (i=1; i<size; i++)
         MPI_Send(message,13,MPI_CHAR,i,type,MPI_COMM_WORLD);
   } else
         MPI_Recv(message,20,MPI_CHAR,0,type,MPI_COMM_WORLD,&status);
   printf( "Message from process =%d : %.13s\n", rank,message);
   MPI_Finalize();
}
```

Figure 9.10 Sample MPI Hello World program.

---

[2] Technically, there are two types of MPI communicators: an *intracommunicator* and an *intercommunicator*. More details can be found in (Gropp, Lusk, and Skjellum 1999).

MPI, standard output is automatically redirected from remote computers to the user's console so the final result will be

```
Message from process =1 : Hello, world
Message from process =0 : Hello, world
Message from process =2 : Hello, world
Message from process =3 : Hello, world
...
```

except that the order of messages might be different—it will depend upon how the processes are scheduled. We will give more details of the MPI routines in Figure 9.10 and other MPI routines later.

The programming model is sometimes called a *single-program-multiple-data* model (SPMD) as each instance of the program (process) will have its own local variables and use different data. The programming model turns out to be very convenient for many applications in which the number of processes needed is fixed. Typically, the number of processes is set to be the same as the number of available processor cores as generally it is less efficient to run multiple processes on a single processor core. It is possible to start processes dynamically from within a process in MPI version 2.

In Figure 9.10, each process will have a copy of the variables named in the source program. That can cause some confusion for the programmer. For example, suppose the source program has a variable `i` declared. Variable `i` will appear in each process, although each is distinct and holds a value determined by the local process.

Setting up the Message Passing Environment. Usually the computers to be used are specified in a file, called a `hostfile` or `machines` file. This file contains the names of the computers and possibly the number of processes that should run on each computer. An implementation-specific algorithm will select computers from the list to run user programs. Users may create their own `machines` file for their program. An example of a `machines` file is

```
coit-grid01.uncc.edu
coit-grid02.uncc.edu
coit-grid03.uncc.edu
coit-grid04.uncc.edu
coit-grid05.uncc.edu
```

Usually, the specified machines are connected together in a cluster through local network. If a `machines` file is not specified, a default `machines` file will be used or it may be that the program will only run on a single computer.

Typically, the MPI implementation uses daemons running on each computer to manage the MPI environment and perform the message passing. These daemons may need to be started before the MPI programs. In LAM-MPI, a specific command called `mpiboot` must be issued by the user. In MPICH-2, the MPI daemons should be already running after the MPI installation but they can be re-started by the user with a similar command, `mpdboot`.

Process Creation and Execution. Using the SPMD model, a single program is written and compiled to include MPI libraries, typically using an MPI script `mpicc`, i.e.,

```
mpicc myProg.c -o myProg
```

`mpicc` typically uses the `cc` compiler and corresponding options (flags) are available. Once compiled, the program can be executed. The MPI version 1 standard does not specify implementation details at all. The actual command to execute the program would depend upon the implementation of MPI, but the common command would be `mpirun` where `-np` option specifies the number of processes. Hence, the command to execute four processes of the program `myProg` would be

```
mpirun -np 4 myProg
```

or

```
mpirun -machinefile myMachines -np 4 myProg
```

if a `machines` file called `myMachines` is specified. Command line arguments for `myProg` can be added after `myProg` and could be accessed by the program in the normal fashion.

The MPI version 2 standard did make certain recommendations regarding the implementation and introduced the `mpiexec` command with certain options. The `-n` option specifies the number of processes. The corresponding commands to the above `mpirun` commands are

```
mpiexec -n 4 myProg
```

or

```
mpiexec -machinefile myMachines -n 4 myProg
```

if the machines file `myMachines` is specified.

Before any MPI routine can be called from within the program, the program must call `MPI_Init()` once (and only once) to initialize the MPI environment. `MPI_Init()` is supplied with the command line arguments that the implementation can use to initialize the environment. The program must call `MPI_Finalize()` to terminate the MPI environment. One typically does not place anything after `MPI_Finalize()` including a `return` and certainly no MPI routine can be called. Therefore, every MPI program has the structure

```
main (int argc, char *argv[]) {
  MPI_Init(&argc, &argv);     /* initialize MPI */

      .
      .
      .

  MPI_Finalize();             /* terminate MPI */
}
```

In most MPI programs, it is necessary to find the rank of a process within a communicator. The rank of a process within a communicator can be found from the routine `MPI_Comm_rank()`. Often, it is also necessary to know how many processes there are in a communicator. This can be found from the routine `MPI_Comm_size()`. A simple program using these routines is

```
main (int argc, char *argv[]) {
  MPI_Init(&argc, &argv);                    /* initialize MPI */
  MPI_Comm_rank(MPI_COMM_WORLD,&myid); /* find no of procs*/
  MPI_Comm_size(MPI_COMM_WORLD,&numprocs); /*find rank*/
  printf("I am %d of %d\n", myid+1, numprocs);
  MPI_Finalize();                            /* terminate MPI */
}
```

All output is redirected to the console, so the output at the console will be

```
I am 3 of 4
I am 1 of 4
I am 4 of 4
I am 2 of 4
```

but in any order, as the order of execution on different computers is indeterminate.

Note that in the above, each process executes the same code. Typically though the master-slave approach is used. Then one might have

```
main (int argc, char *argv[]) {
  MPI_Init(&argc, &argv);
  MPI_Comm_rank(MPI_COMM_WORLD, &myrank); /* find rank */
  if (myrank == 0) master();
  else slave();
  MPI_Finalize();
}
```

where `master()` and `slave()` are procedures to be executed by the master process and slave process, respectively.

Message-Passing Routines.

Data Transfer. One of the principal purposes for the message passing is to send data from one process to another process to continue the distributed computation. The simplest and most fundamental activity is point-to-point message passing in which data is carried in a message that is sent from a process and received at one process. This can be achieved in MPI with an `MPI_Send()` routine at the source and an `MPI_Recv()` routine at the destination as illustrated in Figure 9.11. In this example, the value contained in the variable `x` is sent from process with `rank` = 1 to process with `rank` = 2 and `y` is assigned that value. The arguments in send and receive routines identify the value being sent and other processes. In MPI, there are

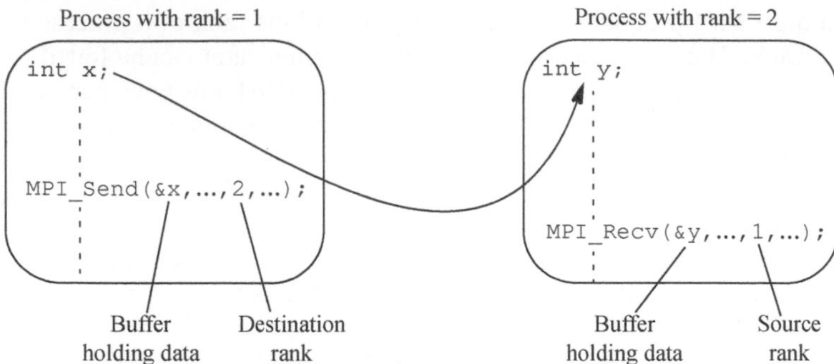

Figure 9.11 MPI point-to-point message passing using MPI_send() and MPI_recv() library calls.

six (formal) parameters in the basic MPI send routine:

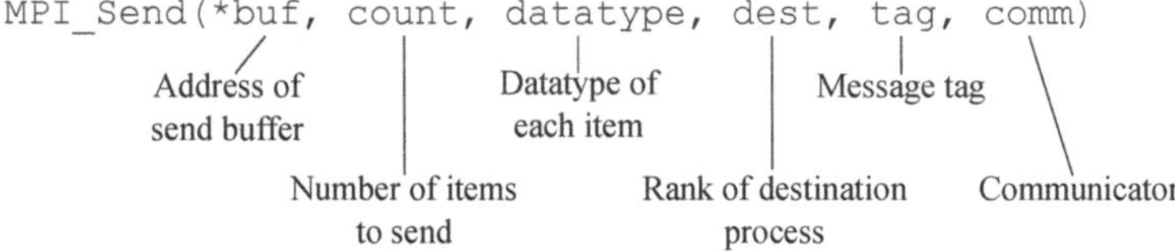

and seven parameters in the basic receive routine:

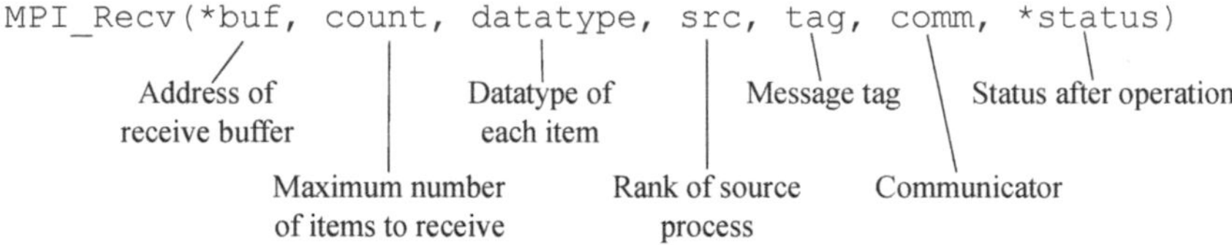

Message tags are used to differentiate between messages being received at different times for different purposes. (Message tags appeared in PVM.) A message tag is an integer that is carried with the message. Only messages with tags specified in the MPI_Recv() will be accepted by the receive routine. It is possible to place a wild card (MPI_ANY_TAG) for the tag in which case the receive will match with any message from the specified process in the communicator. It is also possible to use a wild card for the source rank (MPI_ANY SOURCE), in which case all messages in the communicator to the destination will be accepted.

The semantics of the MPI_Send() and MPI_Recv() are called *blocking*, which means in MPI that each will wait until all its local actions have taken place before returning. After returning, any local variables that were used can be altered without affecting the message transfer. In the case of MPI_Send(), the message may not and probably has not reached its destination when MPI_Send() returns, but the process can continue in the knowledge that the message is safely on its way. MPI_Recv() returns when the message has been received and the data collected. It will cause the process to stall until the message has been received. There are other versions of MPI send and receive routines that have different semantics, see later.

More complex message-passing are possible, and often desirable, in which a message may be sent from one process to multiple other processes (*broadcast*), or messages from multiple processes are sent to one receiving process (*gather*) and the result combined using an arithmetic or logical operation (*reduce*). All the processes in a communicator participate on the collective operation. With the MPI SPMD model, such collective operations are conveniently specified by calling the same routine in each process whether the process is a source of a message or a destination of a message. Then, one parameter in the routine identifies the source as the so-called *root* process.

The broadcast operation is shown in Figure 9.12. In this example, the root is the process with rank 0 although it could be any process. The parameters for the MPI broadcast are

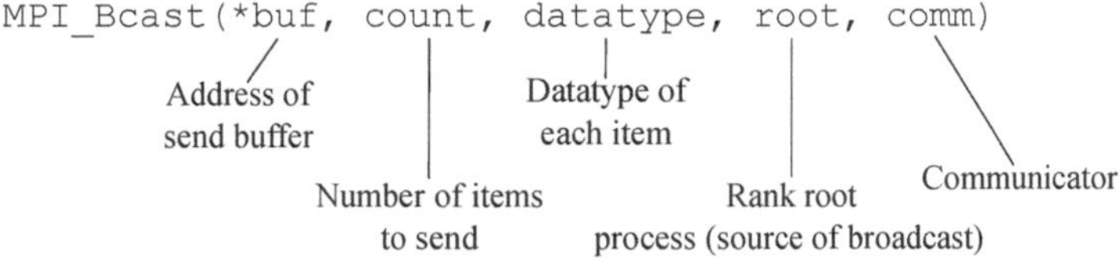

The scatter operation is shown in Figure 9.13. The parameters for the MPI scatter are

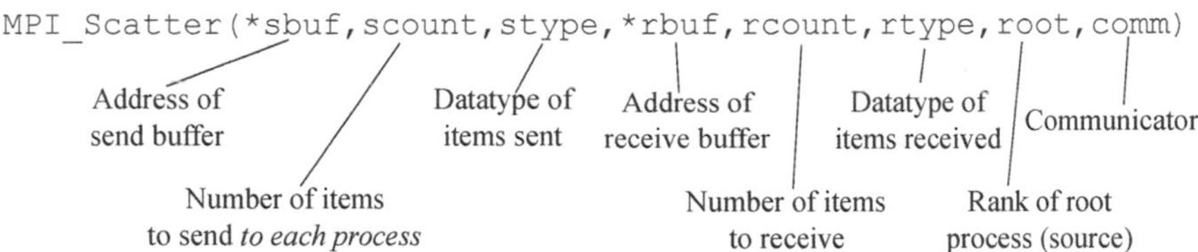

The `scount` parameter refers to the number of items sent to each process. Somewhat strangely, the send and receive types and count are specified separately and could be different. Normally, `scount` would be the same as `rcount`, and `stype` would be the same as `rtype`.

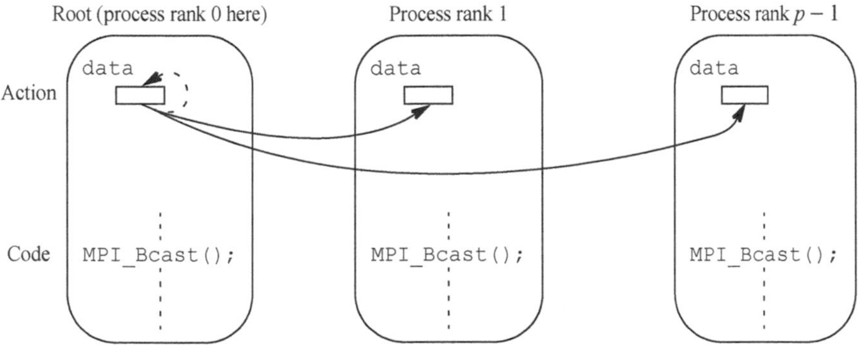

Figure 9.12 MPI broadcast operation.

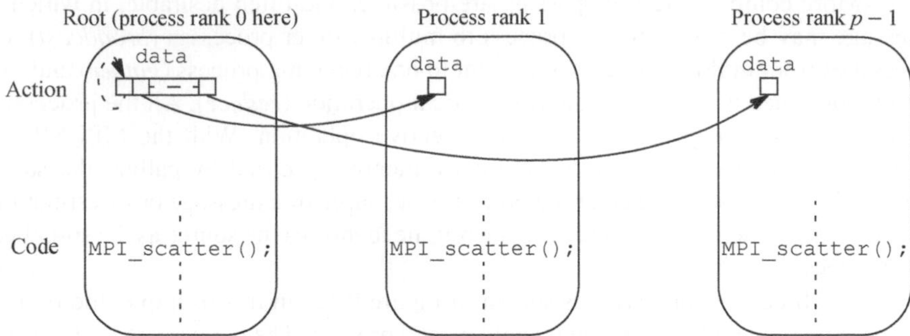

Figure 9.13 Basic MPI scatter operation.

The simplest scatter would be as illustrated in Figure 9.13. One element of an array is sent to different processes. An extension provided in the `MPI_Scatter()` routine would be to send a fixed number of contiguous elements to each process. In the following code, the size of the send buffer is given by `100 * <number of processes>` and 100 contiguous elements are sent to each process:

```
main (int argc, char *argv[]) {
   int size, *sendbuf, recvbuf[100];  /* for each process */
   MPI_Init(&argc, &argv);            /* initialize MPI */
   MPI_Comm_size(MPI_COMM_WORLD, &size);
   sendbuf = (int *)malloc(size*100*sizeof(int));
      :
   MPI_Scatter(sendbuf,100,MPI_INT,recvbuf,100,MPI_INT,0,
         MPI_COMM_WORLD);
      :
   MPI_Finalize();                     /* terminate MPI */
}
```

which is illustrated in Figure 9.14.

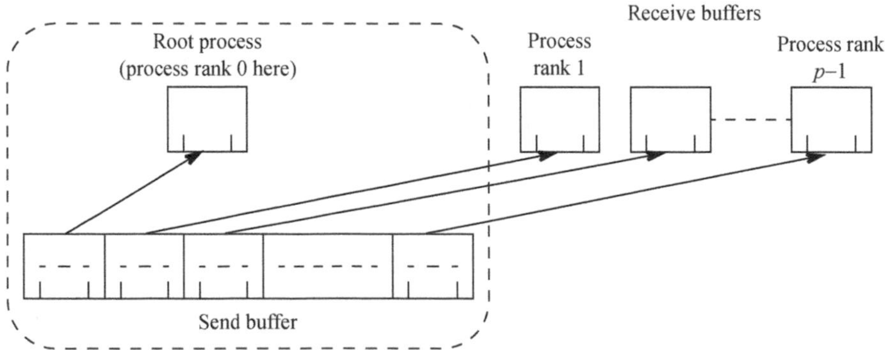

Figure 9.14 Scattering contiguous groups of elements to each process.

The gather operation is shown in Figure 9.15 and is essentially the opposite of scatter. The parameters for the MPI gather are similar to the scatter, i.e.,

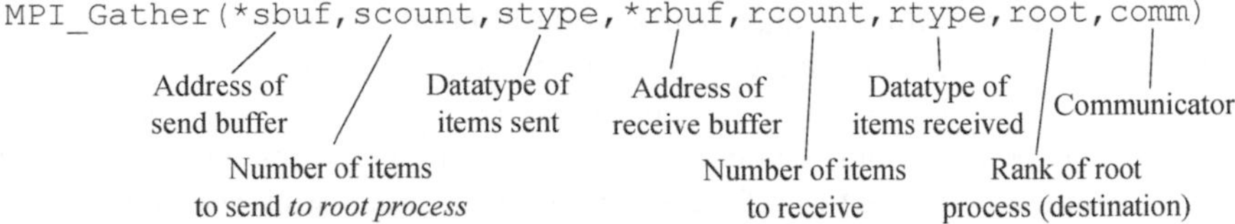

The reduce operation is the gather operation combined with a defined operation on the gathered data. The reduce operation with addition is shown in Figure 9.16. Other pre-defined operations are maximum, minimum, multiplication, and some logical and bit wise operations. Users can define their own operations. All operations are associative and the pre-defined MPI operations are also commutative. The parameters for the MPI reduce are

```
MPI_Reduce(*sendbuf, *recvbuf, count, datatype, op root, comm)
```

Address of send buffer; Address of receive buffer; Number of items to send; Datatype of each item; Operation; Rank of root process (destination); Communicator

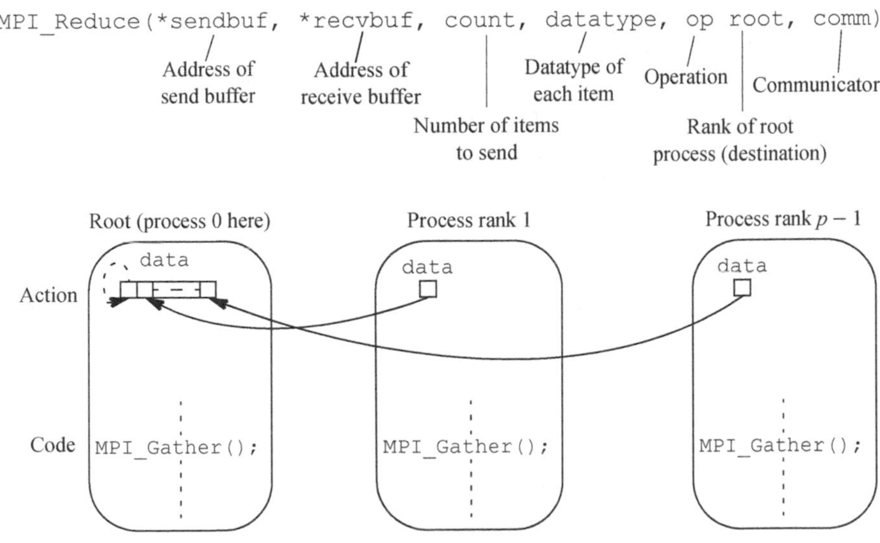

Figure 9.15 Gather operation.

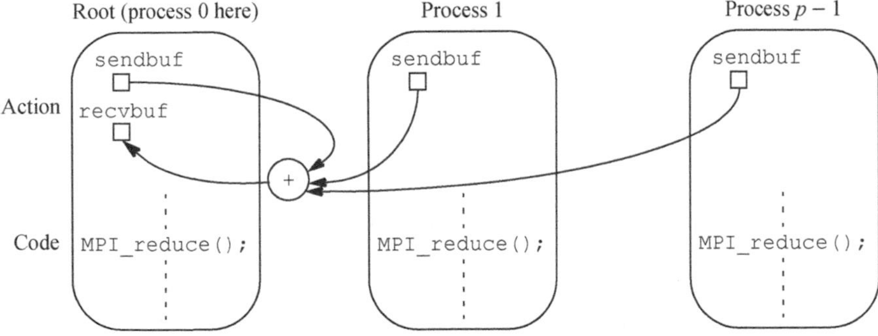

Figure 9.16 Reduce operation (addition).

The actual way that the MPI collective operations are implemented is not defined in the MPI standard except that their effects are equivalent to using a multitude of point-to-point sends and receives. Collective routines would be assumed to be much more efficient than using point-to-point `send/recv` routines. The semantics of the collective message-passing routines (`MPI_Bcast()`, `MPI_Scatter()`, `MPI_Gather()`, and `MPI_Reduce()`) are the same as the semantics of point-to-point communications, i.e., all processes can return when locally complete and do not wait for each other. Those processes involved in sending data can return when the message is on its way in the same way as `MPI_Send()` and those receiving data can return when they individually receive the data in the same way as `MPI_Recv()`.

Synchronization. A related purpose for the message passing is to synchronize processes so the communication continues at a specific point. MPI offers a specific synchronization operation, `MPI_barrier()`, which prevents any process calling the barrier routine returning until all have called it as illustrated in Figure 9.17. This collective routine simply has the name of communicator as its only argument. Synchronization causes all processes to stop to wait for each other and hence can slow the computation down but it may be necessary in some cases to be able to proceed with the rest of the computation.

As mentioned, the `MPI_Send()` does not wait for the message to be received before the process can continue. A synchronous version exists called `MPI_SSend()`, which does not continue until the message has been received. `MPI_Recv()` does not return until the message has been received and can be used with `MPI_SSend()` to effect a synchronous point-to-point message transfer, if desired. Note, it is better to avoid synchronization if possible as it implies that processes are waiting for others.

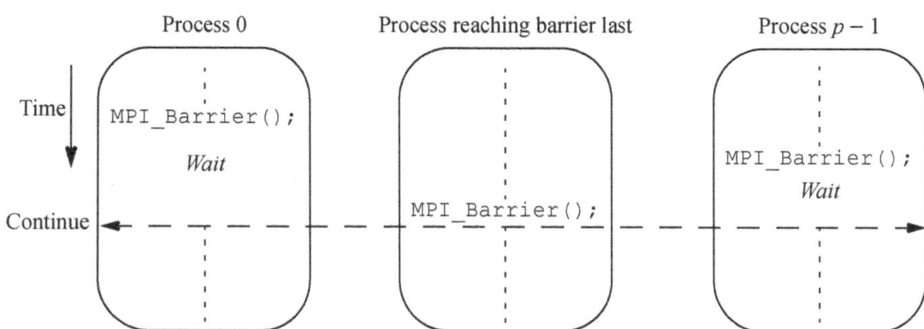

Figure 9.17 Barrier operation.

9.5.4 Grid-Enabled MPI

MPI uses networked computers with IP addresses so it would seem reasonable to extend the MPI application area to include geographically distributed networked computers and there have been several projects to implement MPI across geographically distributed computers connected on the Internet. In fact, MPI alone could be used. Figure 9.18 shows message passing across geographically distributed systems.

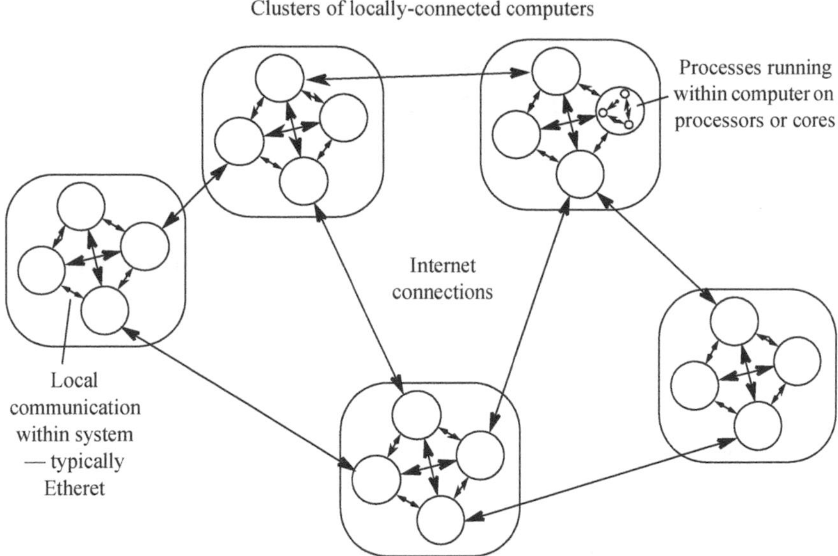

Figure 9.18 Message passing across geograghically distributed computers.

There are messages being sent within a cluster and between clusters at different sites. Within a cluster, the communicating processes may be running on the same computer or may be running on separate computers. On the same computer will result in the highest bandwidth and lowest message latency as most likely shared memory will be used to pass the information in the messages. Between computers in the same cluster will use a local network, typically Ethernet connections, and will incur significant performance reduction. The messages between clusters will use the Internet or a high-speed wide-area network and will incur the worst message passing performance by orders of magnitude.

As mentioned, MPI implementations generally use a file that lists names of all the computers available. If there is a network path and accounts on the machines, we could simply use the default MPI implementation for the whole distributed platform. However, most MPI implementations were designed for cluster configurations and not specifically for a configuration in which some communication would be through the Internet rather than through a local switch. Apart from the extra latency, it is necessary for all processes to be running for them to communicate and it may be difficult to ensure that processes on remote computers will actually be running at the same time. We mentioned this problem in schedulers. Advance reservation addresses the problem.

Some implementations of MPI, for example MPICH-1 uses `fork` to start local processes and `rsh` to start remote processes. The communication method is chosen to suit, for example TCP across a network and possibly using shared memory for a shared memory configuration. MPICH-1 has a very nice internal construction that separates the implementation of MPI routines from the implementation of the communication methods, as illustrated in Figure 9.19. Different implementations of the

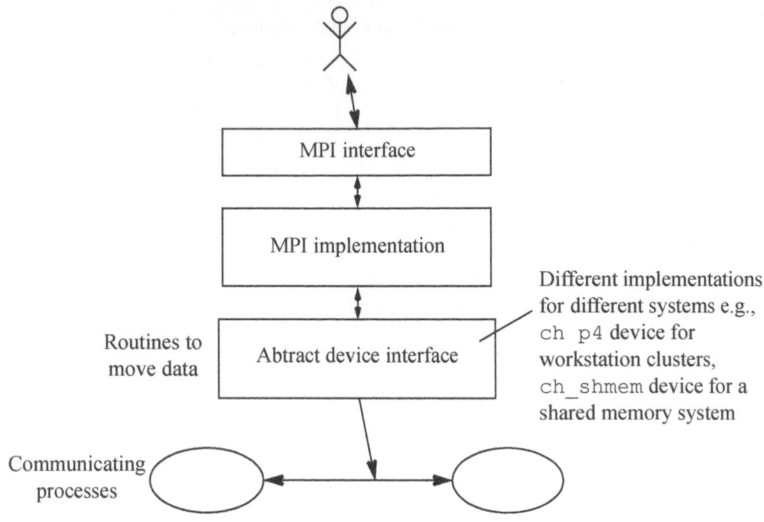

Figure 9.19 MPICH abstract device interface.

abstract device interface are provided for different systems, for example a so-called `ch_p4` device for workstation clusters, and a so-called `ch_shmem` device for a shared memory system. This feature was used by Foster and Karonis (Foster and Karonis 1998) to modify MPICH to operate across a Grid platform, resulting in the version of MPICH called MPICH-G, which included a so-called `globus2` device for Globus version 2. MPICH-G was perhaps the first attempt at porting MPI to a Grid platform and also one of the first projects that directly used the Globus infrastructure and indeed proved the utility of the Globus toolkit.

MPICH-G used all the major components of Globus version 2. It used GSI for authentication and authorization to provide single sign-on and mapping to local accounts. It used GASS (Globus Access to Secondary Storage) to move input files and executable to remote resources and move output back (input and output *staging*). It used GIS (Globus Information System) and MDS (Monitoring and Discovery Service) to obtain information about remote resources. It used GRAM to submit MPI jobs through Globus to remote resources, interfacing with local schedulers. MPICH-G created the appropriate job description file using RSL-1, the job description language used at the time. As pointed out earlier, it is necessary for the communicating processes to be running at the same time. To achieve this, the Globus version 2 component called DUROC (Dynamically Updated Request Online Coallocator) was used to coordinate the start-up of processes and coordinate processes to operate within the single default communicator `MPI_COMM_WORLD`.

MPICH-G is important from a historical perspective as it was the first attempt at using MPI with the Globus infrastructure. Figure 9.20 shows a comparison of using MPI and using MPICH-G in concept. In both cases, after the application is compiled, the same `mpirun` command is invoked and a machines file is used to identify the machines. MPICH-G may also need additional information, for example the specific

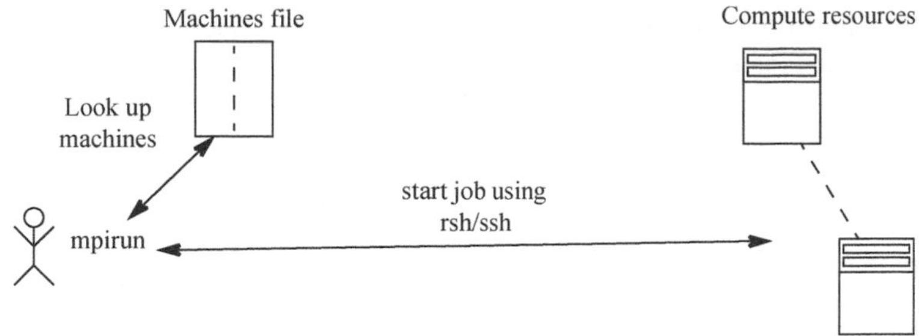

(a) Using MPICH to run a remote job.

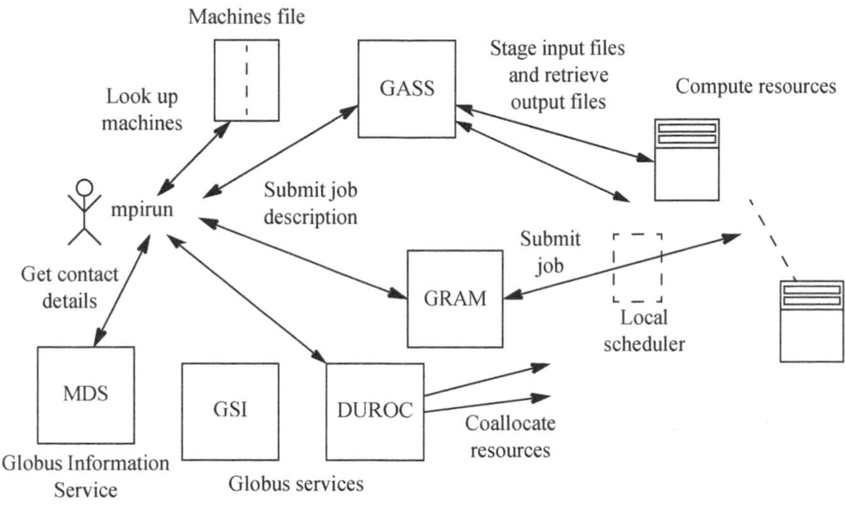

(b) Using MPICH-G to run a remote job.

Figure 9.20 Comparing MPICH and MPICH-G.

job manager name and port number, which could be specified in the machines file or found from a Globus information service (MDS). Other Globus services are also needed as described before. Finally, GRAM is used to submit the job using a job description file created by MPICH-G with `mpirun`. Notice that GSI is used to ensure a secure environment. Originally, MPICH employed the insecure `rsh` where information is passed in plain text although it can be reconfigured to use secure `ssh`. MPICH-G used a communication component called Nexus, which can manage different communication methods. For improved performance, MPICH-G2, a complete re-implementation of MPICH-G, replaced Nexus with its own communication components (Karonis, Toonen, and Foster 2003).

MPICH-G2 implements MPI-version 1.1, which for many applications is sufficient, but there are other issues in its use. It uses Globus version 2. It is usually necessary to work with firewalls. Specific ports must be open. Each site will have a separate independent local job scheduler with many users, and it is important that all MPI processes will be operating at different sites at the same time for them to communicate. The delays in messages in transit are much larger and variable between sites, so often the programmer will have to redesign his/her program to accommodate the indeterminate latencies.

Apart from the early MPICH-G/MPICH-G2 project, there have been other projects for providing tools to run MPI programs across a geographically distributed Grid that does not necessarily use Globus software, for example the Grid-Computing library PACX-MPI (Parallel Computer Extension - MPI) and open-source GridMPI™ project. PACX-MPI can encrypt messages using `ssl`. The GridMPI™ project avoided a security layer altogether for performance. Project investigators argue that geographically distributed high performance computer sites often use their own networks, so that a security is not an issue (GridMPI™). They found that MPI programs scale up to round-trip communication latencies of 20 mS, which represents about 500 miles for 1–10 Gbits/sec networks and the distance between major sites in Japan (where the project was developed) and then MPI programming is feasible.

9.5.5 Grid Enabling MPI Programs

In both GridMPI and PACX-MPI, the idea is not to have to alter the MPI programs at all and simply run MPI programs that would run on a local cluster. That does not preclude modifying the programs to take into account the geographically distributed nature of the compute resources and there have been projects to explore how to modify MPI programs. The most notable aspects for running MPI programs on a Grid are:

- Long and indeterminate communication latency
- Difficulty and effect of global synchronization
- Effect of failures (fault tolerance)
- Heterogeneous computers (usually)
- The need to the communicating processes to be running simultaneously

Mostly, the focus is to ameliorate the effects of latency. One basic concept is to overlap communications with computations while waiting for the communication to complete, which comes directly from regular parallel programming, and is illustrated in Figure 9.21. In this example, process *n* needs to send data through a message to process *m*. The `MPI_Send()` routine in process *n* occurs before the `MPI_Recv()` routine in process *m* in time. After the send routine completes all its local actions, it returns and starts on subsequent computations before the message has been received. The MPI routine `MPI_ISend()` could be used and would return even sooner, before the local actions have taken place, i.e., immediately. Such routines that return immediately and before local actions are complete are called *non-blocking* in MPI. (Routines that return after their local actions are complete are called *blocking* in MPI.) One has to be very careful with non-blocking routines as variables used in the message transfer must not be altered until the local actions have completed.

In Figure 9.21, it is possible for the `MPI_Recv()` to occur before the `MPI_Send()` routine and that case the `MPI_Recv()` routine would wait for the `MPI_Send()` routine. The MPI routine `MPI_Irecv()` could be used and returns

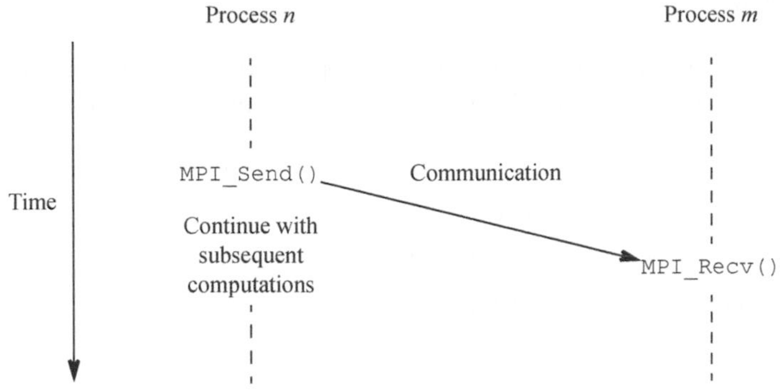

Figure 9.21 Overlapping commmunications with computations (send before receive).

immediately even if the message has not been received and would allow the process to continue with other work while waiting for the message. Then, other MPI routines are used to establish that the message has arrived.

Both the `MPI_Isend()` routine and the `MPI_Irecv()` routine have an additional parameter to identify the particular instance of the routine in the code. This parameter is used in other routines to establish completion of the particular message passing routine. The MPI routine `MPI_Wait()` waits for a particular send or receive routine to complete and then return and is one way to establish that local actions are complete. Another way to establish completion without stalling the process is to use the MPI routine `MPI_Test()`. This routine will test for completion of a non-blocking routine and return setting a Boolean parameter to indicate whether the non-blocking has completed (`true/false`).

The concept of overlapping communication with computation can be extended to allow the processes to be decoupled further. For example, in an algorithm that needs to iterate towards a solution and have to send values obtained to neighboring processes at each iteration, it is possible to buffer the data from multiple messages allowing the source to continue with subsequent iterations. More details of this approach can be found in (Villalobos and Wilkinson 2008).

Other issues in using a Grid include the effects of failures especially on long-running jobs. Most MPI implementations do not handle system failures. If a process fails, normally the whole system will crash. Dealing with potential system failures at the programming level would be very cumbersome, and ideally the system should provide mechanisms for handling system failures. Finally, systems in a Grid environment are very likely to be heterogeneous. Computers of different types are likely in a Grid environment and more so than in a local cluster set up at one time. The seemingly trivial factor that some computers store their data in little endian (least significant byte first) and others store their data in big endian (most significant byte first) needs to be handled.

9.6 SUMMARY

This chapter introduces the types of jobs and applications amenable to porting onto a Grid platform and techniques for porting them. The following are introduced:

- Parameter sweep and its description in a job description language
- Exposing an application as a Web service
- Using Globus and higher level APIs
- Parallel programming with MPI

FURTHER READING

There are many tutorials on MPI. The primary source material can be found in the MPI reference manual (Message Passing Interface Forum 2008) and the Message Passing Interface standard home page. For more information on parallel programming see (Wilkinson and Allen 2005). Grid-enabling application is still a research topic. There have been workshops focusing on this topic such as the Workshop on Grid-Enabling Applications, 15th ACM Mardi Gras Conference, 2008.

BIBLIOGRAPHY

Abramson, D., and J. Komineni. Interprocess communication in GriddLeS: Grid enabling legacy software. Technical report, School of Computer Science and Software Engineering, Monash University.

Afgan, E., and P. Bangalore. 2008. Experiences with developing and deploying Dynamic BLAST. *15th Mardi Gras Conference*, Baton Rouge, Louisiana, USA, Jan. 30.

Bangalore, P. 2007. Experiences building and using UABgrid. Guest lecture, April 3. ITCS 4146/5146 UNC-Charlotte Grid computing course (Spring 2007). http://www.cs.uncc.edu/~abw/ITCS4146S07/index.html

Bangalore, P., and E. Afgan. 2008. G-BLAST: A Grid service for BLAST. *Proc Int Conf. on Grid Computing an Applications*. Las Vegas, Nevada, July 14-17.

BLAST home page http://blast.ncbi.nlm.nih.gov/Blast.cgi

Clark, J., and S. DeRose. 1999. XML path language (XPath) version 1.0 W3C recommendation 16 Nov. http://www.w3.org/TR/xpath

Foster, I., and N. Karonis. 1998. A Grid-enabled MPI: message passing in heterogeneous distributed computing systems. *Proc. Supercomputing 98 (SC98)*, Orlando, FL, November.

Geist, A., A. Beguelin, J. Dongarra, W. Jiang, R. Manchek, and V. Sunderam. 1994. *PVM: Parallel virtual machine*. Cambridge, MA: MIT Press.

GridBus Project. Grid Service Broker: A Grid scheduler for computational and data Grids. University of Melbourne. http://www.gridbus.org/broker/

GriddLeS: Grid enabling legacy software. http://www.csse.monash.edu.au/~davida/griddles/

GridMPI™ http://www.gridmpi.org/index.jsp

Gropp, W., E. Lusk, and A. Skjellum. 1999. *Using MPI portable parallel programming with the Message-Passing Interface.* 2nd ed. Cambridge, MA: MIT Press.

Guan, Z., V. Velusamy, and P. Bangalore 2005. GridDeploy: A toolkit for deploying applications as Grid services. *Int. Conf. on Information Technology Coding and Computing* Las Vegas, Nevada, April 4–6.

Kaiser, H. 2004. The Grid Application Toolkit abstracting the Grid for application programmers. http://www.gridlab.org/WorkPackages/wp-1/Presentations/TheGridApplicationToolkit.ppt

Kaiser, H. 2005. Application development (Current toolkits and approaches for Grid-enabling applications). *SURA Cyberinfrastructure Workshop Series: Grid Technology: The Rough Guide*, Dec. 8-9 Austin, TX. www.sura.org/events/docs/application_toolkits.ppt

Karonis, N., B. Toonen, and I. Foster. 2003. MPICH-G2: A Grid-enabled implementation of the Message Passing Interface. *Journal of Parallel and Distributed Computing* 63 (5): 551–563.

Kielmann, T. 2006. The simple API for Grid applications (SAGA). *17th Global Grid Forum* (GGF17), Tokyo, Japan May 10–12. http://www.ggf.org/GGF17/materials/319/kielman-ggf17.pdf

Krishnan, A. 2005. GridBLAST: a Globus-based high-throughput implementation of BLAST in a Grid computing framework. *Concurrency and Computation: Practice & Experience*, 17 (13): 1607–1623.

Message Passing Interface (MPI) Standard home page. http://www-unix.mcs.anl.gov/mpi/

Message Passing Interface Forum. 2008. MPI: A message-passing interface standard version 2.1. June 23.

Nolan, K. 2008. Approaching the challenge of Grid-enabling applications. *Open Source Grid & Cluster Conference*, Oakland, CA, May 12-16. www.opensourcegridcluster.org/documents/KieranNolan.ppt

OGF23 2008. Working session on parameter sweep. *Open Grid Forum*, June 2–6 Barcelona, Spain. http://www.ogf.org/gf/event_schedule/index.php?id=1294

OpenMPI: Open source high performance computing. http://www.open-mpi.org/

PACX-MPI. Extending MPI for computational Grids. http://www.hlrs.de/organization/amt/projects/pacx-mpi/

Sanjeepan, V., A. Matsunaga, L. Zhu, H. Lam, J. and A. B. Fortes. 2005. A service-oriented, scalable approach to Grid-enabling of legacy scientific applications. *Int. Conference on Web Services (ICWS-2005)*, Orlando, Florida, 553–560.

Snir, M., S. W. Otto, S. Huss-Lederman, D. W. Walker, and J. Dongarra. 1998. *MPI - The complete reference volume 1, The MPI core.* 2nd ed. Cambridge, MA: MIT Press.

Sunderam, V. 1990. PVM: A framework for parallel distributed computing. *Concurrency: Practice & Experience* 2 (4): 315–339.

Villalobos, J. F., and B. Wilkinson. 2008. Latency hiding by redundant processing: A technique for Grid-enabled, iterative, synchronous parallel programs. *15th Mardi Gras Conference*, Baton Rouge, Louisiana, USA, Jan. 30.

van Nieuwpoort, R. Getting started with the Grid application toolkit. http://www.gridlab.org/WorkPackages/wp-1/Doc/JavaGAT-tutorial.pdf

Wilkinson, B., and M. Allen. 2005. *Parallel programming techniques and applications using networked workstations and parallel computers.* 2nd ed. Upper Saddle River NJ: Prentice Hall.

Wilson, G. V. 1995. *Practical parallel programming*. Cambridge, Massachusetts: MIT Press.

Wikipedia. Xpath. http://en.wikipedia.org/wiki/XPathhttp://en.wikipedia.org/wiki/XPath

SELF-ASSESSMENT QUESTIONS

The following questions are multiple choice. Unless otherwise noted, there is only one correct answer for each question.

1. What is meant by Grid enabling an application in the most general sense?

(a) Being able to execute an application on a Grid platform, using the distributed resources available on that platform
(b) Executing a job of a Grid platform
(c) Wrapping an existing application into a service
(d) Adding MPI routines in the application

2. What is meant by parameter sweep?

(a) Executing an application multiple times each time with the arguments specifically incremented by one each time
(b) Executing an application multiple times with the same arguments
(c) Executing an application multiple times with different arguments
(d) Cleaning out the parameters from a computer program

3. What is the XPath expression to select the second `c` element within the second `b` element within the second `a` element of an XML document?

(a) `2a/2b/2c`
(b) `a/b/c[2]`
(c) `a[2]/b[2]/c[2]`
(d) `a2/b2/c2`
(e) `2a2b2c`
(f) None of the other answers

4. What is MPI?

(a) A standard for user-level message passing library routines
(b) An implementation of message passing routines
(c) A portal interface for message passing Grid applications
(d) A portlet interface for message passing Grid applications

5. What is an MPI communicator?

(a) A talkative MPI program
(b) A process that sends messages
(c) An MPI compiler
(d) A way of defining a set of processes that can communicate between themselves

6. In message passing, what is a message tag?

(a) An identifier that identifies the type of data used in a message

(b) An integer included in the message used for message matching
(c) The cost of sending a message
(d) The destination address
(e) Identifies an XML element in a message-passing XML language

7. What type of routine is the `MPI_Recv()` routine?

(a) Synchronous
(b) Blocking
(c) Non-blocking
(d) Asynchronous

PROGRAMMING ASSIGNMENTS

The following assignment requires a Web service environment to complete fully (see Chapters 6 and 7).

9-1 Wrapping an existing program as a Web service — Write a Java program that invokes a command using the `exec` method of `RunTime`, displaying the output. Test with the OS commands such as `ls` and to execute an existing program. Then, write a Web service that calls the Java program and test with a Web service client.

The following assignment requires access to a Grid platform.

9-2 Parameter sweep — The objective of this assignment is to apply the parameter sweep technique to determine whether two images are the same except for scaling. This has practical application. Comparing two images to find a match or similarity where the original images might be of different sizes might occur for example in medical imaging and in security applications (facial images). If the images had the identical scale and resolution one can simply compare the pixels (picture elements) of the two images. With $n \times n$ pixels in each image, this would take $O(n^2)$ operations.

If the scales of each image are different, one simple approach would be to alter the scale of one image to be the same as the other and then compare the images. There needs to be known points of reference in the same place in each image. One image will need to be shifted in x and y directions and rotated to have the points of reference in the same place. Basic scaling and shifting operations are described in Wilkinson and Allen (2005) as follows:

Let the location of each pixel be identified by its x-y coordinates, with the origin at the top left corner of the image (which is the normal place for the origin for digital images). Let x and y be the coordinates a pixel in the original image and x' and y' be the new coordinates of the pixel after the operation.

Scaling: To scale an image by a factor S_x in the x-direction and S_y in the y-direction, the coordinates of each pixel in the object are given by

$$x' = xS_x$$
$$y' = yS_y$$

Shifting: To move a two-dimensional area in an image by Δx in the x-dimension and Δy in the y-dimension, the coordinates of each pixel in the object are changed by

$$x' = x + \Delta x$$
$$y' = y + \Delta y$$

Rotation: To rotate a two-dimensional area in an image through an angle θ about the origin, the coordinates of each pixel in the object are changed by

$$x' = x\cos\theta + y\sin\theta$$
$$y' = -x\sin\theta + y\cos\theta$$

Write a program that can establish whether two images are the same given that their scaling and orientation may be different. The output will be `TRUE` (images represent the same object) or `FALSE` (images do not represent the same object). The variables are S_x, S_y, Δx, Δy, and θ. Scanning through all combinations of these parameters will be computationally extremely time consuming.

Launch the program as a parameter sweep application on a Grid platform and test.

The following assignment requires a cluster/Grid environment with MPI installed.

9-3 MPI — Write an MPI program in C that will contact each available computer and display its name (use `gethostname()`). Add a *ping* test that sends a message to each computer, which responds immediately with a message back. Time the round trip of each ping and display the time with the name of the computer.

Test the program on a locally connected cluster and on geographically connected computers (a Grid platform) and compare the times. Make conclusions.

APPENDIX A
Internet and Networking Basics

In this Appendix, the prerequisite knowledge on basic networking techniques that are necessary for this book and the Grid computing assignments is outlined. We expect that readers already know most of the information in this Appendix, but it is provided for completeness. Self-assessment questions are given at the end of the Appendix.

A.1 INTERNET TECHNOLOGY

Since Grid computing involves Internet protocols and standards, first let us briefly review some well-known Internet interfaces, protocols and communication standards. The standard that establishes basic low level rules for networked computers to communicate and pass data is called TCP/IP (Transmission Control Protocol/Internet Protocol), which was initially developed in the 1970s to provide a unified protocol for packet-switched computer network communications when packet-switched computer networks were gaining interest in the research community.

MAC Addresses. First let us look at how computers can be identified on a network. Computers are physically connected onto a network through a network interface card (NIC), generally an Ethernet connection. Each network interface is given a unique physical 48-bit address called a MAC (Media Access Controller) address, which is assigned during manufacture of the interface. The allocation of addresses is controlled by the IEEE Registration Authority. The MAC address is unalterable. If one replaced the NIC, there would be a different MAC address for the computer through that interface.

IP Addresses. MAC addresses while uniquely identifying the interface and hence the computer through that interface are not really convenient for identifying computers on network. Instead, a number called an IP (Internet Protocol) address or number is used. Each computer is assigned an IP (Internet Protocol) address to identify it for network communication purposes. The IP address is mapped to the NIC MAC address by a software table. The construction of the IP address is arranged for ease of routing to the computer. Also, whereas the MAC address of a particular NIC is unalterable, the IP address assigned to a computer interface can be altered to suit the situation. It is possible for a computer to have more than one NIC, each having a different MAC address and IP address. For example, a front-end computer in cluster usually has one NIC to make an external connection to the Internet and another NIC to make a connection to local compute nodes.

The IP address format (version 4) has 32 bits divided into four 8-bit numbers, each between 0 and 255. For example, a computer interface might be allocated the IP address

129.49.82.1

The mapping of MAC address to IP address can be invoked with the address resolution protocol command `arp -a` (Linux or Windows). A result for a home computer is

```
>arp -a

Interface: 192.168.1.100 --- 0x2
 Internet Address      Physical Address      Type
 192.168.1.1           00-13-10-78-0d-49     dynamic
```

where the physical address is the MAC address. Obtaining more information about the IP configuration can be obtained with the command `ipconfig` (Windows) or `ifconfig` (Linux).

IP Address Formats.

IPv4 Classful Formats (Historical). The *classful* Internet Protocol version (IPv4) formats divides the 32-bit IP address into fields for decoding purposes. It had five formats (classes) identified by the values of the first bits in the address. There were three principal formats, Class A, Class B, and Class C as illustrated in Figure A.1. Class A provided for a large number of local hosts (16,777,216 = 2^{24}) but only 127 networks. Class B provided for 16,384 (2^{14}) networks and a maximum of 65,536 (2^{16}) host computers. Class C provided for 2,097,152 (2^{21}) networks and 256 host computers (no sub-network).

Some addresses were reserved in each class for private networks (class A 10.0.0.0 to 10.255.255.255; class B 172.16.0.0 to 172.32.255.255; class C 192.168.0.0 to 192.168.255.255). Private networks are widely used at home for connecting multiple computers to a single external Internet connection, and for configuring clusters as described in Chapter 1 (Figure 1.1). In Figure 1.1, a switch connects compute nodes together and to a front node. Private IP addresses are used for the

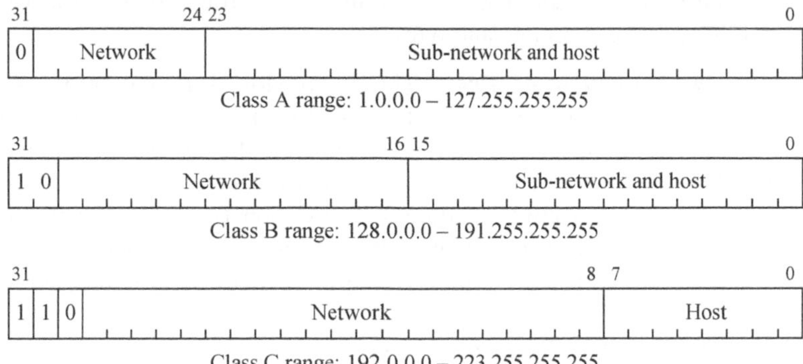

Figure A.1 IPv4 Class A, B, and C formats.

compute nodes, typically Class C addresses for example 192.168.1.100, 192.168.1.101, 192.168.1.102, ...

One pattern within class A (first bit = 0) is reserved for loopback (first 8 bits 01111111). There is also a class for multicast (Class D, first 4 bits 1110), and a reserved class (Class E, first 4 bits 1111).

Classless Inter-Domain Routing (CIDR). In the classful IPv4 format, routers first examine the first bits of the address to determine the class and division of bits in each field. The appropriate masks are used to select the network and host addresses. A problem with the classful approach is the coarse divisions for the host field, either 8 bits, 16 bits or 24 bits (Class C, B, or A). Typically 8 bits is insufficient to specify all the hosts at a site but 16 bits is much too large. Class B would have to be used with many unused patterns. Class A would hardly ever find application. Also routers have to handle the very large number of individual class C addresses.

The Classless Inter-Domain Routing (CIDR) approach was introduced in 1993 to provide greater flexibility and efficiency in allocating network addresses and essentially replaced classful formats from then onwards. Briefly, the format consists of a network field and a host field but the division between network and host can be on any bit boundary. The number of bits for the network is indicated in the CIDR notation by a forward slash after the IP address. For example

129.49.82.1/22

would mean the first 22 bits of the address select the network and the remaining 10 bits select the host. The mask to select the network would have 22 1's followed by 10 0's. CIDR addresses can be aggregated for routing purposes. For example, four /22 addresses could aggregated into one /20 address. The CIDR notation can of course create the same formats as the superseded classful IPv4 formats, i.e., /8 for Class A, /16 for Class B and /24 for Class C.

Blocks of CIDR addresses are allocated by the Internet Assigned Numbers Authority (IANA) to Regional Internet Registries (RIRs) for particular regions of the world. There are currently five regions—North America and parts of the Caribbean,

Europe, Middle East and Central Asia, Asia and the Pacific region, Latin America and parts of the Caribbean region, and Africa.

Addressing Limitations of IPv4. The IP version 4 format provides 256^4 different addresses (2^{32}, over 4 billion addresses) if every pattern were used. That is not actually sufficient to handle the enormous growth in the number of computers on the planet that may be connected to the Internet, and ways are found to work around that, most notably using Network Address Translation (NAT) used with private networks. A NAT maps local IP network addresses to a shared external IP address (differentiating addresses by assigning different port numbers, see later about port numbers).

The IP address format, IP version 6, was introduced as a IETF standard in 1998. It was developed principally to address the addressing limitations of IPv4. IPv6 uses a 128-bit address that provides 2^{128} possible numbers (about 3.4×10^{38}, a very big number!) that will easily be able to handle every computer on the planet for the foreseeable future. The number is divided into eight 16-bit numbers. IPv6 also has a number of other enhancements. Both IPv4 and IPv6 are meant to coexist and operating systems are now designed to handle IPv6. However, IPv6 has had a low adoption partly because the NAT workaround with IPv4 having satisfied private network demand. The greatest adoption has been in Asia.

TCP/IP Layers. Coming back to TCP/IP, TCP/IP has a layered approach in which the top layer is the application layer, then the transport layer, then the Internet layer and finally the link layer. TCP is in the transport layer and packages the information from the application layer into a specified format. TCP is also responsible for making the transfer reliable. If a packet does not get through, it must be re-submitted. The packet is passed down to the Internet layer where the Internet Protocol (IP) resides, which is responsible for determining the route using a routing table, and adds the next destination address to the packet if an intermediate routing site is needed. IP addresses are used at this level. Beneath the IP layer is the link layer, which formats the information to travel along the physical interconnections. Protocols such as ARP (Address Resolution Protocol) reside at this level and used to find the MAC address of a remote computer given the IP address by using a look-up table. The interface driver software launches the packet onto the network. Information is sent along the network in "packets" that contain the data, source and destination addresses, and other information. The addresses at this level are the 48-bit MAC addresses.

Apart from TCP, another Internet protocol at the transport level is UDP (User Datagram Protocol), which is designed for a minimum of protocol actions. It does not guarantee reliability or maintaining the order of packet but offers broadcast and multicast modes. It is used for streaming and some other applications.

Higher-Level Protocols. Higher-level protocols are built on top of TCP/IP and sit at the application layer. Some standard network protocols that are on top of TCP/IP include:

- Telnet — an insecure Internet protocol that allows a connection to another computer on the Internet.

- FTP (File Transmission Protocol) — insecure Internet protocol to exchange files.
- SMTP (Simple Mail Transport Protocol) — Internet protocol for email.
- HTTP (Hypertext Transfer Protocol) — basic protocol of accessing Web pages.
- TLS (Transport Layer Security)/SSL (Secure Socket Layer) Protocol — for making a secure connection.
- SSH (Secure Shell) Protocol — for making a secure connection.

Telnet and FTP are given for historical reference as nowadays they are undesirable because they are insecure and pass all information including important information such as passwords in clear text. They are often not allowed. Secure connections are needed in many computer related activities, including e-business and Grid computing. The SSL protocol (Secure Socket Layer) was proposed by Netscape Communications and has been adopted widely. The Transport Layer Security protocol (TLS) was subsequently proposed with no major differences to SSL. We describe the SSL protocol in Chapter 4, Section 4.5.3. It uses public key cryptography which is also described in Chapter 4 and used for necessary secure mechanisms in Grid computing. SSH was also mentioned in Chapter 4 for making secure connections (Section 4.5.2).

Secure protocols can use SSH or SSL to maintain a secure connection. The SFTP (Secure File Transfer) protocol generally uses SSH for the secure connection. It can be invoked by a client such as FillZilla with URL protocol `sftp://` or with the Linux command `sftp`. The protocol HTTPS (i.e., `https://`) is using HTTP over a SSL connection and is very common for use with a secure Web site. It is possible to perform FTP over SSL (so-called FTPS) for secure file transfers but that is not common. More common is SFTP. As described in Chapter 6, Grid computing uses the secure GridFTP protocol that operates under the GSI security envelope and has features for increased performance.

A network service (server) is an implementation of a protocol with a set of capabilities. Examples of network servers include FTP servers and HTTP servers. The specific protocol implemented will be "on top" of the TCP/IP protocol.

Ports. In addition to an IP address that identifies the host computer, a port number is used to identify the process (application) that will handle the request with a given protocol. Port numbers are divided into:

- Well-known ports
- Ephemeral ports

Well-known port numbers employ ports 0 to 1023. They are used for processes that handle standard protocols and requests such as illustrated in Figure A.2. The Internet Assigned Numbers Authority (IANA) is responsible for assigning port numbers to protocols.

Ephemeral ports (short-lived ports) are allocated to client applications. Applications need to use different port numbers to those already used by other applications. Ephemeral ports have now been divided into two types:

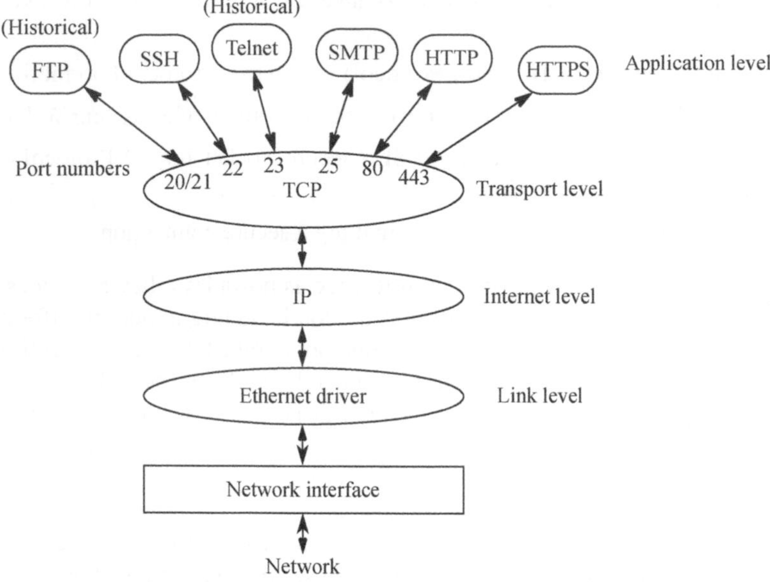

Figure A.2 TCP/IP stack and port numbers.

- Registered ports
- Dynamic and/or private ports

Registered ephemeral ports are from 1024 to 49151. Software companies can register ports in the registered port range. The Internet Corporation for Assigning Names and Numbers, ICANN, is responsible for this registration. A listing can be found at Wikipedia ("List of TCP and UDP port numbers"). Dynamic and/or private ephemeral ports are from 49152 to 65535 and are not used for any prescribed application and could be reasonably used for any purpose. The allocation of port numbers being used for processes running on a system can be found with the `netstat` command (Linux or Windows). Note that both Linux and Windows use ports; ports are a universal concept. The design of the protocols is also not tied to a specific operating system.

There are many "unofficial" but de facto ports. Port 8080 is the alternative port to 80 for HTTP servers. The Tomcat server used for Web services typically uses port 8080. We use port 8080 for the GridSphere Grid course portal (as this portal is run on top of Tomcat). The default port for the secure Globus container is 8443, although any port can be used. We sometimes use port 8440 for the Globus container for the command-line assignments to run Grid jobs so as not to conflict with GridSphere. Port usage is either to avoid conflict with existing applications or personal preference and can be altered. Of course, everyone who is using the services needs to know the port numbers. Ports have to be open through firewalls, which can be a significant issue for a Grid computing environment. Table A.1 shows some of the ports used by Globus 4 installations. More information on a Globus installation can be found in Appendix D.

TABLE A.1 SOME PORTS USED BY GLOBUS 4 GRID COMPUTING INSTALLATIONS

Component	Port (all TCP)
Container (for services such as GRAM, etc.)	8080 (HTTP) or 8443 (HTTPS)
GridFTP	2811 plus ephemeral ports for data
MyProxy server	7512
Grid-enabled SSH	22
Gridsphere/Tomcat	8080 (HTTP) or 8443 (HTTPS), possibly change to 8440
Client	Ephemeral port

World Wide Web. The Internet was transformed into the World Wide Web that we know today by two key developments—the HTTP (Hypertext Transfer Protocol) protocol for accessing Web pages and the HTML (HyperText Markup Language) used to specify how Web pages should display—coupled with easy to use Web browsers. The transformation started in early 1990s when HTTP was developed at CERN and the Mosaic browser developed at NCSA. Tim Berners-Lee at CERN is credited with the birth of the World Wide Web by his pioneering work on HTTP. A driving force behind the work is ease of use for world-wide collaboration.

Domain Names. For ease of use by humans, domain names are used rather than IP addresses. For example, coit-grid01.uncc.edu is the name of one server used for Grid computing in the Department of Computer Science at UNC-Charlotte. Its IP address is 152.15.98.24. The conversion between an IP address and its name done by a domain name server (DNS), a distributed domain name database. The domain name identifies the computer (server).

URL (Uniform Resource Locator). A URL (Uniform Resource Locator) is a way of identifying and accessing a Web page. The general format is shown below

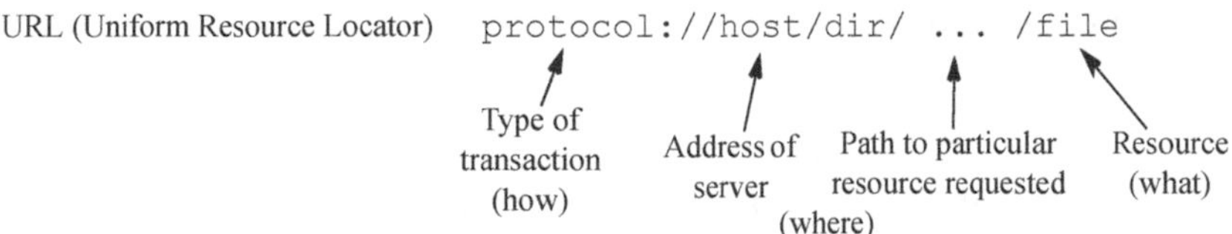

Knuckles (2001) describes URLs as having a "how" part, a "where" part and a "what" part. The "how" is the protocol to be used, such as HTTP or SFTP. The "where" part is the name of the server, and the "what" part is of the name of the resource that is requested. In our example, the where is the name of the server plus the path to the resource.

There are about 74 official and unofficial protocols that can be used in URLs. Protocols include previously mentioned protocols such as the ubiquitous `http` and `https`. An example is the home page for this book, which is

```
http://www.cs.uncc.edu/~abw/GridComputingBook/index.html
```

The particular Web page being selected is the file called `index.html` within the directory called `GridComputingBook`, within the home directory of `abw`. A tilde (~) indicates the home directory of the user.

Browsers are usually set to look for a file called `index.html` by default so that the URL could simply be written as

```
http://www.cs.uncc.edu/~abw/GridComputingBook
```

Browser are also usually configured to respond to other protocols.

The protocol `file` is typically applied for accessing a file on one's computer, hence the format

```
file://host/dir/ ... /file_name
```

In that case, `host` becomes `localhost` (the name of the machine that one is issuing commands from) and could be omitted, resulting in the somewhat strange construction of three `/`'s

```
file:///dir/ ... /file_name
```

This construction is seen in Chapter 2 when specifying a file to be transferred between the local computer and a remote host (Section 2.3.1), e.g.,

```
globus-url-copy gsiftp://www.coit-grid02.uncc.edu/~abw/hello \
                 file:///home/abw/
```

The protocol for the remote host in Globus is `gsiftp`. The backslash symbol (\) is the continuation symbol, needed if a command crosses onto a new line.

If the server is using the default port number for the application/protocol, the port number does not need to be specified. If it not using the default port number, it would need to be specified by adding it after the server name, separated by a colon, i.e.,

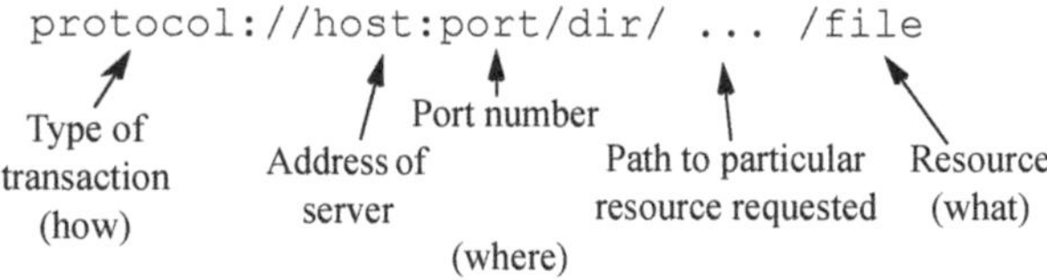

An example of this construction is the URL of our Grid course portal, `http://coit-grid02.uncc.edu:8080/gridsphere`. Port 8080 is specified because the default port for `http` is 80.

FURTHER READING

Williams (2001) is an excellent clear book covering networking.

BIBLIOGRAPHY

Knuckes, C. D. 2001. *Introduction to interactive programming on the Internet using HTML & JavaScript*. New York: John Wiley & Sons.

Wikipedia. List of TCP and UDP port numbers. http://en.wikipedia.org/wiki/List_of_TCP_and_UDP_port_numbers

Williams, R. 2001. *Computer systems architecture: A networking approach.* Harlow England: Addison-Wesley.

SELF-ASSESSMENT QUESTIONS

The following questions are multiple choice. Unless otherwise noted, there is only one correct answer for each question.

1. TCP/IP is

(a) A communications standard for computers to interconnect and exchange data
(b) A browser software package like Internet Explorer
(c) A special form of directory
(d) An international standards committee

2. What is the name of the file stored with the URL given as `www.cs.uncc.edu/~abw/ITCS4146F08` and where is it stored, given that `ITCS4146F08` is a directory?

(a) `ITCS4146F08` in the directory `abw`
(b) `index` in the directory `abw/ITCS4146`
(c) `index` in the directory `abw/public/ITCS4146`
(d) The first file in alphabetical order in the directory `abw/public`
(e) None of the other answers

3. In the command:

```
globus-url-copy gsiftp://www.coit-grid02.uncc.edu/~abw/hello \
        file:///home/abw/
```

why are there three forward slashes (three `/`s) in the file URL?

(a) The general form of file URL specifies three forward slashes
(b) The general form of file URL is `file://host/path` and when host is `localhost`, it can be omitted and so one is left with three forward slashes
(c) It is a mistake. There should be two forward slashes
(d) It indicates that `home` is a second-level directory

APPENDIX B
Linux and Windows Command-Line Interfaces

In this Appendix, the prerequisite knowledge on Linux and Windows commands that are necessary for this book and Grid computing assignments is outlined. We expect that most readers already know the information provided in this Appendix, but it is provided for completeness and for reference. Self-assessment questions are given at the end of the Appendix.

B.1 LINUX SYSTEM AND COMMANDS

Most Grid systems use a Linux platform and it is often necessary to interact with such systems on the command line. Some of the Grid computing assignments we have described also require basic Linux commands and an understanding of the Linux platform. The following information should be useful for conducting the assignments that require Linux command line tools. We shall assume that the local computer is a Windows system.

Login. Remote login to a Linux server from a Windows system is accomplished though an `ssh` client such as PuTTY, which can be obtained from:

http://www.chiark.greenend.org.uk/~sgtatham/putty/download.html

Figure B.1 shows the PuTTY window. The entered host name can be saved for future

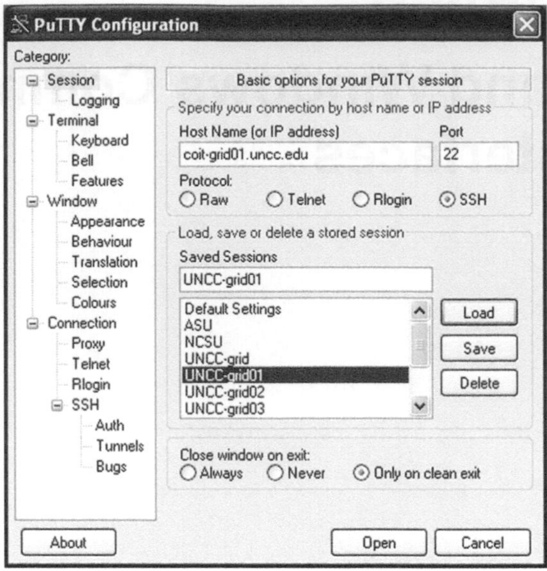

Figure B.1 PuTTY client window.

use with the save button and retrieved with the load button. The connection sequence will generally consist of a prompt for the username and a prompt for the user's password (i.e., a password authentication sequence as described in Chapter 4, Section 4.1.2).

Often one needs to transfer files between `localhost` and a remote computer. There are Windows GUI SFTP clients such as FileZilla. FileZilla can be obtained from

http://filezilla-project.org/

There are other clients that can be used. For example, WinSCP is a combined `sftp` and `ssh` client for Windows, available from:

http://winscp.net/eng/index.php

and has the advantage that it can be used for both making an `ssh` connection and for transferring file between `localhost` and a remote computer with convenient drag-and-drop actions.

Shell. shell is the term used to describe a Linux command interpreter (program) that the user interacts with to issue commands. The most common shell is `bash` (Bourne-Again shell, which is based upon the Bourne shell `sh`) and `tcsh` (an extension of the C shell, `csh`). `bash` is the default and most commonly used shell. The minor differences in shells are only in the built-in commands and in the files they read on start-up. We shall assume `bash`.

Once logged in, the shell will be invoked and will wait for a user command. The shell prompt is the symbol that indicates that the shell is waiting for a user command. The default `bash` shell prompt is the symbol `$`. Usually the configuration is set up so that the prompt is preceded with a server name and current directory, e.g.:

```
[bwilkinson@coit-grid01 ~]$
```

In this example, the user's account is `bwilkinson` on server `coit-grid01` and ~ indicates the home directory. If necessary, the specific shell can be invoked either by using `exec` command, e.g., the command `exec tcsh`, or by simply typing the name of the shell, e.g., `tsch`. The `exec` command is defined to replace the current shell instead of creating new process.

Built-In Shell Commands. The shell has built-in commands that supplement the Linux commands. For example, the `cd` command to change directories is a built-in shell command. `exec` is a built-in command. The command `echo` exists as a built-in command and as a Linux command.

Linux File Structure. The Linux file structure is hierarchical with the topmost level being the root directory. A forward slash (/) is used to indicate a directory (as opposed to a backslash (\) used in Windows). Some key directories in the root directory that a user might come across are:

`home`	Contains the user's home directories and many services
`bin`	Contains binaries (command) the user can use
`sbin`	As `bin` but not intended for the user. Used by Linux.
`usr`	Contains in subdirectories commands and libraries for normal use
`dev`	Device files for disk drives, input/output interfaces, etc.
`etc`	Configuration files for the machine
`tmp`	Temporary files (not preserved between program invocations)

Commands can be held in either `/bin` or `/usr/local/bin`.

Input/Output Streams. The default input device is the user's keyboard. Information from the default input device is referred to as the standard input stream and called `stdin`. The default output device is the user's display console. Information to the default output device is referred to as the standard output stream and called `stdout`. The default standard error stream is for logging error messages and is called `stderr`. The default stderr device is the user's display. `stdout`, `stdin`, and `stderr` reside in `/dev`. Also in `/dev` is `/dev/null` — a special "file" that discards everything that is written to it!

Users. The user called *superuser* (more commonly called *root*) has complete access to all resources and accounts. It can perform administrative tasks such as creating a new user account and modify system configurations. The regular users do not have such privileges—they usually have full access to their own account but limited access elsewhere. Access privileges in Linux are attached to files and directories. Three types of file access may be granted:

- *Read* permission — the user is allowed to read the contents of the file.
- *Write* permission — the user is allowed to modify the contents of the file.
- *Execute* permission — the user is allowed to execute the file. This permission would only be used with executable programs.

The permissions are also applied to directories. Read permission on a directory would allow the files within the directory to be listed. Write permissions on a directory would allow files within the directory to be deleted or renamed.

Files and directories are assigned an *owner*. The user that creates the files or directory is the owner by default. Files can also be assigned a *group* ownership, which is useful for collaborative projects to share files. Group accounts are established by superuser with the command `groupadd`. Users can be members of named group accounts. The user can issue the command `groups` to find which groups they are members of.

Permissions are assigned to:

- Owner
- Group
- Other

Other refers to any user other than the owner or group. One bit is needed for each permission—*read*, *write*, and *execute*. Coupled with *owner*, *group*, and *other*, this leads to nine file/directory permission bits. The permission bits can be displayed with the `ls -l` command. For example

```
[abw@coit-grid03 abw]$ ls -l
total 28
-rwxrw-r--     1 abw    mpi       9749 Oct  1 15:12 a.out
-rw-rw-r--     1 abw    mpi       3216 Oct  1 14:45 bitmap.c
-rw-rw-r--     1 abw    mpi       1499 Oct  1 14:47 bmptest.c
-rw-rw-r--     1 abw    mpi       3570 Oct  1 13:09 mandelbrot.c
drwxrwxr-x     2 abw    mpi       4096 Sep 11 11:22 mpi_lab
```

Here `d` stands for directory. The access permissions are read access (`r`), write access (`w`), and execute access (`x`), for user, group, and other, respectively. The group above is called `mpi`. The permissions can be set with the `chmod` command. For example, to grant read access to file `bmptest.c` to members of group `mpi`, one can issue the command:

```
chmod g+r bmptest.c
```

The Linux permissions arrangement is generally workable although as described it does not allow a file/directory to be accessed by multiple named groups. It does allow users to be members of multiple groups. Files/directories only have one group owner.

Linux Command Structure. Linux commands typically consist of the command name, and a list of *options* (sometimes called *flags* or *switches*) possibly with arguments. The options are preceded with a hyphen and are traditionally a single letter, but can be more letters and may have alternative full names. They are case sensitive and are used to modify or select a specific action within the command or provide additional required information. There are sometimes a very large number of options, which are not listed here. A common notation to indicate optional command-line parameters in documentation is to use `[]`. In the following, we will give the command and options in bold and name any argument for the option in italics. There is sometimes an option (`-help` or `--help`) that will describe command and list all flags available for the command and their purpose. Alternatively, one can use the command

man *command*

to display an on-line reference manual for the *command*.

Commands can be quite long. If it is needed to go onto next line, the \ (backslash) is used at the end of a line that will be continued on the next line. The backslash symbol must immediately precede a newline character.

Commonly Used Commands for Grid Computing.

Changing Password. Once you have logged on, you may need to change your password as your account may have been set up with a password that was sent to you in an insecure way (say by email). The command is

passwd Change user password

which will prompt for the current password and then the new password.

Viewing and Navigating Directories. The following are frequently used commands for viewing and navigating directories.

ls -a	List files in current directory. **-a** says list all files including hidden files (those starting with a period)
pwd	Print the full path of the current directory
cd *directory*	Change user's directory to that specified, or if none specified, to user's home directory
cd ..	Move up one directory

Creating Directories and Files. Files are usually created using an editor. The standard Linux editor is called vi. One simple editor that may be more convenient for the casual user and part of most Linux installations is called nano. To invoke nano, type

```
nano [filename]
```

Figure B.2 shows the nano window with the contents of an XML file called `test1.xml`. One simply types into the window to change the contents the file. Use control-O to save and control-X to exit. (Available commands are given at the bottom of the window.)

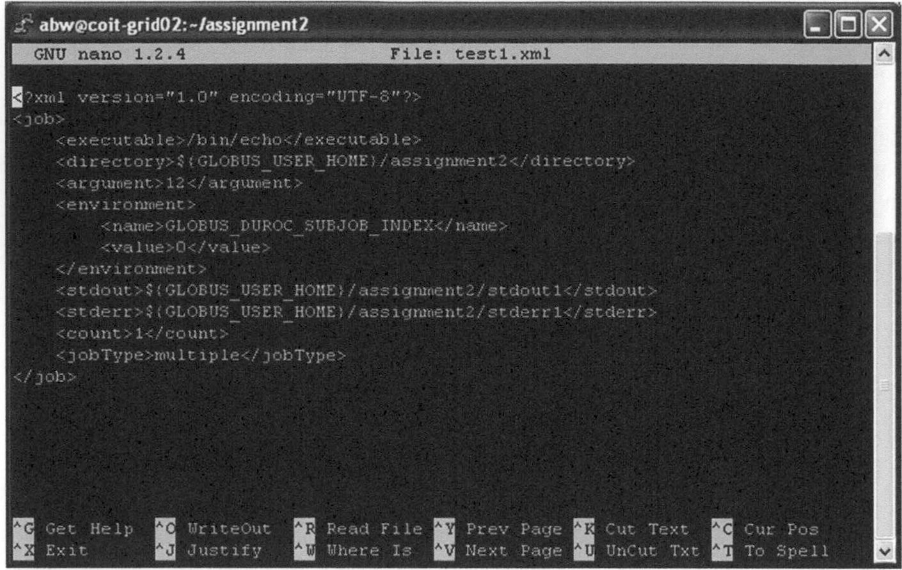

Figure B.2 Nano editor window with a file displayed.

Manipulating Directories and Files. Common commands used for manipulating directories and files include

`mkdir directories`	Create one or more directories
`cat files`	Read file(s) and send to standout output, i.e., display them.
`cp file1 file2`	Copy *file1* to *file2*
`cp file1 directory`	Copy *file1* to *directory*
`mv sources destination`	Move or rename files and directories
`rm files`	Delete one of more files
`rmdir directories`	Delete one of more directories. Must be empty.
`rmdir -r directories`	Remove non-empty directories (recursive remove)

Switching User. Occasionally in Grid computing, you need to switch to another user and execute a command as that user. The command for this is `sudo` (superuser do):

sudo -u *user command*	Execute command as specified user

For `sudo` to work, you have to be authorized to act for the specified user and for the specified command. This is done by being listed with allowed commands in the privileged `sudo` configuration file `/etc/sudoers`. Users will be prompted for a password, and subsequently every five minutes.

The Globus installation documentation strongly recommends creating a user in addition to the regular users called `globus` to start and stop the Globus container and for other administrative tasks. One can execute start the globus container as user `globus` with

```
sudo -u globus globus-start-container
```

Appendix D describes setting up and using the `globus` user as part of the installation process.

Environment Variables. Each Linux process operates in a software environment. *Environment variables* define values that control system behavior in the environment. Each shell has a set of *environment variables*. For example, `HOME` specifies the home directory. The value of the environment variable is obtained by prefixing it with a `$` symbol (a shell facility). For example, `$HOME` returns the path to the user's home directory. One can find the current value of an environment variable with the `echo` command. For example

```
echo $HOME
```

will display the path to the user's home directory.

Creating Environment Variables. Environment variables can be created for specific uses. A common environment variable for the Java environment is `JAVA_HOME`, which specifies the location of the Java JDK. An environment variable can be set on the command line with:

variable=value	Sets `variable` to `value`

That would only set the variable for the current shell. For global meaning, it can be followed by the command:

export *variable*	Make shell variable have global meaning (built-in `bash` command)

or the two commands combined with

export *variable=value*

For example

```
export GLOBUS_LOCATION=/usr/local/gt4
```

sets the Globus environment variable `GLOBUS_LOCATION` to the path `/usr/local/gt4`.

Command(s) can be placed in a start-up script such as `bash_profile` so that they take effect on subsequent login sessions. After editing the start-up script, rather than re-login, one can use the command

source *file*	Read and execute contents of file (built-in `bash` command)

where `file` is the start-up script.

`PATH` ***Variable.*** An important environment variable is called `PATH`, which specifies a list of directories to look in for each command. The directories are separated in the list by colons. (They are separated by semicolons in Windows as a colon would indicate a drive letter.) The home directory in a path is indicated with a period. Suppose one wants to add the path `$HOME/myProgs/myProg1` to the current path. One can add it with the construction

```
PATH = $PATH:$HOME/myProgs/myProg1
```

Usually the current directory is not put in the path for security considerations. Then to execute any program on the command line that is in your current directory, one would have to precede the name of the program with `./`.

Ports in Use. Sometimes, one needs to know the ports in use so that any new port usage does not conflict with existing ports. The Linux (and Windows) command for that is `netstat`, which has various options to display connections and routing tables, for example

netstat -a	Display all entries
netstat -t	Display tcp connections only
netstat -r	Display routing tables

B.2 WINDOWS COMMAND LINE ENVIRONMENT

Grid computing assignments, such as creating Web services in an Apache Axis/Tomcat environment, creating WSRF services using Globus core environment, or creating customized JSR portlets in a Gridsphere portal environment, can be done on a personal Windows computer. The necessary software is available for download. Tasks can be done through a Windows command line or a mixture of command line actions and Windows GUI drag-and-drop actions. As described in the Preface, there are significant advantages in using a personal computer when possible rather than using a centralized server.

Windows Command-Line Interface (Console). The Windows command-line interface (console) is found by first going to `start` (bottom left corner of display) and selecting `Run`, which will bring up a window where programs can be started by typing in their name. The command-line interpreter, `cmd.exe`, can be started this way by simply typing `cmd`. The command-line interpreter provides a console window for issuing commands. These commands are limited and derived from the 1970s MSDOS PC operating system. Drives are given letters. Typically the first hard drive would be C. Paths start with the drive followed by a colon. As mentioned earlier, paths in Windows use the backslash (\) as a path delimiter rather than the forward slash (/) used in Linux. The command prompt is usually the path to the current working directory followed by a greater than symbol, e.g., `C:\>`, although it can be altered by modifying the `PROMPT` environment variable. Some commands are similar to Linux, for example

`cd` *`directory`*	Change user's directory to that specified, or if none specified, to user's home directory
`cd ..`	Move up one directory

but some are different, for example

`dir`	List files in current directory

Environment Variables. Once software has been downloaded from the Internet for Grid computing assignments, it is installed usually by simply moving the unpacked executable to the local disk (C:), possibly placing it in the `Program File` directory. Then, typically one needs to set up the environment variables used by the software to identify locations for itself and for other software. For example, if the Java Development Kit (JDK) is being installed, its location needs to be given in the `JAVA_HOME` environment variable. Also the `PATH` system variable has to include the path to the installed software. Mostly, this does not occur automatically when installing open-source software by downloading it from the Internet. However, setting and editing environment variables in Windows is particularly easy. Go to `start` and right-click `My Computer`. Go down menu and click on `Properties`. Click on the `Advanced` tab. Click on the `Environment Variables` tab. Click new, edit, or delete as appropriate. There are two sets of environment variables, `User variables` and `System variables` as illustrated in Figure B.3. Normally, the variables to set or edit are system variables, which will apply across all users. But be careful not to alter system variables that could affect the operating system adversely. The separator in the `PATH` variable is a semicolon.

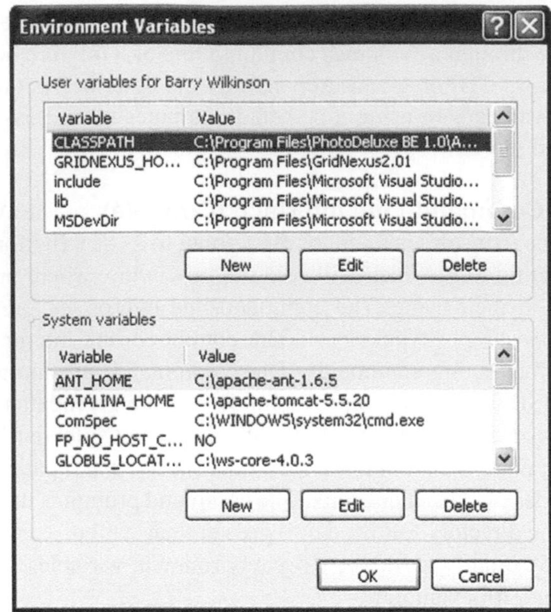

Figure B.3 Windows user and system environment variables.

The value of an environment variable is obtained by surrounding the variable name with `%`'s (rather than preceding the name with a `$` symbol as in Linux). Hence, to obtain the contents of the `PATH` variable, one would type

```
echo %PATH%
```

Environment variables can be set on the command line with the command

`set variable=value` Sets `variable` to `value`

That would only set the variable temporarily. The easiest way to set the variable permanently is to use the environment variable window shown in Figure B.3.

BIBLIOGRAPHY

Siever, E., S. Figgins, and A. Weber. 2003. *Linux in a nutshell: A desktop quick reference*. 4th ed. Sebastopol, CA: O'Reilly Media, Inc.

SELF-ASSESSMENT QUESTIONS

The following questions refer to Linux. The questions are multiple choice. Unless otherwise noted, there is only one correct answer for each question.

1. In the shell prompt

```
[abw@coit-grid02 ~/assignment2]$
```

what does the symbol ~ mean?

(a) Nothing. It is optional
(b) Default port number
(c) Path from the current directory
(d) Path to the user's home directory

2. When one executes the following commands

```
$ bash
```

and

```
$ exec bash
```

what is the difference in the results?

(a) Nothing
(b) `exec` will not create a new process but will replace shell
(c) `exec` will create a new process and will not replace shell
(d) `exec` will execute command but will return to original shell afterwards
(e) None of the other answers

3. What is the symbol \ (backslash) used for on a Linux command line?

(a) To ignore the remaining part of the line
(b) To specify a path of an executable
(c) To allow a command line to stretch over to the next line
(d) To specify a URL

4. What is the purpose of using the `netstat -t` command?

(a) To see which TCP ports are in use
(b) To make a TCP connection
(c) To see the status of the Internet
(d) To alter the TCP port to use

5. In Linux, what command makes a directory?

(a) `mkdry`
(b) `mkdir`
(c) `make_dir`
(d) `cp`

6. If a command line argument starts with the symbol, `$`, what can we infer about the argument?

(a) The argument refers to dollars
(b) The argument name starts with the `$` symbol

(c) The argument is an environment variable
(d) The argument is a file

7. What command would you use to make a secure connection to a remote computer?

(a) `telnet`
(b) `ssh`
(c) `ftp`
(d) `sec_con`

APPENDIX C
XML Markup Language

In this Appendix, markup languages are described, in particular XML, which is used in components of the Grid infrastructure and occurs in several places in the book. XML is used in job description languages in Chapter 2. It is used in Chapter 5 in authorization/authentication statements. Web services described in Chapters 6 and 7 rely on XML for their flexibility. XML is used to represent information in information services in Chapter 7. It appears in Chapter 8 in both workflow editors for describing workflows internally and in portals for describing portlet layout. It appears in Chapter 9 in defining parameter sweep in job description languages. For completeness, we shall begin with markup languages in general and HTML.

C.1 MARKUP LANGUAGES

Markup languages provide a way of describing information in a regular format that can be read by computers or humans. They use *tags* that identify and delineate information. These tags enable the information to be interpreted and processed by computer programs in a platform-independent way. The concept of markup languages developed in the 1960s and 1970s, notably with IBM's markup language called Generalized Markup Language (GML) developed in the 1960s. The markup language called the Standard Generalized Markup Language (SGML) was introduced by IBM in 1974 and ratified as an International Organization for Standardization (ISO) specification for a markup language in 1986. SGML uses the symbols < and > as the first and last character of the tags. Pairs of tags surrounded the information, a begin tag `<tag_name>` and a matching end tag `</tag_name>`, where `tag_name` is the name of the tag, as shown below

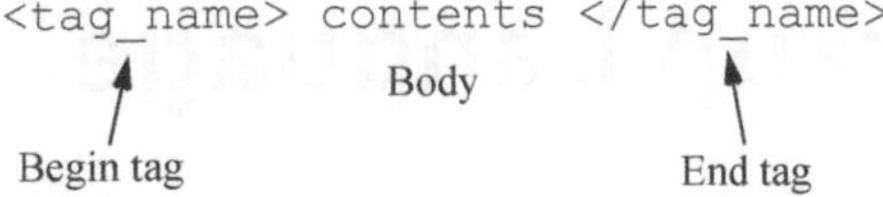

The term *element* is the name given to the construction (start tag, end tag and the contents between the start tag and end tag). Note the / in the end tag. Elements can be nested. Tags in SGML do not need to be in pairs if just the presence of a tag is required. In SGML, the actual syntax and tag names (element names) are defined by Document Type Definitions (DTDs), which are associated with the SGML document, either embedded into the document or held in a separate file. Hence, one can devise many languages based upon SGML. Different markup languages use specific names for their tags for specific purposes.

SGML, as such, is not now widely used as the underlying markup language, having been taken over by XML (see later), but SGML forms the basis for HTML and XML.

C.2 HYPERTEXT MARKUP LANGUAGE (HTML)

HyperText Markup Language (HTML) is a markup language specifically designed for Web pages. It was introduced as part of the initial development of the World Wide Web by Berners-Lee in the early 1990s. It is based upon SGML (tags with angle brackets), but the names of the tags are predefined with specific meanings. "Hypertext" refers to the text's ability to link to other documents. "Markup" in this context refers to providing information to tell a browser how to display page and other things. An HTML document consists of a head section and a body section as shown in Figure C.1. The head section is used mainly for meta-information about the HTML

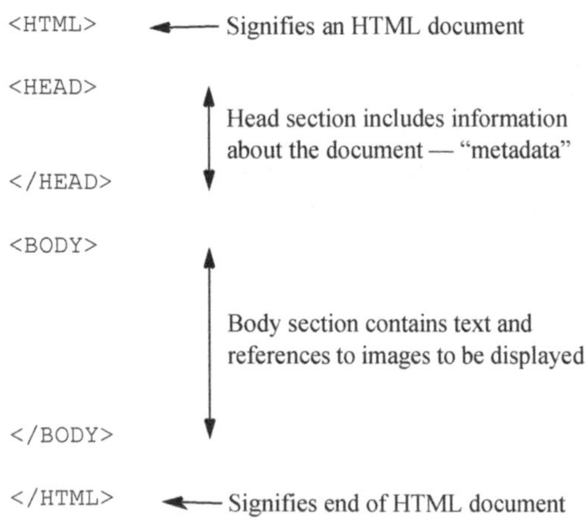

Figure C.1 HTML page format.

document.[1] What is intended to be displayed on the Web page is placed in the body. If we just put text in there, it would simply be displayed. Usually, we want to format the text is some fashion, for example make some text larger or in bold. HTML tags can specify such details, for example `<B>` to start bold text, `</B>` to end bold text, `<I>` to start italic text, `</I>` to end italic text, etc. To display some text in italics, the construction is

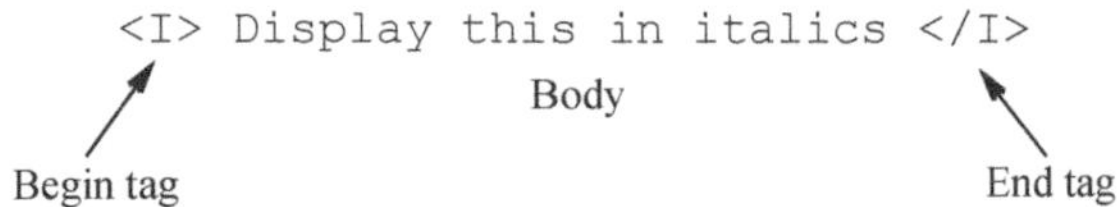

Some tags in the original HTML do not have to form pairs. For example, line break tag `<BR>` and new paragraph tag `<P>` would not naturally need more than saying when to introduce. In HTML 2.0 introduced in 1995, the paragraph tag can optionally include an end tag (i.e., `<P> ... </P>`) although break (`<BR>`) was still a singleton. This practice continued in HTML 4.0 introduced in 1997, which remains the standard as of 2009. (HTML 5 is in draft form.) XHML 1.0, introduced in 2000, is a version of HTML that conforms to XML syntax. XML requires all elements to be closed except for a special case of an empty body, which as a special notation `<tag_name/>` (see Section C.3 on XML).

Attributes. Many HTML tags can have *attributes* that specify something about the body between tag pair. For example

```
<FONT COLOR=red SIZE=3 FACE=Times>
This text is displayed in red in Times font, about 12 pt.
</FONT>
```

Figure C.2 shows an HTML page with bold italic text and also text attributes. The concept of attributes is very important and will appear as we discuss XML. There are many features in HTML not mentioned here such as being able to specify numbered and bulleted lists, tables with defined numbers of rows and columns, and images. A good introduction to HTML is by Knuckles (2001). Knuckles also gives an introduction to JavaScript, a language that can be used within HTML pages to provide much more functionality in defining what will be displayed and provides language constructs to do computations. JavaScript provides the ability to specify data entry formats through text field, buttons, pop-up menus, etc. and a means of acting upon the data that is input, for example to alter what is displayed. We refer the reader to Knuckles for more information on JavaScript.

HTML is somewhat unstructured and written usually by humans even though it is meant to be processed by computer. However, HTML can be dynamically generated through the Java Server Pages (JSP) technology, see Chapter 8.

---

[1] It is possible to have JavaScript code located in the head section that writes information in the body for display, or is a JavaScript routine used in body.

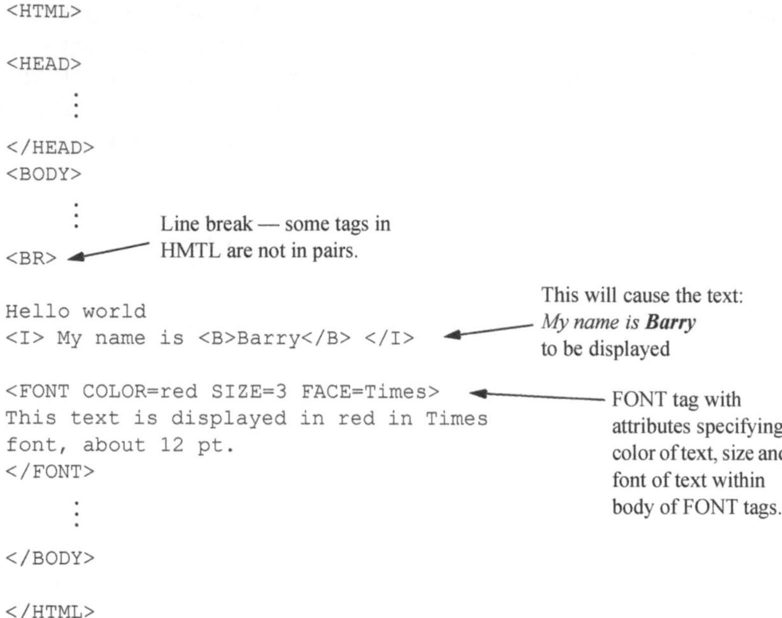

Figure C.2 Examples of HTML text tags and attributes.

C.3 EXTENSIBLE MARKUP LANGUAGE (XML)

Extensible Markup Language (XML) is a very important standard markup language for constructing other languages. It was ratified in 1998 and developed to represent textual information in a structured manner that could be read and interpreted by a computer. It is a foundation for Web services which are the basis of Grid computing. Some key aspects of XML include:

- Names of tags can be defined broadly at will (but will be defined, to be explained).
- Tags can have attributes.
- Tags can be nested.
- Tags are used in pairs to delineate information and make it easy to process.

XML is a development from SGML. Just as SGML could be used to create specific markup languages such as HTML, XML can be used to create markup languages. The names of the tags and their meaning in XML are not predefined and unique languages can be invented for a particular purpose. How XML tags are defined for a particular XML language will be described later. Table C.1 gives XML languages encountered in the book.

Generation of XML Documents. XML documents can be created by humans or automatically by computer. A sample XML document is the purchase

TABLE C.1 XML LANGUAGES IN BOOK

XML language	Chapter
RSL-2 (Resource Specification Language version 2)/ JDD (Job Description Document)	Chapter 2
JSDL (Job Submission Description Language)	Chapter 2
SAML (Security Assertion Markup Language)	Chapter 5
WSDL (Web Service Description Language)	Chapter 6
WSDD (Web Service Deployment Descriptor)	Chapter 6
GWSDL (Grid Web Service Definition Language) — Historical	Chapter 7
JXPL — GridNexus workflow language	Chapter 8
JSR 168 Portlet deployment descriptor file	Chapter 8
XPML (eXtensible Parameter Modeling Language) — Gridbus resource broker	Chapter 9

order shown in Figure C.3, which we shall use to explain concepts. This figure is based closely upon Graham et al. (2005) and uses a hypothetical XML language to specify the purchase order. A date is given as an attribute. Dates used in XML documents often use or require the ISO 8601 date format yyyy-mm-dd. This particular date format will sort in chronological order. The purchase order might be written directly by a human or more likely created by software from information entered by a human on an on-line form. Either way, the purchase order is likely to be processed by a computer. An example of an XML document that is likely to be created by humans in Grid computing is an XML job description language document. Another example of an XML document that could be generated either by humans or more likely by a computer is a WSDL document describing a Web service interface. Even when XML documents are generated by a computer program, programmers still have to be able to read these documents to fully understand what is going on.

Attributes. Attributes can occur within the start tag to provide additional information relating to the element. The attribute values are required to be written with quotation marks (single or double), i.e., they appear as strings. The use of attributes rather than providing the information in the body is XML language-specific and will be dependent upon the application area. In many cases, the information could have been provided either as an attribute or in the body. The XML language designer chooses the way for the specific information. For example, suppose a tag `<order>` is used to define a purchase order and another tag `<bar_code>` is used to define the product and the number of products needs to be defined. It could be defined as an attribute called say `number`, as in

```
<order number = "10">
    <barCode> 1234 </barCode>
</order>
```

```
<purchaseOrder id="53912" submitted="2009-01-01">
 <billTo>
  <name>Department of Computer Science</name>
  <company>The University of North Carolina at Charlotte</company>
  <street>9201 University City Blvd</street>
  <city>Charlotte</city>
  <state>NC</state>
  <postalCode>28223</postalCode>
  <country>USA</country>
 </billTo>
<shipTo>
  <name>Department of Computer Science</name>
  <company>The University of North Carolina at Charlotte</company>
  <street>9201 University City Blvd</street>
  <city>Charlotte</city>
  <state>NC</state>
  <postalCode>28223</postalCode>
  <country>USA</country>
 </shipTo>
 <order>
  <item barCode="45623881">
   <description>computer system model 1234 </description>
   <quantity>3</quantity>
   <price>1234.78</price>
  </item>
  <item barCode="36877780">
   <description>Laser printer type 2666 </description>
   <quantity>3</quantity>
   <price>1234.78</price>
  </item>
 </order>
</purchaseOrder>
```

Figure C.3 Sample XML document.

or alternatively with another tag say called <number>, as in

```
<order>
     <barCode> 1234 </barCode>
     <number> 10 </number>
</order>
```

Notice an attribute will apply to everything within the body. A common use of attributes in XML languages we come across is to give a name to an element that can then be referred to elsewhere in the document, see for example in WSDL. In WSDL, attributes are used for several other purposes. The type of a message is an attribute. Attributes are used to declare namepaces that apply to the element (see later), an approach that comes from the underlying XML syntax.

It is possible to have an empty body. This would only make sense if there were attributes that provide information. For example, we could have

```
<order barCode = "1234" number = "10"> </order>
```

Because an empty body

```
<tagName></tagName>
```

occurs frequently, there is a short form of this construction, which is:

```
<tagName/>
```

which breaks from having all tags in pairs. An example with attributes would be

```
<order barCode = "1234" number = "10" />
```

Document Structure. Formally, an XML document consists of an optional prolog, and a root element. The prolog includes instructions to specify how to process the XML document. These are called *processing instructions*. Processing instructions are identified by the construct

```
<? ... ?>
```

that is, start with `<?` and end with `?>`. Processing instructions include meta-information and comments. Meta-information is information about the document rather than the specific information being carried by the document. One processing instruction identifies the document as a XML document, e.g.,

```
<?xml version="1.0" encoding="UTF-8"?>
```

This processing instruction says that the document uses XML version 1.0 and that the encoding is UTF-8 (Unicode Transformation Format-8). The UTF-8 format is a subset of the Unicode format that includes all of the ASCII character and uses 8 bits. Other encodings are possible, including the full Unicode encoding for international applications. However, UTF-8 is usually sufficient.

Comments can be placed within the prolog and have the same form as in HTML, i.e.,

```
<!-- this is a comment  -->
```

The root element is the outermost element that follows the prolog and contains contents of document. Other elements are within root element and elements can be nested.

Tags. Tag names are part of the particular XML language and mean certain things just as reserved words are used to mean certain things in a particular programming language. There are restrictions in the construction of the names of XML tags. Names are case sensitive and must start with a letter. They can include a hyphen, underscore, and colon, but the use of a colon might be confusing to humans as it is also used to separate a prefix in namespaces (see below). We shall look at how the names are established but first let us look at the concept of namespaces.

Namespace Mechanism. The namespace mechanism provides a way of distinguishing tags of the same name and is used widely within XML documents. It particularly addresses the problem of combining XML documents when different documents use the same tag names to mean different things. With the namespace mechanism, the tag name is combined with an additional namespace identifier to qualify it. The fully qualified name is given by namespace identifier and original XML tag name. Namespace identifiers use URIs (Uniform Resource Identifiers), a Web naming mechanism that includes URLs (Uniform Resource Locators), email addresses (e.g., `mailto:abw@uncc.edu`) and globally unique names called Uniform Resource Names. A URL is typically used as a namespace identifier. Whatever is used within the possibilities of URIs, it does not need to refer to a physical resource. It is just a name and what it might refer to, be it a Web page or other resource, it is not used as such. It could be a URL that refers to a read-me document for documentation purposes, but that is not required.

Rather than simply concatenating the namespace identifier with the local name in the document, which could be very long, the namespace identifier is associated with a short name that becomes a prefix to the local name. The association of prefix with namespace is defined in an attribute of the element that the namespace applies. For example, suppose we wish to use the URL `http://www.cs.uncc.edu/~abw/ns` as the namespace to be referred to by the prefix `pons` (purchase order namespace). The attribute in the element is given as

```
xmlns:pons="http://www.cs.uncc.edu/~abw/ns"
```

Suppose now the whole purchase order document shown in Figure C.3 is to use this namespace and the prefix `pons`, then the document would have the form

```
<pons:purchaseOrder id="53912" submitted="2009-01-01"
    xmlns:pons="http://www.cs.uncc.edu/~abw/ns">
    <pons:billTo>
        <pons:name>Department of Computer Science</pons:name>
        ⋮
    </pons:billTo>
        ⋮
</pons:purchaseOrder>
```

i.e., every tag would have the namespace prefix.

The namespace can be applied to every tag automatically if that is required without having to write a prefix, by using the attribute with

```
xmlns ="http://www.cs.uncc.edu/~abw/ns"
```

Then, the document would have the form

```
<purchaseOrder id="53912" submitted="2009-01-01"
    xmlns ="http://www.cs.uncc.edu/~abw/ns">
    <billTo>
        <name>Department of Computer Science</name>
        ⋮
    </billTo>
        ⋮
</purchaseOrder>
```

The namespace becomes the default namespace, unless overridden by a prefix. There can be more than one namespace attribute, each associated with a different prefix and the different prefixes used with different tags in the element. Namespaces declared in an element can be used in all nested elements.

Defining the Tags. So far, we have not said what tag names and attribute names are legal in a document and how the tags are associated with a particular meaning. Variables with their names are declared in a normal computer program before being used. One way to define what tags and attributes are legal in a particular XML document is to describe them within the document in so-called Document Type Definitions (DTD), which is somewhat similar in concept to declaring variables in a computer program. DTD was used earlier with the Standard Generalized Markup language (SGML). However, this approach has serious limitations in terms of system integration and flexibility and is not currently used. In fact, they are not allowed in SOAP messages (the messaging protocol used for XML documents, see Chapter 6).

XML Schemas. An alternative much more powerful approach to DTDs is to define legal tags and attributes in another XML document and associate that document with the XML document either explicitly or by common agreement. The XML document describing the tags and attributes is called an XML *schema*. XML schemas, being expressed in XML, provide a flexible way of handling legal names and structures. They can handles namespaces and have the notation of data types, i.e., they can associate specific allowed datatypes for attributes and within elements. Each XML language will have its own schema.

At first, it seems a circular argument that a document has an associated XML schema document. How is that XML schema document defined? It would need its own schema, and that schema would need its own schema and so on. The answer is that we use a *Schema Definition Language* (XSD) with fixed definitions of the tags and structure. XSD is the underlying XML schema or all XML languages. It includes primitive data types. The `element` element in XSD is used to define the structure of an element in the XML language. The element name is defined by the attribute of `element` called `name`.

As a very simple example of an XML schema, consider the purchase order shown in Figure C.3. This purchase order has `id` and `date` as attributes and has three main elements, billTo, shipTo, and order. The elements billTo and shipTo both consist of addresses. Each address consists of the elements name, company, street, city, state, postalCode, and country. The order is described in the element order and consists of items being ordered. Each item consists of a product description (description), a quantity (quantity) and a price (price). In addition, the item part number is given as an attribute (barCode).

A minimal schema for such purchase orders is shown in Figure C.4. The namespace prefix xsd is a convention for the XSD namespace, that is

```
<xsd:schema xmlns:xsd="http://www.w3.org/2001/XMLSchema">
<xsd:element name="purchaseOrder" type="PurchaseOrderType"/
>
<xsd:complexType name="PurchaseOrderType">
 <xsd:sequence>
  <xsd:element name="billTo" type="Address"/>
  <xsd:element name="shipTo" type="Address"/>
  <xsd:element name="items" type="Items"/>
 </xsd:sequence>
 <xsd:attribute name="id" type="xsd:string"/>
 <xsd:attribute name="date" type="xsd:date"/>
</xsd:complexType>

<xsd:complexType name="Address">
 <xsd:sequence>
  <xsd:element name="name" type="xsd:string"/>
  <xsd:element name="company" type="xsd:string"/>
  <xsd:element name="street" type="xsd:string"/>
  <xsd:element name="city" type="xsd:string"/>
  <xsd:element name="state" type="xsd:string"/>
  <xsd:element name="postalCode" type="xsd:string"/>
  <xsd:element name="country" type="xsd:string"/>
 </xsd:sequence>
</xsd:complexType>

<xsd:complexType name="Items">
 <xsd:sequence>
  <xsd:element name="item" minOccurs="0" maxOccurs="unbounded">
   <xsd:complexType>
    <xsd:sequence>
     <xsd:element name="description" type="xsd:string"/>
     <xsd:element name="quantity" type="xsd:positiveInteger"/>
     <xsd:element name="price" type="xsd:decimal"/>
    </xsd:sequence>
    <xsd:attribute name="barCode" type="xsd:string" use="required"/>
   </xsd:complexType>
  </xsd:element>
 </xsd:sequence>
</xsd:complexType>

</xsd:schema>
```

Figure C.4 Simple XML schema for a purchase order.

```
xmlns:xsd="http://www.w3.org/2001/XMLSchema"
```

`xsd` is attached to the element names and data types that come from XSD. In the body of the schema, first the overall purchase order structure is defined, consisting of a `billTo` element, a `shipTo` element and a `items` element. The purchase order attributes `id` and `date` also have to be defined. Two of the elements, `billTo` and `shipTo`, refer to addresses and are given the type called `address`, which refers to an element called `address` and given below in the schema. The `order` element refers to items. The format of `items` is also given below. The format of an `item` element is then described. The names of elements that appear in a purchase order also appear as such in the schema and are defined there. The types may be primitive types such as integers, string, etc., which are predefined words in XSD. The types may be complex types that are constructed of simpler types and are then given arbitrary names.

This example is extremely simple and omits many powerful features of XSD such as providing for limits for what is allowed in the XML document. We do see one constraining feature in Figure C.4—the `items` element can have zero or more items, which is defined by

```
<xsd:element name="item" minOccurs="0" maxOccurs="unbounded">
```

The other elements in the purchase order must be present. The schema can include comments.

Associating a Document with a Schema. Once a schema exists, it can be associated with XML documents, or *instances* of documents that use the schema. The association of a instance of an XML language to the particular schema is done by attributes within the instance document, as illustrated in Figure C.5. The overall relationship, which begins with the root language XSD, is illustrated in Figure C.6.

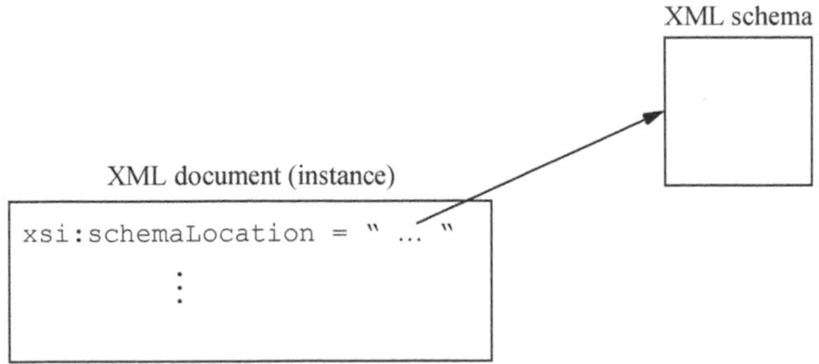

Figure C.5 Associating a schema with an instance of an XML document.

Schema Validation. The XML schema approach allows documents that are supposed to conform to XML syntax and a specific schema to be checked against the schema using a XML/schema preprocessing step prior to actually processing the document. There are two formal aspects in this preprocessing:

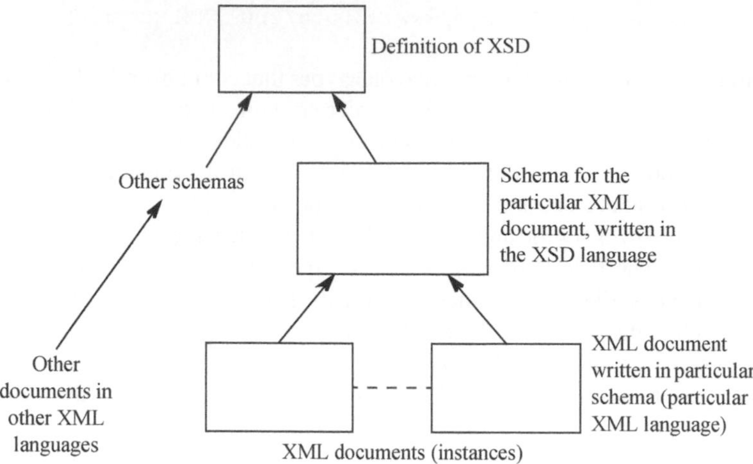

Figure C.6 Structure of XML documents.

- *Well-formedness* — document conforms to the rules of XML syntax
- *Validity* — checks that document again the specific schema

Rules that are checked for well-formedness include that tags have matching end tags and nesting is correct. Rules that are checked for validity include checking that the tag names exist in the schema and are used correctly, and the use of the attributes correspond to the schema. XML parsers exist to perform well-formedness and validity.

C.4 XML LANGUAGE EXAMPLES

Table C.1 lists XML languages that appear in the main chapters of the book for particular purposes. Figures C.7, C.8, and C.9 show some sample XML documents in three of these languages. In Figure C.7, the XML document uses the language called

```
<?xml version="1.0" encoding="UTF-8"?>
<job>
    <executable>/bin/echo</executable>
    <directory>${GLOBUS_USER_HOME}</directory>
    <argument>abc</argument>
    <environment>
        <name>GLOBUS_DUROC_SUBJOB_INDEX</name>
            <value>0</value>
    </environment>
    <stdout>${GLOBUS_USER_HOME}/stdout</stdout>
    <stderr>${GLOBUS_USER_HOME}/stderr</stderr>
    <count>1</count>
    <jobType>multiple</jobType>
</job>
```

Figure C.7 Globus 4.0 job description document (JDD).

```
?xml version="1.0" encoding="UTF-8"?>
<deployment name="defaultServerConfig"
 xmlns="http://xml.apache.org/axis/wsdd/"
 xmlns:java="http://xml.apache.org/axis/wsdd/providers/java"
 xmlns:xsd="http://www.w3.org/2001/XMLSchema">
 <service
  name="services/MyService" provider="Handler"
        use="literal" style="document">
  <parameter name="className" value=
   "edu.cs.uncc.abw.services.MyService"/>
  <wsdlFile>
    /schema/services/MyService_instance/My_service.wsdl
  </wsdlFile>
  <parameter name="allowedMethods" value="*"/>
  <parameter name="handlerClass"
   value="org.globus.axis.providers.RPCProvider"/>
  <parameter name="scope" value="Application"/>
  <parameter name="providers" value="GetRPProvider"/>
  <parameter name="loadOnStartup" value="true"/>
 </service>
</deployment>
```

Figure C.8 Web service deployment descriptor file (WSDD language).

Job Description Document (JDD), based upon the earlier *Resource Specification Language* version 2 (RSL-2). JDD is used in Globus 4.0 to describe the job to be submitted through the Globus Grid Resource Allocation Manager (GRAM) to a Grid platform. Here one sees the job specified by the job tag (`<job>` tag), and the executable within the job specified (`<executable>` tag) together with the arguments for the executable, input and output files and a environment variable. Notice, namespaces are not used in this document as it is not expected that the document would be combined with others in different languages. Also, there is no linkage to the language schema. That will be implied by its usage. In Figure C.8, we see a Web service deployment descriptor file written in the XML language called *Web Service Deployment Descriptor* (WSDD) for use in a Globus 4.0 environment. This file provides information on how the service is deployed into the service container. It requires certain information including where the service will be found within the Web service container, the location of the service class file, and the location of the WSDL file. More information on the WSDD construction can be found at (Sotomayor and Childers 2006). In Figure C.9, we see the first part and last part of a Web service description file using the XML language called *Web Services Description Language* (WSDL). Some names are declared in parts not shown. Both WSDD and WSDL documents use namespaces and show the linkage to the schemas.

FURTHER READING

Several books describe XML usually as part of describing Web services, although as we saw in the last section there are many applications for XML in addition to Web

```
<?xml version="1.0" encoding="UTF-8"?>
<definitions
  name="MathService"
  targetNamespace="http://www.globus.org/namespaces/examples/core/
      MathService_instance"
    xmlns="http://schemas.xmlsoap.org/wsdl/"
    xmlns:tns="http://www.globus.org/namespaces/examples/core/
      MathService_instance"
    xmlns:wsdl="http://schemas.xmlsoap.org/wsdl/"
    xmlns:wsrp="http://docs.oasis-open.org/wsrf/2004/06/
      wsrf-WS-ResourceProperties-1.2-draft-01.xsd"
    xmlns:wsrpw="http://docs.oasis-open.org/wsrf/2004/06/
      wsrf-WS-ResourceProperties-1.2-draft-01.wsdl"
    xmlns:wsdlpp="http://www.globus.org/namespaces/2004/10/
      WSDLPreprocessor"
    xmlns:xsd="http://www.w3.org/2001/XMLSchema">

      :

<portType name="MathPortType"
  wsdlpp:extends="wsrpw:GetResourceProperty"
  wsrp:ResourceProperties="tns:MathResourceProperties">
    <operation name="add">
      <input message="tns:AddInputMessage"/>
      <output message="tns:AddOutputMessage"/>
    </operation>

    <operation name="subtract">
      <input message="tns:SubtractInputMessage"/>
      <output message="tns:SubtractOutputMessage"/>
    </operation>

    <operation name="getValueRP">
      <input message="tns:GetValueRPInputMessage"/>
      <output message="tns:GetValueRPOutputMessage"/>
    </operation>
</portType>
</definitions>
```


Figure C.9 Web service description file (WSDL language).

services. The book by Graham et al. (2005) is thorough but very large (792 pages) for cursory reading. The purchase order example is derived from this book. The book by Benz with Durant (2003) is also good but again very large at 945 pages. XML reference and tutorial materials can be found on-line for example:

- WW3C consortium home page: http://www.w3.org/XML/
- W3Schools XML Tutorial: http://www.w3schools.com/xml/

BIBLIOGRAPHY

Benz B., with J. R. Durant. 2003. *XML programming bible*. New York, NY: Wiley Publishing, Inc.

Graham, S, D. Davis, S. Simeonov, G. Daniels, P. Brittenham, Y. Nakamura, P. Fremantle, D. König, and C. Zentner. 2005. *Building Web services with Java: Making sense of XML, SOAP, WSDL, and UDDI.* 2nd ed. Indianapolis, Indiana: SAMS publishing.

Knuckles, C. D. 2001. *Introduction to interactive programming on the Internet using HTML and JavaScript*. New York: John Wiley & Sons.

Sotomayor, B., and L. Childers. 2006. *Globus toolkit 4: Programming Java services*. San Francisco: Morgan Kaufmann.

Wikipedia. Standard Generalized Markup Language. http://en.wikipedia.org/wiki/SGML

SELF-ASSESSMENT QUESTIONS

The following questions are multiple choice. Unless otherwise noted, there is only one correct answer for each question.

1. What is an XML schema?

(a) A way to make XML documents secure
(b) A way of defining XML tags
(c) A way of encoding data for transmission
(d) None of the other answers

2. What is the difference between how are tags defined in HTML and how are tags defined in XML?

(a) No difference
(b) In both HTML and XML, they are pre-defined
(c) In XML, they are pre-defined.
(d) In HTML, they are pre-defined; in XML, they can be defined arbitrarily to create a particular XML language

3. A pair of XML tags with empty contents:

```
<tag_name></tag_name>
```

can be written as

```
<tag_name/>
```

Under what circumstances would such a construction make sense?

(a) Never
(b) When there are attributes
(c) When there are no attributes
(d) Under any conditions

APPENDIX D
Globus Installation Tutorial
by Jeremy Villalobos

The Globus documentation is thorough but it can be hard to use by novice users. To create a working Globus Grid node, one needs to follow instructions from several installation tutorials available at http://www.globus.org. This Appendix is made so that the reader can have a Grid node up and running with basic Globus services. The instructions are meant to get you started. Users can then look to the Globus website for more advance settings.

D.1 OVERALL STEPS

The minimum steps needed in order to have a working Grid node are:

1 Install Globus
2 Set sudo
3 Add environmental variables
4 Create a simple CA
5 Create host certificate
6 Create user certificate
7 Test basic tools

D.2 SOFTWARE REQUIREMENTS

To install Globus, you need an SQL server, Ant, and Java. Also, Perl is needed for GRAM jobs and the xinetd service is necessary for GridFTP services.

- **SQL server** — a database management program. Two open source choices are MySql and PostgreSql.
- **Ant** — a compiling and deploying tool. A programmer can create Ant scripts that can be used to compile, test, and deploy applications, which are often written in Java language.
- **Perl** — an interpreted language. Also needed are the Perl archive modules perl-Archive-Tar and perl-Archive-Zip.
- **xinetd** — an operating system service. The program listens on all ports and depending on the port number of the request, it dispatches the appropriate program.
- **cc** — a C compiler (for source installation).

If you can, use the package manager to get these packages. Type:

```
$ apt-get install postgres-server perl xinetd #Debian,Ubuntu
$ yum install postgres-server perl xinetd    #Red-Hat,Fedora,SuSE
```

It is recommended to use Sun's Java virtual machine–there is a GNU virtual machine that is known to cause errors. Also, we recommend installing Ant manually instead of using repositories. To check the Java version, type:

```
$ java -version
```

The returned text should be:

```
java version "1.6.0_07
Java(TM) SE Runtime Environment (build 1.6.0_07-b06)
Java HotSpot(TM) 64-Bit Server VM (build 10.0-b23, mixed mode)
```

If you get:

```
java version "1.6.0_0"
OpenJDK  Runtime Environment (build 1.6.0_0-b11)
OpenJDK 64-Bit Server VM (build 1.6.0_0-b11, mixed mode)
```

you are using an open-source version that may have incompatibility issues. Get Java JDK from http://java.sun.com and Ant from http://ant.apache.org. Setting up the binaries is easy—just unpack the tar.gz files and place the folder at /usr/local/ or a location of your choosing. Then, set the following environmental variable in the file /etc/profile:

```
JAVA_HOME=/usr/local/java
ANT_HOME=/usr/local/ant
PATH=$JAVA_HOME/bin:$ANT_HOME/bin:$PATH
export JAVA_HOME ANT_HOME PATH
```

Make sure JAVA_HOME points to the JDK and not the JRE.
To test the installation, as any user source the /etc/profile file:

```
$ source /etc/profile
```

and test the installation:

```
$ java -version
$ which javac
```

D.3 FIREWALL REQUIREMENTS

Most problems in the installation and configuration for the Globus package can be attributed to firewalls. Globus uses ports that are closed by default on firewalls. Here is a list of ports you should open on your firewall for Globus 4:

- GSIFTP: 2811
- GRAM: 8443
- Ephemeral ports: high port range.

The ephemeral ports are used to communicate job status and to transfer files with gsiftp. The ports are usually in the high ranges (30,000+), but it is recommended to set a range for these port so that you can make sure you have those ports open on the firewall. The port ranged is set with the environmental variable GLOBUS_TCP_PORT_RANGE. For example, if you want to open a range such as 50000 to 50100, you set the range this way:

```
$ export GLOBUS_TCP_PORT_RANGE=50000,50100
```

You should also check within your institution if there are other firewalls between your Globus server and the Internet.

D.4 INSTALL GLOBUS PACKAGE

Go to http://www.globus.org/toolkit/downloads/ and select the latest stable version. You have the choice to download a package manager file for your distribution, or you can download the source code. On this tutorial, we will use the source code because we can cover all the Linux distributions, Unix, and Mac OS X at one time. If you decide to use a package manager file, just check to make sure the package manager script accomplished the steps we mentioned in this section.

1 Create the Globus User

Create user globus–this will be the user that controls the Globus tools and authenticates and validates use of the resources. It is not recommended to use root to manage the Globus tools. Do the following steps as root:

```
$ useradd globus
```

2 Create Directory

Create a directory for Globus software and set ownership:

```
$ cd /usr/local                          #or some other path
$ mkdir globus-4.0.7
$ chown -R globus.globus globus-4.0.7
```

Put the tar.gz source package in another directory such as /home/globus and unpack it as globus user:

```
$ tar -zxvf gt4.0.7-all-source-installer.tar.gz
```

Go into the new directory and configure the package. At this point we will refer to the globus path, which is /usr/local/globus-4.0.7, as $GLOBUS_LOCATION:

```
$ export GLOBUS_LOCATION=/usr/local/globus-4.0.7
```

3 Configure and Compile the Code

Configure and compile software:

```
$ ./configure -prefix=$GLOBUS_LOCATION
$ make
$ make install
```

During the make install command, the script performs a check for Perl modules that will cause hard-to-find errors when using the tools. Install the necessary missing Perl modules using a package manager if necessary. If you encounter any other error, look at the error message. Most errors are related to package conflicts, absence, or incompatibility. Resolve these errors using the distribution's package manager.

4 Add Globus to the System Environment

Add GLOBUS_LOCATION variable to /etc/profile. Also, add the executable to the machine's path, and source some scripts that set other variables.

```
/etc/profile
....
GLOBUS_LOCATION=/usr/local/globus-4.0.7
GLOBUS_TCP_PORT_RANGE=50000,50100
PATH=$PATH:$GLOBUS_LOCATION:$GLOBUS_LOCATION/bin:$GLOBUS_LOCATION/etc
. $GLOBUS_LOCATION/etc/globus-user-env.sh
. $GLOBUS_LOCATION/etc/globus-devel-env.sh
export GLOBUS_LOCATION GLOBUS_TCP_PORT_RANGE PATH
```

5 Modify the Sudo File

Sudo allows the administrator to give users special privileges to run commands that only root can run. Globus needs root privileges to be able to sudo into other accounts. Globus will sudo into another user account when it runs a file transfer or job submission job. Globus reads from the file gridmap-file in order to map a certificate to a user name. This process provides more accountability for the user's actions since it is known at all times who issues a command or file transfer. It also provides privacy and security to users because a Linux permission system would not allow one user account to access another account's processes and data without explicit permission.

Modify the sudo file using the command visudo as root user. As root user type:

```
$ visudo
```

To edit the file press `i` key. Add the text below to that file:

```
globus ALL=(ALL) NOPASSWD: /usr/local/globus-
4.0.7/libexec/globus-gridmap-and-execute -g /etc/grid-
security/grid-mapfile /usr/local/globus-4.0.7/libexec/globus-
job-manager-script.pl *

globus ALL=(ALL) NOPASSWD: /usr/local/globus-
4.0.7/libexec/globus-gridmap-and-execute -g /etc/grid-
security/grid-mapfile /usr/local/globus-4.0.7/libexec/globus-
gram-local-proxy-tool *
```

To save and exit type `ESC` key and `SHIFT` + `z` twice. Alternatively, you can press `ESC` and then give the command `:exit`. The command is shown in the shell's left-bottom corner.

6 Create a Certificate Authority (CA)

There are multiple methods to build your trust infrastructure to be used by Globus. We will explain how the simplest out-of-the-box certificate authority is created.

Run this command as globus user, and follow the instructions:

```
$ $GLOBUS_LOCATION/setup/globus/setup-simple-ca
```

Then run this command as root user:

```
$ $GLOBUS_LOCATION/setup/globus_simple_ca_Hash_setup/setup-gsi  -default
```

This will create a certificate authority. The certificate authority as explained in Chapter 4 gives a higher level of trust to the public keys that users and servers exchange. Now you can request host certificates (for a server) and user certificates (for a person).

7 Get a Host Certificate

Whenever a secure communication is established, the participating entities have to mutually authenticate themselves. If a user submits a job to a server, not only does the user need to authenticate itself to the server, but also the server needs to prove it is the expected server. This provides security against spoofing and DNS poisoning attacks. As root type:

```
$ grid-cert-request -host serverrequestingcertificate.com
```

where serverrequestingcertificate.com is the name of the server. The command will put a certificate request at /etc/grid-security/hostcert_request.pem. Then switch to user globus and type:

```
$ grid-ca-sign -in /etc/grid-security/hostcert_request.pem \
-out /tmp/hostsigned.pem
```

Switch to root again and move hostsigned.pem file.

```
$ mv hostsigned.pem /etc/grid-security/hostcert.pem
```

There are several commands one can use to transfer files. There is GridFTP, Reliable File Transfer, Replica Location Service, and others. We will go over the configuration for GridFTP. The reader is referred to the Globus website for information in configuring the other file transfer services.

8 Set up GridFTP

Make sure the following entries are present in /etc/services file.:

- gsiftp 2811/tcp # GSI FTP
- gsiftp 2811/udp # GSI FTP

Some Linux distributions come with the services already included. If not, add the lines to your file. Add a file called gsiftp to /etc/xinet.d and put the following text in it:

```
$ cd /etc/xinet.d
$ touch gsiftp
$ vi gsiftp
```

```
service gsiftp
{
instances = 1000
socket_type = stream
protocol = tcp
wait = no
user = root
env += GLOBUS_LOCATION=/usr/local/globus-4.0.7
env += LD_LIBRARY_PATH=/usr/local/globus-4.0.7/lib
env += GLOBUS_TCP_PORT_RANGE=50000,50100
server = /usr/local/globus-4.0.7/sbin/globus-gridftp-server
server_args = -i
log_on_success += DURATION
log_on_failure += USERID
nice = 10
disable = no
}
```

Try to understand what the code is doing and make sure the variables are correct for your system.

To have xinet.d start up during system boot, type:

```
$ chkconfig -level 345 xinetd  on
```

To check the service is on, type:

```
$ chkconfig -list xinetd
```

Start the service:

```
$ /etc/init.d/xinetd start
```

9 Set up GSI Gatekeeper

The GSI Gatekeeper is a pre-Web service part of the Globus toolkit. You can use GRAM to submit jobs instead and we will not include the GSI gatekeeper as part of the installation for simplicity.

10 Set up GRAM

In order to receive jobs, the Grid node needs a container. There is no script provided by Globus that will start a container during boot time. First make sure you use your host certificate for the container. On the command line, type:

```
$ cd /etc/grid-security
$ cp hostcert.pem containercert.pem
$ cp hostkey.pem containerkey.pem
$ chown globus.globus containercert.pem containerkey.pem
```

You also need to have created the grid-mapfile in /etc/grid-security/:

```
$ touch /etc/grid-security/grid-mapfile
```

To start a container as globus, type the following:

```
$ globus-start-container
```

You can shut down the container using the command:

```
$ globus-stop-container
```

Since the command returns an error and does not shut down the container, killing the Java process may be needed.

This configuration does not include setting up Reliable File Transfer, which is not necessary for a basic installation.

11 Get a User Certificate

Each user, just like each server needs to have a certificate to authenticate itself to other entities. We will use the user griduser to refer to the unprivileged user accessing Grid resources in your Globus Grid. In order to create a certificate, type:

```
$ grid-cert-request
```

The wizard will require you to set a passphrase. The wizard will create three files at ~/.globus/. The files are:

- usercert.pem
- usercert_request.pem
- userkey.pem

The file usercer.pem is empty at the moment because the certificate has not been signed by the certificate authority. Copy the usercert_request.pem file to a directory where user globus has write permission (such as /tmp) and switch to user globus. Then, sign the certificate with the command:

```
$ globus-ca-sign -in usercert_request.pem -out usercert.pem
```

Turn back to the griduser and move the signed certificate to ~/.globus/usercert.pem:

```
$ cp /tmp/usercert.pem ~/.globus/usercert.pem
```

You can verify the process was done correctly by typing as griduser:

```
$ grid-proxy-init -verify
```

12 Add a User to grid-mapfile

Every user that wants to use a resource must show a certificate. If the certificate is allowed to work on this host, the certificate must be mapped to a local user. To map a user to a host you need two pieces of information:

1. First the local user name. For example user1.
2. Then you need the subject from the certificate. This can be read out of the certificate or you can print it with the command:

```
$ grid-cert-info -subject
```

Suppose your subject is:

```
/O=Grid/OU=GlobusTest/OU=simpleCA-grid02.domain.edu/OU=domain.edu/CN=User
```

then type the two pieces of information into the following command:

```
$ grid-mapfile-add-entry -dn \
"/O=Grid/OU=GlobusTest/OU=simpleCA-grid.domain.edu/OU=domain.edu/CN=User" \
-ln user1
```

Now the user can use services from this host.

D.5 TESTING

D.5.1 Submit a Test Job

As a user who owns a certificate, create a test file called test.xml. Put the following text in it:

```
<?xml version="1.0" encoding="UTF-8"?>
<job>
    <executable>/bin/hostname</executable>
    <directory>${GLOBUS_USER_HOME}</directory>
    <environment>
        <name>GLOBUS_DUROC_SUBJOB_INDEX</name>
        <value>0</value>
    </environment>
    <stdout>${GLOBUS_USER_HOME}/stdout</stdout>
    <stderr>${GLOBUS_USER_HOME}/stderr</stderr>
    <count>1</count>
    <jobType>multiple</jobType>
</job>
```

The file will run /bin/hostname on the server, and output the result in a file called stdout, which will be placed in the user's home directory. The error pipe will be put in the stderr file.

Run the command:

```
$ grid-proxy-init
```

The command will create a proxy. It will require your passphrase.

Then submit the job with the command:

```
$ globusrun-ws -submit -F localhost -f test.xml
```

The output should be:

```
Submitting job...Done.
Job ID: uuid:8fdec044-6b6f-11dd-bf92-0014220920ef
Termination time: 08/17/2008 08:44 GMT
Current job state: Active
Current job state: CleanUp
Current job state: Done
Destroying job...Done.
```

and the files should have the correct output. Use the flag -dbg to get more verbose output in case you need to troubleshoot.

D.5.2 Submit a Test File Transfer

To test GridFTP, create a file as user:

```
$ touch testfile.txt
```

Then transfer the file using the command:

```
$ globus-url-copy file:///home/user/testfile.txt \
gsiftp://localhost/home/user/file_transfered.txt
```

There should be a new file named file_transfered and no errors. If you get an error when you do this from another server, a firewall problem along the route is the most likely cause.

Congratulations, you have a basic Globus installation working !

There are other services provided by Globus, but this installation is a good starting point.

D.6 TROUBLESHOOTING

D.6.1 Error Log Files Location

The Web service GRAM component will output the errors to stdout and stderr. The output can be redirected:

```
$ globus-start-container > container.out 2> container.err
```

or you can put both on one file with

```
$ globus-start-container 1>&2 container.log
```

D.6.2 Connection Problems

The most common problem is to have the firewall ports open, but still have the connection timing out. Follow these steps to resolve the problem:

1 Verify your local firewall has the necessary ports open.

2 Check with the institution's system administrator that there are no other firewalls between your server and the Internet. If there is another firewall, try to get them to open the ports or to add your Globus server to the DMZ (DeMilitarized Zone).

3 Check that the port is open from another server on the Internet. There are several Web services that can do this. One of them is http://www.yougetsignal.com/tools/open-ports/.

If the port is open to the Internet, but the problem persists, the problem is likely not related to the firewall.

D.6.3 Using a Globus Server within a NAT

Globus is not designed to work inside a NAT network. However, there are a few workarounds that can help in using Globus on a non-production NAT network. The main problem is that Globus performs a domain verification technique to prevent spoofing. But if you have a host inside a NAT, it may report its address as being 192.168 ... which the other host will not trust. One solution to get this to work is to specify the expected subject from the server that is inside the NAT. You can do this on globusrun-ws with the -subject-authz option. Example:

```
$ globusrun-ws -submit -F localhost -f test.xml -subject-authz "servers subject"
```

This can become handy if you are just testing the globus software and do not have a production server with a WAN address.

D.6.4 Security Error Problems

Most security errors may be related to confusions about when to use root and when to use globus user. To prevent these common errors, we specified which user you should be when performing important steps. Under most basic setups, root should NOT have a .globus directory in its home directory. This will create security errors–this could happen if you request a non-host certificate, if you use Java Cog kits setups as root, and also if you use OGCE portal package as root.

BIBLIOGRAPHY

GT4 Admin Guide. http://www.globus.org/toolkit/docs/4.0/admin/docbook/index.html

SELF-ASSESSMENT QUESTIONS

1. If after setting up the Grid, submitting a job fails with the error message "connection refused," and the job was submitted from another host, the most likely cause is:

(a) The software was not configured correctly, check sudoers.
(b) SimpleCA may be causing the error
(c) The firewall may not be configured correctly

2. Assuming we have a Grid configured exactly as suggested by this document and a job is submitted. Select one likely port that would be used to give your client program the status on the job's progress:

(a) 50133
(b) 33030
(c) 22004
(d) None of the other answers

3. Which of the following is NOT necessary to check before testing GRAM?

(a) The user doing the testing has a certificate
(b) The user issues the command from his home directory
(c) A GRAM container is running
(d) The user was added to the grid-mapfile file

4. What command should be used to make sure certificates are being created properly?

(a) `grid_cert_verify`
(b) `grid_cert-init -verify`
(c) `grid-proxy-init -verify`
(d) `globusrun-ws`

GLOSSARY

access control list (ACL). A list that defines what types of access are allowed for each object, usually files or directories, by the user or group of users.

advance reservation. Requesting actions to occur at a time in the future. In the context of job schedulers, requesting a job to start at some time in the future.

AES (Advanced Encryption Standard). An encryption algorithm proposed in 2001. Replaces the DES as a government standard.

Apache Axis (Apache eXtensible Interaction System). A Java-based software environment to support Web services, implementing SOAP. From the Apache Software Foundation.

Apache Tomcat. Servlet engine supporting Java servlets and JavaServer pages. From the Apache Software Foundation.

array job. A group of instances of the same job with different arguments.

authentication. The process of deciding whether a particular identity is who he/she (or it) says he/she (or it) is.

authorization. The process of deciding whether a particular identity can access a particular resource.

backfilling. In a job scheduler—the mechanism in which lower priority jobs are run while waiting to run a high priority job, to utilize the compute resource fully.

Beowulf cluster. A computing cluster constructed using off-the-shelf computers and other commodity components and software.

binding (SOAP binding). The mapping to concrete SOAP exchanges and specific protocols.

BLAST (Basic Local Alignment Search Tool). A software tool is to find statistical matches between gene sequences. An input sequence is compared against a database of known sequences.

block encryption algorithm. An encryption algorithm in which the data is divided into sections and the encryption process is performed on each section using the same key.

building. The process of compiling a group of the software components into an executable form.

certificate. A digital document belonging to the user, which validates a public key as belonging to the user.

certificate authority (CA). A trusted third party that certifies that the information in the certificate is indeed correct, in particular that the public key does in fact belong to the user or resource named on the certificate.

checkpointing. A mechanism to enable a program to restart at a point during its execution rather than at the beginning. Information is stored at regular intervals during the execution of the program to enable the program to be restarted at the last checkpoint.

ciphertext. Encrypted information.

ClassAd. A Condor mechanism that enables jobs to be matched with resources (machines) according to job and machine characteristics.

client proxy. Another name for client stub.

client stub. A component (Java class) between a client and the network responsible for converting client requests into SOAP messages and converting incoming SOAP messages into data for the client.

cloud computing. A business model in which services are provided on servers that can be accessed through the Internet.

cluster computing. A group of computers connected together through a local network to create a high-performance computing platform.

CoG (Commodity Grid) Kits. APIs using commodity software technologies to provide a higher-level interface to Grid components than the base Grid middleware. Conceived and used in the Globus project.

Community Authorization Service (CAS). Component developed by the Globus community to provide an authorization service in a Globus environment using proxy certificates.

Condor. Software developed at the University of Wisconsin—Madison to use networked computers for high performance computing. Also used as a job scheduler.

credentials. Private key and signed certificate of end-entity (user or resource).

cross-certifying. Process in which the certificates of a pair of certificate authorities are signed by each other rather than by themselves.

cryptography. The theory and mechanisms for encrypting and decrypting information.

Cyclic Redundancy Check (CRC) word. Data pattern attached to a file and computed from the file contents. Alterations to the pattern can indicate errors in storage or after transmission.

DAGMan (Directed Acyclic Graph Manager). Condor's meta-scheduler that provides the ability to specify dependencies between Condor jobs so that jobs start in a certain order.

data confidentiality. Assurance that information transfers are protected against eavesdroppers.

data integrity. Assurance that a message was not modified in transit (intentionally or by accident).

decryption. The process of returning encrypted information back to its original form.

delegation. The process of giving authority to another identity, usually a computer/process, to act on your behalf.

deploying (a service). Placing files in the correct locations in a container, after which the service can be accessed by clients.

DES (Data Encryption Standard). An early encryption standard developed by IBM in the 1970s originally to encrypt unclassified U.S. government documents.

Diffie-Hellman key exchange. An algorithm developed by Diffie, Hellman, and Merkle which enables two parties each having a private key to obtain each other's key in a secure fashion.

digital signature. An encrypted message digest to provide a way of achieving authentication and data integrity.

dispatcher. In a job scheduler, the component that takes jobs from the job queue and sends them to computer resources.

distinguished name. The owner's identification following the naming convention of X.500 namespaces and used in X.509 certificates to identify the owner uniquely. Uses a concatenation of attributes.

DMTF (Distributed Management Task Force). A group created in 1992 for IT systems management infrastructure.

DRM (Distributed Resource Management). Used to describe a Grid computing job scheduler and the like.

DRMAA (Distributed Resource Management Application). A standard API specification for the submission and control of jobs to DRMs.

dynamic content. Information on a Web page that can be altered during viewing of page or when the page is refreshed. Technologies to achieve this include JavaScript and JSP.

e-Business. Using a Grid-like infrastructure to improve business models and practices, sharing corporate computing resources and databases.

e-Infrastructure. A Grid-like research infrastructure.

e-Science. Describes conducting scientific research using distributed networks and resources of a Grid infrastructure.

encryption. A process which involves modifying information into patterns that cannot be recognized and understood except by intended authorized parties.

end-entity. A term used with digital certificates to indicate its owner as given in the subject line of the certificate—a user or a computing resource.

enterprise Grids. Grids formed within an organization for collaboration.

EPR (EndPoint Reference). The way that a service and associated resource is identified in WSRF. Consists of an XML document that includes the address of the service given as a URI. If the EPR refers to a resource, the resource identification number (key) is included in the EPR.

GAT (Grid Application Toolkit). A toolkit of higher-level Grid middleware APIs not tied to specific Grid software, co-funded by the European Commission under the Fifth Framework Programme. Completed in 2005.

gateway. A grid portal.

GGF (Globus Grid Forum). The central forum for discussing and developing Grid computing standards. Globus Grid Forum became the Open Grid Forum in 2006.

Globus core. The common runtime environment of Globus that hosts services. Has different implementations (C, Java, and Python).

Globus toolkit. A software toolkit that provides a suite of software tools (APIs) for building a Grid environment.

GRAM (Grid Resource Allocation Management). The software component in Globus that processes job submission requests. The GRAM is a job manager that is only able to submit jobs locally through the fork mechanism and must interface to a job scheduler otherwise.

Grid-enabling. Being able to create an application that uses the distributed resources available on a Grid platform effectively. Not simply executing the application on a single local or remote resource.

Grid Security Infrastructure (GSI). The security component of the Globus toolkit.

Grid service. The term introduced by OGSI to describe an extended Web service that conforms to the OGSI standard. Now used more broadly to describe a Web service that conforms to Grid computing infrastructure standards.

GridFTP. A non-WSRF component of Globus used for high-speed file transfers. Provides for large data transfers, secure transfers, fast transfers, reliable transfers, and third-party transfers.

gridmap file. A file used in Globus that maps the user's distinguished name to the corresponding local account username.

GridNexus. A Grid computing workflow editor with a graphical interface developed at the University of North Carolina–Wilmington. Based upon the Ptolemy II workflow editor.

GridSphere. A Grid computing portal framework developed by a collaborative group (primary group: University of California, San Diego, Poznan Supercomputing and Networking Center, and Albert Einstein Institute).

GridWay. A Grid computing meta-scheduler developed by the Distributed Systems Architecture Research Group from Universidad Complutense de Madrid. It interfaces to Globus components.

GWSL (Grid Web Service Definition Language). A modified WSDL language that enables state to be specified in OGSI. Now obsolete.

HTML (HyperText Markup Language). A language used to describe how to display Web pages and connections to other pages elsewhere through *hypertext* links.

index service. A local service registry holding information on resources.

interface definition/description language (IDL). A language that enables a service interface to be described in a manner independent of any particular programming language.

Internet engineering task force (IETF). A standardization body formed in 1985 for Internet standards, including the previously developed TCP/IP protocol.

Java servlets. Java objects that receive requests from Web clients and generate responses. The Servlet interface in javax.servlet package defines required methods that must be implemented for client-servlet interaction.

JDD (Job Description Document). An XML job description language used in Globus version 4. Simplified syntax to RSL-2 language.

JSDL (Job Submission Description Language). A job description language developed as a GGF standard.

JSP (JavaServer Pages). A technology from Sun Microsystems for creating dynamic Web content through Java servlets.

JSR (Java Specification Request) 168. A Java portlet specification released in October 2003. Also called Java Portlet Specification version 1.0.

JSR 286 Portlets Specification v2.0. An updated version of JSR 168.

JXPL. An XML language used in the GridNexus workflow editor for describing the workflow.

key (cryptographic). A number used with a cryptographic algorithm to encrypt or decrypt information.

key (WSRF). A resource identifier for a WS-resource.

LDAP (Lightweight Directory Access Protocol). A network protocol used to access network-accessible databases.

marshalling. The process of converting source data into a SOAP message.

MD5 (Message-Digest 5). Cryptographic hash functions introduced by Rivest in 1991.

MDS (Monitoring and Discovering System). A collection of components in Globus that can be used to identify remote resources in the Grid for running jobs and storing information. In Globus version 4, these components consist of index service, trigger service, and WebMDS servlet.

message digest. A data pattern computed from the contents of a message such that if the message is altered, the computed data pattern will change with high probability.

message-level protocol. A protocol for messages in which specific portions of the message contents can be encrypted rather than the whole message (c.f. with transport-level protocols).

meta-scheduler. A job scheduler that assigns jobs across a distributed computing environment such as a Grid.

MPI (Message-Passing Interface). A standard specification for message-passing APIs. Does not define the implementation.

mutual authentication. The process when two parties both satisfy themselves that the other party is indeed the person whom they claim to be.

MyProxy. A software tool developed by the National Center for Supercomputing Applications, University of Illinois for managing credentials. Allows users to retrieve credentials upon request. Includes a credential repository and a certificate authority.

namespace. A way of distinguishing tags of the same name in an XML document by assigning a namespace to specific tags using tag prefixes.

non-repudiation. The process in which a communicating party cannot deny being involved in the communication.

OASIS (Organization for the Advancement of Structured Information Standards). A standardization consortium that focuses on structured information standards including the XML markup language.

OGF (Open Grid Forum). The central forum for discussing and developing Grid computing standards.

OGSA (Open Grid Services Architecture). A standard proposed by Foster et al. (2002) and introduced by the Open Grid Forum to define standard mechanisms for creating, naming, and discovering service instances and address architectural issues relating to interoperable services for Grid computing.

OGSI (Open Grid Services Infrastructure). The first attempt by the Grid computing community to specify how OGSA could be implemented. Now obsolete.

on-demand computing. Providing access to distributed computing resources when requested by clients and paid for by the clients when used.

OpenSSL toolkit. An open source software package that implements the SSL/TSL protocols and provides features for creating certificate authorities and for signing certificates.

parallel programming. Programming multiple processors or computers to operate collectively to solve a program.

parameter sweep. Running the same job with different input parameters. Can be done on different computers at the same time.

partner Grids. Grids set up between collaborative organizations or institutions.

passphrase. Similar to a password but can be very long and incorporate complete sentences with spaces. A properly chosen long passphrase makes it more secure.

password-based authentication. An authentication method where users enter their user name and password that were established for their account.

PKI (Public Key Infrastructure). Describes the structure that binds users to their public keys in public key cryptography.

portal (Grid). A user-friendly interface to a Grid computing environment deployed on the Web.

portlet. Presentation-layer software component for an area within a portal to provide specific functionality.

portType. The abstract definition of the service in WSDL version 1.1. Specifies the messages that must be sent to the service and the messages that must be returned, in abstract terms, that is, as inputs and outputs.

private key. A key that the owner keeps secret in cryptography. One of the two keys used in public key cryptography held by the owner and kept secret.

proxy certificate (proxy). A digital certificate that carries the authority of the user to act on its behalf.

public key. One of the two keys used in public key cryptography, made available to all.

PURSe (Portal-based User Registration Service). Grid computing software for user registration and credential management. Used in Grid portals.

PVM (Parallel Virtual Machine). Suite of message-passing libraries developed at Oak Ridge National Laboratories in the 1980s.

RC2. An encryption algorithm designed by R. Rivest.

RC4. An encryption algorithm designed by R. Rivest.

registration authority (RA). A component that acts for a certificate authority for some management functions, including processing user requests, confirming their identity and entering their information into the certificate repository database.

remote procedure call. A mechanism that allows invoking a procedure on a remote computer and getting back the results from the procedure.

resource broker. A job scheduling component that can negotiate for resources. The term is also used to describe a powerful scheduler that can optimize resource performance.

resource home. A component in WSRF/GT4 that provides resource management functions including locating the resources. Different resource home classes for different configurations.

resource properties. Name given to data items in a WS-resource.

RFT (Reliable File Transfer) service. A WSRF service front-end to GridFTP.

RLS (Replica Location Service). Part of the data management component in Globus that handles replicated data.

RSA. A public key algorithm developed by Rivest, Shamir, and Adleman.

RSL-1 (Resources Specification Language version 1). (Historical) Job description meta-language used in Globus version 1 and version 2.

RSL-2 (Resources Specification Language version 2). (Historical) An XML language version of the Globus job description language RSL-1, used in Globus version 3.

SAGA (Simple API for Grid Applications). Standardized toolkit of higher-level Grid middleware APIs not tied to specific Grid software.

SAML (Security Assertion Markup Language). A framework that embodies both an XML language for making assertions for authentication and authorization decisions and a request-response protocol for such assertions.

science portal. A Grid portal with an emphasis for use in the science domain.

server stub. A component (Java class) between a server and the network responsible for converting server requests/responses into SOAP messages and converting incoming SOAP messages into data for the server.

service container. A software hosting environment for services.

service registry. A component that provides a means of locating the service. Used in a service-oriented architecture.

service-oriented architecture. An architecture centered around service providers offering services and clients that access the service providers. The services are published in a service registry and the clients use this registry to locate the service and then bind with the service provider to invoke the service.

SGE (Sun Grid Engine). A job scheduler provided by SUN Microsystems for distributed compute resources. Open source version is called Grid Engine. Supported commercial version is called N1 Grid Engine (N1GE).

SHA (Secure Hash Algorithms, SHA-1 and SHA-2). Cryptographic hash functions.

simpleCA. An implementation of a certificate authority part of the Globus toolkit. Based upon OpenSSL certificate authority.

single sign-on. The process of logging onto a computing environment and being able to access resources within the environment without further user manual authentication.

SOAP. A communication protocol for passing XML documents, standardized by W3C organization.

SSH (Secure Shell). A protocol to make a secure connection between two computers and to exchange data in a secure fashion. Originally developed by Tatu Ylönen in 1995. Uses public key cryptography.

SSL (Secure Sockets Layer) protocol. A network protocol for making a secure connection that uses a public key cryptographic algorithm. SSL can be used with the HTTP protocol to create HTTPS.

staging (a file). Arranging that files are moved to where they are needed. Input staging refers to input files. Output staging refers to output files.

stateful Web service. A Web service that has the ability to remember information from one invocation to the next and between invocations by different clients.

stateless Web service. A Web service that does not remember information from one invocation to the next.

static content. Information on the Web page displayed without changing under varying circumstances.

streaming. Sending a stream of data from one location to another location as it is generated.

symmetric key cryptography. A cryptographic method that uses the same key to encrypt and to decrypt the information.

transport-level protocol. A protocol for messages in which the whole message is encrypted rather than portions of the message (c.f. with message-level protocols).

trigger service. A Globus 4 component that responds to specific conditions occurring within the Grid environment. Subscribes to a source of WSRF information such as an index service to be notified of changes.

TSL (Transport-Layer Security). A protocol based upon, and very similar to, SSL (secure sockets layer protocol) version 3, which provides for complete encrypted message transfers.

tunneling. In a SSH program, a secure connection forged through an insecure network by encrypting the messages with the SSH protocol and using the SSH port.

UDDI (Universal Description, Discovery and Integration). A discovery mechanism for Web services introduced in 2001 and standardized through OASIS.

UNICORE (UNiform Interface to COmputing REsources). European Grid computing project. Initially funded by the German ministry for education and research (BMBF) and continued with other European funding.

unmarshalling. The process of converting a SOAP message into data for the destination.

unsigned certificate. A certificate that does not have a digital signature and hence is unvalidated.

utility computing. Similar to on-demand computing—providing access to distributed computing resources when requested by clients—in a similar way as utilities such as electrical, gas, and water are metered.

virtual organization. A group of people, both geographically and organizationally distributed, working together on a problem, sharing computers and other resources. Term may also include the shared resources.

W3C (World Wide Web Consortium). A standardization body that works on standardization of Web-related technologies including XML. Founded by Tim Berners-Lee.

Web service. A software component that provides actions invoked through standard Internet protocols and addressing. Uses the XML language WSDL for defining the service interface.

Web service container. A software environment for Web services that provides the communication components to and from Web services.

WebMDS (Web Monitoring and Discovering System). A servlet that provides a Web-based interface to display XML-based information such as resource property information. Front-end to index services.

workflow. The description of a job execution sequence consisting of a group of jobs with inter-dependencies.

WS-Addressing. A specification for how to address Web services in WSRF. Defines the endpoint reference.

WS-BaseFaults. A specification for how to report faults in WSRF.

WS-BPEL (Web Services Business Process Execution Language). An OASIS standard introduced in 2004 for constructing Web service workflows in business applications.

WSDD (Web Service Deployment Descriptor). An XML language for describing how to deploy a Web service.

WSDL (Web Service Description Language). An XML interface definition language for describing a Web service interface.

WSFL (Web Services Flow Language). An XML language introduced by IBM for constructing Web service workflows in business applications. Superseded by WS-BPEL.

WS-Notification. Collection of specifications in WSRF that specifies how to configure services as notification producers or consumers.

WS-Resource. A Web service and an associated resource in WSRF.

WS-ResourceLifetime. A specification to manage resource lifetimes in WSRF.

WS-ResourceProperties. A specification for how resource properties are defined and accessed in WSRF.

WSRF (WS-Resource Framework). A specification proposed in 2004 to describe how to make a Web service stateful and other features in OGSA. Ratified by OASIS.

WSRP (Web Services for Remote Portlets). An OASIS standard for defining a Web service interface for interacting with "presentation-oriented Web services."

WS-Security. A standard that provides mechanisms for secure SOAP message exchanges including using of XML signature for message confidentiality and XML encryption for message integrity.

WS-ServiceGroup. A specification for how to group services or WS-Resources together in WSRF.

X.509 certificate. A certificate format defined by International Telecommunications Union (ITU).

XLANG. An XML business Web service orchestration language proposed by Microsoft. An extension of WSDL. Superseded by WS-BPEL.

XML (eXtensible Markup Language). A markup language for document exchange and processing in a platform-neutral way. Used to create specific XML languages for specific purposes.

XML encryption. Encryption of SOAP messages containing XML documents that allows part of the SOAP message to be encrypted and made confidential.

XML signature. A W3C standard for signing parts of a document, typically an XML document.

XPath. A language for selecting an XML element in an XML document. Not an XML language.

XPML (eXtensible Parametric Modeling Language). An XML language used in the Gridbus resource broker.

ANSWERS TO SELF-ASSESSMENT QUESTIONS

The following are the answers to the multiple-choice self-assessment questions that are given at the end of each chapter. Unless otherwise noted, there is only one correct answer for each question.

CHAPTER 1

1 (c)
2 (d)
3 (a), (b), (c), and (f)
4 (d)
5 (b)

CHAPTER 2

1 (b)
2 (c)
3 (a)
4 (b)
5 (a)
6 (c)
7 (b)
8 (c)
9 (d)
10 (a)
11 (c)
12 (b)

CHAPTER 3

1 (b)
2 (b)
3 (d)
4 (a)
5 (b)
6 (d)
7 (a)
8 (a)
9 (b) and (c)
10 (c)
11 (c)
12 (c)

CHAPTER 4

1 (d)
2 (d)
3 (b)
4 (c)
5 (a), (b), and (c)
6 (a) and (b)
7 (a)
8 (c), (e)
9 (b)
10 (b)
11 (b)
12 (a)
13 (b) and (d)
14 (c), (e), and (f)
15 (a) and (d)
16 (a) and (b)
17 (c)
18 (d)
19 (b)

CHAPTER 5

1 (c)

2 (c)
3 (b)
4 (d)
5 (b)
6 (a), (b), and (e)
7 (b)
8 (a)
9 (b)
10 (d)
11 (b)

CHAPTER 6

1 (b)
2 (a)
3 (c)
4 (d)
5 (a)
6 (a)
7 (c)
8 (a)
9 (b)
10 (c)
11 (b)
12 (a)
13 (d)

CHAPTER 7

1 (a)
2 (b)
3 (a)
4 (c)
5 (d)
6 (c)
7 (d)
8 (c)
9 (a)
10 (c)
11 (d)

CHAPTER 8

1 (b)
2 (c)
3 (c)
4 (b)
5 (d)
6 (b)
7 (a)
8 (c)
9 (d)
10 (b)
11 (a)
12 (b)
13 (d)
14 (f)
15 (a)
16 (c)

CHAPTER 9

1 (a)
2 (c)
3 (c)
4 (a)
5 (d)
6 (b)
7 (b)

APPENDIX A

1 (a)
2 (c)
3 (b)

APPENDIX B

1 (d)
2 (b)
3 (c)
4 (a)
5 (b)
6 (c)
7 (b)

APPENDIX C

1 (b)
2 (d)
3 (b)

APPENDIX D

1 (c)
2 (a)
3 (b)
4 (c)

INDEX

T

U

V

W

X